T0276846

Modern Nanoparticles Technology

Modern Nanoparticles Technology

Edited by **Mindy Adams**

New York

Published by NY Research Press,
23 West, 55th Street, Suite 816,
New York, NY 10019, USA
www.nyresearchpress.com

Modern Nanoparticles Technology
Edited by Mindy Adams

International Standard Book Number: 978-1-63238-332-7 (Hardback)

Printed in the United States of America.

Contents

Preface   VII

Properties and Applications   1

Chapter 1   **Thermal Property Measurement of Al_2O_3-Water Nanofluids**   3
Fei Duan

Chapter 2   **Thermal Effects on the Ferromagnetic Resonance in Polymer Composites with Magnetic Nanoparticles Fillers**   25
Mirosław R. Dudek, Nikos Guskos and Marcin Kośmider

Chapter 3   **Magnetic Nanoparticles: Its Effect on Cellular Behaviour and Potential Applications**   39
Hon-Man Liu and Jong-Kai Hsiao

Chapter 4   **Self-Assembly of Nanoparticles at Solid and Liquid Surfaces**   55
Peter Siffalovic, Eva Majkova, Matej Jergel, Karol Vegso, Martin Weis and Stefan Luby

Chapter 5   **Nanofluids**   81
Wei Yu, Huaqing Xie and Lifei Chen

Chapter 6   **Nanoparticle Dynamics in Polymer Melts**   103
Giovanni Filippone and Domenico Acierno

Chapter 7   **Dielectric and Transport Properties of Thin Films Deposited from Sols with Silicon Nanoparticles**   123
Nickolay N. Kononov and Sergey G. Dorofeev

Chapter 8   **View on the Magnetic Properties of Nanoparticles**
            Co_m (m=6,8,10,12,14) and Co_6O_n (n=1-9)   157
            Jelena Tamulienė, Rimas Vaišnoras,
            Goncal Badenes and Mindaugas L. Balevičius

Chapter 9   **Organic Semiconductor Nanoparticle Film:**
            Preparation and Application   187
            Xinjun Xu and Lidong Li

Chapter 10 **Magnetic Properties and Size Effects**
            of Spin-1/2 and Spin-1 Models of Core-
            Surface Nanoparticles in Different Type Lattices   203
            Orhan Yalçın, Rıza Erdem and Zafer Demir

Chapter 11 **Thermal Conductivity of Nanoparticles Filled Polymers**   223
            Hassan Ebadi-Dehaghani and Monireh Nazempour

            Permissions

            List of Contributors

Preface

This book was inspired by the evolution of our times; to answer the curiosity of inquisitive minds. Many developments have occurred across the globe in the recent past which has transformed the progress in the field.

Nanoparticles have major applications in biomedical, electronic and other fields. A broad overview of nanoparticle materials has been presented in this book. Nanoparticles have made a significant impact in the scientific domain in the past few years. The properties of numerous conventional materials changed when shaped from nanoparticles. Nanoparticles have a larger surface area per weight as compared to larger molecules which makes it more reactive and effective than other molecules. This book discusses the inherent properties of nanoparticles and their applications in various fields. It includes a number of well researched studies which will help students and experts interested in the field of nanotechnology.

This book was developed from a mere concept to drafts to chapters and finally compiled together as a complete text to benefit the readers across all nations. To ensure the quality of the content we instilled two significant steps in our procedure. The first was to appoint an editorial team that would verify the data and statistics provided in the book and also select the most appropriate and valuable contributions from the plentiful contributions we received from authors worldwide. The next step was to appoint an expert of the topic as the Editor-in-Chief, who would head the project and finally make the necessary amendments and modifications to make the text reader-friendly. I was then commissioned to examine all the material to present the topics in the most comprehensible and productive format.

I would like to take this opportunity to thank all the contributing authors who were supportive enough to contribute their time and knowledge to this project. I also wish to convey my regards to my family who have been extremely supportive during the entire project.

Editor

Properties and Applications

Thermal Property Measurement of Al$_2$O$_3$-Water Nanofluids

Fei Duan

School of Mechanical and Aerospace Engineering, Nanyang Technological University
Singapore

1. Introduction

Fluids have been applied in the cooling in the most important industries including microelectronics, manufacturing, metrology, etc. With increasing thermal loads that require advances in cooling the new higher power output devices with faster speeds and smaller feature, the conventional heat transfer fluids, such as water, engine oil, ethylene glycol, etc., demonstrate the relative low heat transfer performance. The use of solid particles as an additive suspended in the base fluid is a potential alternative technique for the heat transfer enhancement, i.e. thermal conductivity of metallic or nonmetallic solids might have two orders of magnitude higher than the conventional fluids. The enhancement of thermal conductivity of conventional fluids with the suspension of solid particles, such as micrometer-sized particles, has been well known for more than 100 years (Choi, 1995). However, the conventional micrometer-sized particle liquid suspensions require high concentrations (>10%) of particles to achieve such an enhancement. Because they have the rheological and stability problems such as sedimentation, erosion, fouling, and pressure drop in flow channels, the fluids with the micrometer-sized particle have not been of interest for practical applications. The recent advance in materials technology has made it possible to produce nanometer-sized particles that can overcome these above problems. The innovative fluids suspended with nanometer-sized solid particles can change the transport and thermal properties of the base fluid, and make the fluid stable.

Modern nanotechnology can produce materials with average particle sizes below 50 nm. All solid nanoparticles with high thermal conductivity can be used as additives of nanofluids. These nanoparticles that have been usually used in the nanofluids include: metallic particles (Cu, Al, Fe, Au, Ag, etc.), and nonmetallic particles (Al$_2$O$_3$, CuO, Fe$_3$O$_4$, TiO$_2$, SiC, carbon nanotube, etc.). The base media of nanofluids are usually water, oil, acetone, decene, ethylene glycol, etc. (Li et al., 2009). A 40% increase in thermal conductivity was found in the Cu oil-based nanofluids with 0.3% volume concentration, while the Al$_2$O$_3$ water-based nanofluids exhibited a 29% enhancement of thermal conductivity for the 5% volume concentration nanofluids (Eastman et al., 1997).

The Al$_2$O$_3$ nanoparticles were selected to prepare the water-based nanofluids in this study due to their chemical stability. Preparation of nanofluids is the key step in the use of nanoparticles for stable nanofluids. Two kinds of methods have been employed in producing nanofluids: the single-step method and the two-step method. The single-step method is a process combining the preparation of nanoparticles with the synthesis of nanofluids, for which the nanoparticles are directly prepared by the physical vapor deposition technique or the liquid chemical method (Choi, 1995; Eastman et al., 1997). The processes of drying, storage, transportation, and dispersion of nanoparticles can be avoided, so the aggregation of nanoparticles is minimized and the stability of fluids is increased. But a disadvantage of the method is that only low vapor pressure fluids are compatible with the process. It limits the applications of the method. The two-step method for preparing nanofluids is a process by dispersing nanoparticles into base liquids. Eastman et al. (1997), Lee et al. (1999), and Wang et al. (1999) used this method to produce the Al_2O_3 nanofluids. Nanoparticles used in the method are firstly produced as a dry powder by inert gas condensation, chemical vapor deposition, mechanical alloying, or the other suitable techniques before the nano-sized powder is then dispersed into a fluid in the second processing step. This step-by-step method isolates the preparation of the nanofluids from the preparation of nanoparticles. As a result, aggregation of nanoparticles may take place in both the steps, especially in the process of drying, storage, and transportation of nanoparticles. The aggregation would not only result in the settlement and clogging, but also affect the thermal properties. The techniques such as ultrasonic agitation or the addition of surfactant into the fluids are often used to minimize particle aggregation and improve dispersion behavior. Since nanopowder synthesis techniques have already been commercialized, there are potential economic advantages in using the two-step synthesis method. But an important problem that needs to be solved is the stabilization of the suspension to be prepared.

Nanofluids are a new class of solid-liquid composite materials consisting of solid nanoparticles, with sizes typically in the order of 1 - 100 nm, suspending in a heat transfer liquid. Nanofluids are expected to have superior properties compared to conventional heat transfer fluids. The much larger relative surface area of nanoparticles should not only significantly improve heat transfer capabilities (Xie et al., 2001), but also increase the stability of the suspensions. In addition, nanofluids can improve abrasion-related properties as compared to the conventional solid/fluid mixtures. Successful applications of nanofluids would support the current trend toward component miniaturization by enabling the design of smaller but higher-power heat exchanger systems (Keblinski et al., 2005). The thermal properties including thermal conductivity, viscosity, and surface tension have been investigated.

1.1 Thermal conductivity of nanofluids

Since the model reported by Maxwell (1892), the classical models have been derived by Hamilton & Crosser (1962), Bruggeman (1935), and Xuan & Li (2000) for predicting the effective thermal conductivity of a continuum mixture with the assumed well-dispersed solid particles in the base fluid. The Maxwell model was developed to determine the effective thermal conductivity of liquid-solid suspensions for a low volumetric concentration of spherical particles. This model is applicable to statistically homogeneous low volume fraction liquid-solid suspensions with randomly dispersed and uniform spherical particles in size. For non-spherical particles, the thermal conductivity of the nanofluids depends not only on the volume fraction of the particles, but also on the shape of the particles. Hamilton & Crosser (1962) modified the Maxwell model to determine the effective thermal conductivity of nonspherical particles by applying a shape factor for the effective thermal conductivity of two-component mixtures. The Hamilton-Crosser model considers the nanoparticle aggregation. For spherical particles, the Hamilton-Crosser model reduces to the Maxwell model. In the Bruggeman model, the mean field approach is used to analyze the interactions among the randomly distributed particles (Bruggeman, 1935). The model by Xuan & Li (2000) is not specified for any particular shape of particles. However, the classical models were found to be unable to predict the anomalously high thermal conductivity of nanofluids. This might be because these models do not include the effects of particle size, interfacial layer at the particle/liquid interface, and the Brownian motion of particles (Jang & Choi, 2004; Keblinski et al., 2002; Wang et al., 1999; Yu & Choi, 2003). Recently, Yu & Choi (2003) proposed a modified Maxwell model to account for the effect of the nano-layer by replacing the thermal conductivity of solid particles with the modified thermal conductivity of particles, which is based on the so called effective medium theory (Schwartz et al., 1995). The model can predict the presence of thin nano-layers less than 10 nm in thickness. Yu & Choi (2004) proposed a modified Hamilton-Crosser model to include the particle-liquid interfacial layer for nonspherical particles. The model can predict the thermal conductivity of the carbon nanotube-in-oil nanofluids reasonably well. However, it fails to predict the nonlinear behavior of the effective thermal conductivity of general oxide and metal based nanofluids. Xue (2003) presented a model for the effective thermal conductivity of nanofluids considering the effect of the interface between the solid particles and the base fluid based on the Maxwell model and the average polarization theory. Xue (2003) demonstrated that the model predictions were in a good agreement with the experiments of the nanotube oil-based nanofluids at high thermal conductivity and nonlinearity. However, Yu & Choi (2004) found that the predicted values from the model by Xue are inaccurate by using two incorrect parameters, as same as the finding of Kim et al. (2004). Xue & Xu (2005) obtained an equation for the effective thermal conductivity based on the Bruggeman model (Bruggeman, 1935). The equation takes account of the effect of interfacial shells by replacing the thermal conductivity of nanoparticles with the assumed value of the "complex nanoparticles", which introduces interfacial shells between the nanoparticles and the base fluids. The model can explain the size dependence of the thermal conductivity of nanofluids (Xuan & Li, 2000). Xie et al. (2001) considered the interfacial nano-layer with the linear thermal conductivity distribution and proposed an effective thermal conductivity model to account for the effects of nano-layer thickness, nanoparticles size, volume fraction, and thermal conductivities of fluids, and nanoparticles. They claimed that the calculated values could agree well with some available experimental data.

Temperature is one of the important factors influencing the thermal conductivity of nanofluids (Das et al., 2003; Li & Peterson, 2006; Yang & Han, 2006). Xuan et al. (2003) considered the Brownian motion of suspended nanoparticles on the basis of the Maxwell model. The prediction from the model is in an agreement with the experiment results, especially when the effect of nanoparticle aggregation is taken into account. But the model may be not accurate for the second term in the equation. Wang et al. (2003) proposed a fractal model for predicting the thermal conductivity of nanofluids based on the effective medium approximation and the fractal theory, developed firstly by Mandelbrot (1982). It can describe the disorder and stochastic process of clustering and polarization of nanoparticles within the mesoscale limit. A comprehensive model considering a large enhancement of thermal conductivity in nanofluids and its strong temperature dependence was deduced from the Stokes-Einstein formula by Kumar et al. (2004). The thermal conductivity enhancement takes into account of the Brownian motion of the particles. However, the validity of the model in the molecular size regime has to be explored and it may not be suitable for a large concentration of the particles where interactions of particles become important. Bhattacharya et al. (2004) developed a technique to compute the effective thermal conductivity of a nanofluid using the Brownian motion simulation. They combined the liquid conductivity and particle conductivity. The model showed a good agreement of the thermal conductivity of nanofluids. Jang & Choi (2004) combined four modes of energy transport in the nanofluids, collision between base fluid molecules, thermal diffusion of nanoparticles in fluids, collision between nanoparticles due to the Brownian motion, and thermal interaction of dynamic nanoparticles with the base fluid molecules in their model, which considered the effects of concentration, temperature, and particle size. The predictions from this model agree with the experimental data of Lee et al. (1999) and Eastman et al. (2001). However, it may not be suitable in the high temperature since the Brownian motion effect was neglected. Prasher et al. (2005) proposed that the convection caused by the Brownian motion of nanoparticles is primarily responsible for the enhancement in the effective thermal conductivity of nanofluids. By introducing a general correlation for the heat transfer coefficient, they modified the Maxwell model by including the convection of the liquid near the particles due to the Brownian motion. The result showed that the model matched well with the experimental data under different fluid temperature in a certain range. A model for nanofluids, which takes account the effects of particle size, particle volume fraction and temperature dependence as well as properties of the base fluid and the particle subject to the Brownian motion, developed by Koo & Kleinstreuer (2004).

Although many models have been proposed, no theoretical models are available for predicting the thermal conductivity of nanofluids universally up to now. More experimental data are required. Such data should include more studies of the effects of size and shape of the nanoparticles, the interfacial contact resistance between nanoparticles and base fluids, the temperature dependence, the effect of the Brownian motion, or the effect of clustering of particles.

Experimental works have been reported on the thermal conductivity of nanofluids. The main techniques are the transient hot wire (THW) method (Kestin & Wakeham, 1978), the temperature oscillation technique (Wang et al., 1999), and the steady-state parallel-plate method (Das et al., 2003). Among them, the TWH method has been used most extensively. Since most nanofluids are electrically conductive, a modified hot-wire cell with an electrical system was proposed by Nagasaka & Nagashima (1981). The advantage of the method is its almost complete elimination of the effect of natural convection. The measuring principle of the THW technique is based on the calculation of the transient temperature field around a thin hot wire as a line source. A constant current is supplied to the wire to raise its temperature. The heat dissipated in the wire increases the temperature of the wire as well as that of the nanofluids. This temperature rise depends on the thermal conductivity of the nanofluids in which the hot wire is at the center. Therefore, the thermal conductivity value of the fluid can be determined. The oscillation method was proposed by Roetzel et al. (1990) and further developed by Czarnetzki & Roetzel (1995). In principle, the thermal diffusivity of a fluid can be measured very accurately by considering amplitude attenuating of thermal oscillation from the boundary to the center of the fluid. However, for direct measurement of thermal conductivity one has to consider the influence of the reference materials as well. Since the defects of the reference materials might bring out the uncertainty in the thermal conductivity measurement, a direct evaluation of the thermal conductivity of the fluid is less accurate. The apparatus for the steady-state parallel-plate method can be constructed on the basis of the design by Challoner & Powell (1956). The steady-state parallel-plate method needs to measure the temperature increase accurately in each thermocouple (Das et al., 2003). The difference in temperature readings needs to be minimized when the thermocouples are at the same temperature. In this method, it has to follow the assumption that there is no heat loss from the fluid to the surrounding. As a result, guard heaters would be applied to maintain a constant temperature in the fluid. However, it is challenging to control the conditions in which no heat radiated to the surrounding from the fluid. Thus, the TWH method was selected for this study.

1.2 Viscosity of nanofluids

Viscosity of nanofluids is an important parameter in the fluid transporting. However, the data collected showed that no theoretical models (Batchelor, 1977; Brinkman, 1952; Einstein, 1906; Frankel & Acrivos, 1967; Graham, 1981; Lundgren, 1967) succeed in predicting the viscosity of nanofluids accurately until now. A few theoretical models were used to estimate particle suspension viscosities. Almost all the formulae were derived from the pioneering work of Einstein (1906), which is based on the assumption of a linearly viscous fluid containing the dilute, suspended, and spherical particles. The Einstein formula is found to be valid for relatively low particle volume fractions less than 0.01. Beyond this value, it underestimates the effective viscosity of the mixture. Later, many works have been devoted to the "correction" of his formula. Brinkman (1952) has extended the Einstein formula for use with moderate particle concentration. Lundgren (1967) proposed an equation under the form of a Taylor series. Batchelor (1977) considered the effect of the Brownian motion of particles on the bulk stress of an approximately isotropic suspension of rigid and spherical particles. Graham (1981) generalized the work of Frankel & Acrivos (1967), but the correlation was presented for low concentrations. Almost no model mentioned could predict the viscosity of nanofluids in a wide range of nanoparticle volume fraction so far. According to these correlations the effective viscosity depends only on the viscosity of the base fluid and the concentration of the particles, whereas the experimental studies show that the temperature, the particle diameter, and the kind of nanoparticle can also affect the effective viscosity of a nanofluid. A good understanding of the rheological properties and flow behavior of nanofluids is necessary before nanofluids can be commercialized in the heat transfer applications. These factors influencing the viscosity include concentration, size of nanoparticles, temperature of nanofluids, shear rate, etc. Thus, more thorough investigations should be carried out on the viscosity of nanofluids.

In the measurement, the rotational rheometer, the piston-type rheometer, and the capillary viscometer are the most popular tools used to measure the viscosity of nanofluids. Rotational rheomters use the method that the torque required to turn an object in a fluid is a function of the viscosity (Chandrasekar et al., 2010). The relative rotation determines the shear stress under different rates. The advantage of this type of measurement is it is not affected by the flow rate of the fluids. The operation is simple and high repeatable. The piston type rheometer is based on the Couette flow inside a cylindrical chamber (Nguyen et al., 2007). It composes the magnetic coils installed inside a sensor body. These coils are used to generate a magnetically-induced force on a cylindrical piston that moves back and forth over a very small distance, imposing shear stress on the liquid. By powering the coils with a constant force alternatively, the elapsed time corresponding to a round trip of the piston can then be measured. Since the measurement of the piston motion is in two directions, variations due to gravity or flow forces are minimized. Because of the very small mass of the piston, the induced magnetic force would exceed any disturbances due to vibrations. However, the piston type viscometer is that the duration of the heating phase necessary to raise the fluid sample temperature is relative long, especially under the elevated temperature condition, some base fluids may be evaporated. The capillary viscometer is introduced in U-shaped arms (Li et al., 2007). The capillary viscometer is submerged in a glass water tank. A water tank is maintained at a prescribed constant temperature for the capillary viscometer by the water circulation. The vertical angle of the viscometer is accurately controlled with a special tripod. Li et al. (2007) pointed out that the capillary tube diameter may influence the apparent viscosity and result in inaccurate in the nanofluids at higher nanoparticle mass fractions, especially at a lower temperature. In addition, nanoparticles might stain at the inner wall of the bore. Because of the narrow diameter, cleaning is difficult if the nanoparticles are left. In our study, we adopted the rotational rheometer to measure the viscosity of nanofluids because of its simplicity and repeatability.

1.3 Surface tension of nanofluids

Interfacial properties such as surface tension play an important role for the fluids having a free surface, however, the studies of the interfacial properties of nanofluids are limited. An understanding of nanofluid properties is essential so that we can optimize the usage of nanofluids and understand their limitations. The temperature dependence of surface tension of the liquid is crucial in the bubble or droplet formation. Wasan & Nikolov (2003) studied the spreading of nanofluids on solid surfaces and found that the existence of nanoparticles near the liquid/solid contact line can improve its spreading. Vafaei et al. (2009) investigated the effects of size and concentration of nanoparticles on the effective gas-liquid surface tension of the aqueous solutions of the bismuth telluride nanoparticles. Kumar & Milanova (2009) found that the single-walled carbon nanotube suspensions in a boiling environment can extend the saturated boiling regime and postpone catastrophic failure of the materials even further than that previously reported if the surface tension of the nanofluids is carefully controlled. The surface tension of a liquid strongly depends on the presence of contaminants or dispersion agents such as surfactants.

Pendant droplet analysis is a convenient way to measure surface tension of fluids. It is assumed that the droplet is symmetric and the drop is not in motion. The advantage of the technique is that the calibration is straightforward, only based on the optical magnification. This can lead to a high accuracy. Another advantage is that the cleanliness requirement is not high. Surface tension is determined by fitting the shape of the droplet to the Young-Laplace equation which relates surface tension to droplet shape. Pantzali et al. (2009) used the pendant droplet method to measure the surface tension of the CuO water-based nanofluids. The other common method to measure surface tension is the capillary method (Golubovic et al., 2009). The main component of the device is a capillary tube in which the liquids would show a significant rise with a meniscus due to the surface tension in order to balance the gravity force. The disadvantage of the capillary method is that cleaning is difficult if the nanoparticles are left in the small diameter capillary. Thus, the pendant drop technique was selected in this study.

In sum, the reported thermal property measurement are scattered, and lack of agreement with the models. It might be due to various factors such as the measuring technique, particle size, base fluid, volume fraction of nanoparticles in fluids, temperature, etc. The lack of reliable experimental data is one of the main reasons for no universal theoretical or empirical models. Therefore, we investigated the thermal properties of the Al_2O_3 water-based nanofluids. The thermal conductivity, viscosity, and surface tension were measured. The effects of particle volume fraction, temperature and particle size were discussed at the end of experiments.

2. Experimental procedure

2.1 Preparation of Al_2O_3 nanofluids

As discussed by Kwek et al. (2010), different sizes of the Al_2O_3 nanoparticles and the surfactant, Cetyltrimethylammonium Bromide (CTAB), were purchased from Sigma Nanoamor and Aldrich respectively. During the experiments, we dispersed the Al_2O_3 nanoparticles with an average diameter of 25 nm and particle density of 3.7 g/cm^3 into 100 ml of the de-ionized water to prepare the different volume concentrations (1%, 2%, 3%, 4%, and 5%). Oxide-particle volume concentrations are normally below 5% in order to maintain moderate viscosity increases. To investigate the particle size effect on the thermal conductivity and viscosity, additional four sets of nanofluids each with a constant volume concentration of 5% but with different particle sizes (10 nm, 35 nm, 80 nm and 150 nm) were prepared. Sample preparation is carried out by using a sensitive mass balance with an accuracy of 0.1 mg. The volume fraction of the powder is calculated from the weight of dry powder using the density provided by the supplier and the total volume of the suspension.

$$vol\% = \frac{m/\rho}{100ml\text{water} + m/\rho} \tag{1}$$

where m and ρ are the mass and density of the Al_2O_3 nanoparticles respectively.

The surfactant, CTAB with the density is 1.3115 g/cm^3 at volume percentage of around 0.01-0.02 can stabilize the nanofluids (Sakamoto et al., 2002). The amount of 0.01 vol % CTAB was added into the Al_2O_3 water-based nanofluids to keep the nanoparticles well dispersed in the base fluid, water.

The nanofluid was then stirred by a magnetic stirrer for 8 hours before undergoing ultrasonicfication process (Fisher Scientific Model 500) for one and a half hours. This is to ensure uniform dispersion of nanoparticles and also to prevent the nanoparticles from the aggregation in the nanofluids.

2.2 Thermal conductivity measurement of Al_2O_3 nanofluids

In the study, we adopted the THW technique for measuring thermal conductivity, as shown in Fig. 1. The setup consists of a direct current (DC) power supply, a Wheatstone bridge

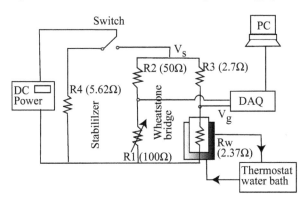

Fig. 1. Schematics of the THW setup (Kwek et al., 2010).

circuit, and a thin platinum wire surrounded by a circular nanofluid container, which is maintained by a thermostat bath. The DC power supply provides a constant voltage source to the Wheatstone bridge circuit at a constant rate to allow a uniform increment of temperature with respect to time. As the resistors used in this experiment have low values of resistance, V_s is adjusted to a value of between 0 to 2.5V. A data acquisition unit (Yokogawa Electric Corporation, DaqMaster MW100) is applied to capture the readings, recorded in a computer. The voltage supplied by the stabilizer, the voltage supplied in the Wheatstone circuit, the voltage for the platinum wire and the voltage across bridge (V_g) can be monitored during the experiments. The main experimental cell is a part of the Wheatstone bridge circuit since the wire is used as one arm of the bridge circuit. Teflon spray is used for coating a platinum (Pt) wire to act as an electric insulation because the Al_2O_3 nanofluids are electrically conductive. The Pt wire has good resistance as a function of temperature over a wide temperature range. The resistance-temperature coefficient of the Pt wire is 0.0039092 °C (Bentley, 1984). The Pt wire of 100 μm in diameter and 180 mm in length was used in the hot-wire cell whose electric resistance was measured. The dimensions of the nanofluid container are chosen to be sufficiently large to be considered as infinite in comparison with the diameter of the Pt wire. The volume and diameter of the nanofluid container are 100 ml and 30 mm respectively.

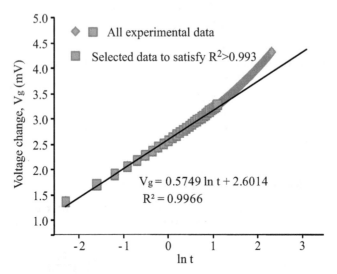

Fig. 2. V_g as a function of (ln t) with the linear fitting curve.

To investigate the effect of temperature from 15 to 55 °C on the thermal conductivities of the nanofluids, the nanofluid container was enclosed with an acrylic container connected to a thermostat bath. Different temperatures of nanofluids can be reached during the measurement process. The nanofluid temperature was monitored with a thermocouple. In the measurement of the thermal conductivity of the Al_2O_3 nanofluids, the cylindrical shaped nanofluid container was filled with 100 ml of the Al_2O_3 water-based nanofluid. The required temperature was set at the thermostat to maintain a uniform temperature in the nanofluid. Then the DC power source was switched on with the input voltage (V_s) being adjusted to 0.5 V while the switch in the circuit remained on the stabilizer resistor (R4 in Fig. 1) circuit. Thereafter, the switch was turned to the Wheatstone bridge circuit and V_g (Fig. 1) was balanced by adjusting manually the variable resistor in circuit. Once there was no voltage change, the circuit was considered as being balanced. Again, it was switched back to the stabilizer resistor circuit and input voltage V_s was then set to the desired value of 2.0 V before the switch was set back to the Wheatstone bridge circuit. The unbalanced voltage change (V_g) occurring in the hot wire was recorded for 10 seconds in the computer via a data acquisition unit. The input voltage to the circuit was also recorded for each run. This measured unbalanced voltage over the natural logarithm of time was plotted in Fig. 2 by using Equation (2) (Kwek et al., 2010). The thermal conductivity is then calculated from the slope and intersect.

$$V_g = \frac{R_3}{(R_3 + R_w)^2}(\beta R_w)\frac{V_s q}{4\pi k}\left(\ln t + \ln \frac{4\alpha}{a^2 C}\right)$$ (2)

where V_g can be obtained directly from the Wheatstone bridge circuit, V_s is the voltage supplied, R_w is the known resistance of the Pt wire, R_3 is the resistance along same branch of Wheatstone circuit, q is the heat rate per unit length, α is the thermal diffusivity of the surrounding medium, β is the resistance-temperature coefficient of the wire, k is the thermal conductivity to be determined, and $C = \exp(0.5772)$.

Figure 2 shows a sample of the unbalanced voltage (V_g) as a function of the natural logarithm of time. The best fitting with $R_2 > 0.993$ was applied to determine the thermal conductivity. The average thermal conductivity was then determined.

Before the experiments of nanofluids, the THW setup was calibrated with the de-ionized water, the procedure was as same as the experimental process for measuring the thermal conductivity of nanofluids. The calibration showed that the accuracy of the measured thermal conductivity values is in ±2% from the documental data.

2.3 Viscosity measurement of Al$_2$O$_3$ nanofluids

Fig. 3. The image of the controlled shear rate rheometer (Contraves LS 40).

As shown in Fig. 3, the controlled shear rate rheometer (Contraves LS 40) was applied to measure the viscosity of the Al$_2$O$_3$ nanofluids. The rheometer has a cup and bob geometry. The bob is connected to the spindle drive while the cup is mounted onto the rheometer. As the cup is rotated, the viscous drag of the fluid against the spindle is measured by the deflection of the torsion wire. The cup and bob geometry requires a sample volume of around 5 ml, hence, the temperature equilibrium can be achieved quickly within 5 minutes. The spindle type and speed combination would produce satisfactory results when the applied torque is up to 100% of the maximum permissible torque. In the measurement, the cup was placed onto the rheometer while the bob was inserted into the top shaft. The nanofluids were then transferred to the cup in preventing any bubbles forming. Afterwards, the bob was lowered down until it was completely inserted into the cup and immersed in the nanofluids. The lever knob was then adjusted until the bob and cup were concentric. After the measuring settings such as the minimum and maximum shear rates were set, the experiment was run. The viscosity as a function of the shear rate was plotted.

For the temperature effect, the rheological property of the nanofluids was measured by the viscometer with the thermostat, which controls temperature in Fig. 3. The viscosity measurement was started at 15 °C, and temperature was gradually increased to 55 °C at an interval of 10 °C. The nanofluid temperature was also measured by using a thermocouple. All the viscosity measurements were recorded at steady state conditions.

Before the measurement of nanofluids, the viscometer was calibrated with the de-ionized water, having an error within ± 1%.

2.4 Surface tension measurement of Al_2O_3 nanofluids

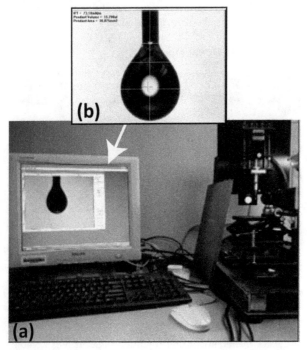

Fig. 4. Surface tension measurement for Al_2O_3 water-based nanofluids, (a) FTA 200 system; (b) a pendant droplet of the fluid for measurement.

The surface tension of the Al_2O_3 water-based nanofluids under different volume concentrations was measured with First Ten Angstroms (FTA) 200, illustrated in Fig. 4a. The precision syringe pumps (KD Scientific Inc., USA) was used to drive the Al_2O_3 water-based nanofluids to form a pendant droplet as shown in Fig. 4b. An epi-fluorescent inverted microscope with a filter set (Nikon B-2A, excitation filter for 450 - 490 nm, dichroic mirror for 505 nm and emission filter for 520 nm) was used to monitor the hanging droplet . A sensitive interline transfer CCD camera (HiSense MKII, Dantec Dynamics, Denmark) was employed for recording the droplet shape.

n the experiments, the Al_2O_3 nanofluids with a certain volume concentration were filled into the syringe, which was held at the loading platform as shown in Fig. 4a. Once a pendant nanofluid droplet was formed, the image of droplet was taken. The surface tension, the droplet volume, and the surface area were then computed.

The calibration was conducted with the de-ionized water before the surface tension of nanofluid was measured. It was found that the surface tension of pure water was 72.93 ± 1.01 mN/m at room temperature. The value is very close to the standard value at 71.97 mN/m (Vargaftik et al., 1983).

3. Results and discussion

3.1 Thermal conductivity of Al₂O₃ nanofluids

3.1.1 Effect of volume concentration on thermal conductivity

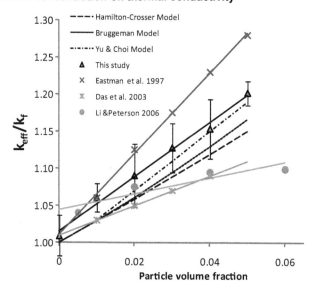

Fig. 5. Thermal conductivity enhancement as a function of volume concentrations of Al_2O_3 water-based nanofluids at 25 °C.

Each of the experimental data represents the average of six measurements at a specific concentration under room temperature. As shown in Fig. 5, the effective thermal conductivity ratio (k_{eff}/k_f) of the nanofluids is plotted as a function of nanoparticle volume fraction for a series of the Al_2O_3 nanofluids prepared from 25 nm Al_2O_3 powders and measured at 25 °C. k_{eff} is the measured thermal conductivity of the nanofluids and k_f is the thermal conductivity of pure water. Figure 5 also illustrates the data reported by Eastman et al. (1997) (33 nm), Das et al. (2003) (38.4 nm), Li & Peterson (2006) (36 nm), and the prediction from the Hamilton-Crosser model (Hamilton & Crosser, 1962), Bruggeman model (Bruggeman, 1935), and the modified model by Yu & Choi (2003). Direct quantitative comparisons are not possible in this case as the particle size used by the other researchers differs from this experimental results (25 nm). It can be noted that the previous experimental results, the predicted thermal conductivity, and the measured values in the study increase with an increase of nanoparticle concentration in a distinct linear fashion. However, the slopes are not same. From our experimental results, it is found that a small volume percentage at 1 - 5% addition of the Al_2O_3 nanoparticles in the water significantly increases the effective thermal conductivity of the Al_2O_3 water-based nanofluids by 6 to 20% respectively. If we disregard the minor differences in the particles size, clear discrepancies were found between the previous experimental data and ours on the amount of enhancement in Fig. 5. This difference may be caused by the various factors such as the different particle preparation, the particle source, or even the measurement technique. Up to now, there are no standard guidelines on the preparation of nanofluids such as the amount and type of surfactant added, the time duration for ultrasonification process, the measurement method and procedures, and the size and shape of nanoparticles in use. All these might add up to account for the difference in the experimental data.

By comparing the percentage difference in the effective thermal conductivity ratio with the measured values, our data are more consistent with the predicted values of the Yu & Choi correlation than those of the other correlations, especially at a high volume concentration where the percentage difference at 0.04 and 0.05 volume fraction is around 0.4 % and 1 % respectively. Thus, the conventional models underestimate the thermal conductivity enhancement when compared against the measured values. The reason may be that the present proposed models did not take into account the additional mechanisms such as the interfacial layer, the Brownian motion, the size and the shape of nanoparticles, and the nanoparticle aggregation. At this stage, most of these aforementioned mechanisms are neither well established nor well understood. Therefore, more experimental works are required before the concrete conclusions can be inferred from the thermal behavior of nanofluids.

3.1.2 Effect of temperature on thermal conductivity

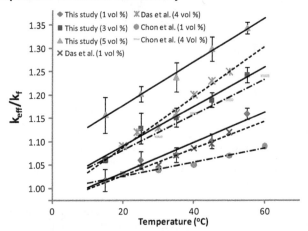

Fig. 6. Temperature dependence of thermal conductivity enhancement for the Al_2O_3 water-based nanofluids.

The effective thermal conductivity ratio (k_{eff}/k_f) is expressed with a reference of the measured value of water at the related temperatures. The measurement was made for the Al_2O_3 water-based nanofluids with the given particle concentrations at different temperatures. Figure 6 shows the enhancement of thermal conductivity of Al_2O_3 nanofluids with temperature. There is a considerable increase in the enhancement from 15 to 55 °C in the nanofluids of 1 vol %, 3 vol %, and 5 vol %. With 1 vol % particles at about 15 °C, the enhancement is only about 1.7 %, but about 16 % at 55 °C. The present measurement shows that a higher enhancement can be achieved in the nanofluid having small volume ratio of nanoparticles in the fluids at a higher temperature. The measurement of 3 vol % and 5 vol % nanofluids shown in Fig. 6 demonstrates the enhancement goes from 6 % to 24 % and 15 % to 34 % respectively as a function of temperature from 15 to 55 °C. The average rate of enhancement in these cases is higher compared with that of 1 vol % nanofluids. The increasing slope of the fitted line of the 1 vol %, 3 vol %, or 5 vol % nanofluids has a gradient of 0.003575, 0.0045 or 0.00475 respectively. Thus it can be said that the enhancement of thermal conductivity with increases of temperature depends on the concentration of nanoparticles. The above trends are also explained by the experimental results of Das et al. (2003) (38.4 nm) and Chon et al. (2005) (47 nm) in Fig. 6. From the data of Das et al., the increasing rates are 0.002 and 0.005 for 1 vol % and 4 vol %, whereas the results of Chon et al. show the increasing rates of 0.001 and 0.003 for the nanofluids at 1 vol % and 4 vol %. The increasing trends observed are quite similar.

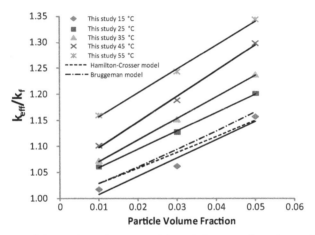

Fig. 7. Enhancement of thermal conductivity of the Al₂O₃ water-based nanofluids against particles concentration and comparison with models.

Figure 7 shows that there is a close agreement between the measured thermal conductivity and the Hamilton-Crosser and the Bruggeman models at 15 °C. However, this agreement is only at the low temperature. At higher temperature, the experiments of the Al₂O₃ water-based nanofluids disagree with the models. It is suggested that the present models cannot reflect on the effective conductivity with temperature. Das et al. (2003) stated that the main mechanism of the thermal conductivity enhancement in nanofluids can be thought as the stochastic motion of nanoparticles, and that the Brownian motion would depend on the fluid temperature. This enhancement in our experiments can be supported by the results of Das et al. (2003) and Chon et al. (2005). Their data have the maximum enhancements of 25 % and 19 % for 4 vol % at 55 °C whereas the Hamilton-Crosser model (Hamilton & Crosser, 1962) and the Bruggeman model (Bruggeman, 1935) predict only 12 % and 13 %, regardless of the temperature effect. At the low temperature, the Brownian motion was less significant. Thus the present results indicate that it is possible to have a threshold temperature at which the effective thermal conductivity of nanofluids starts deviating from that of the usual suspension and the enhancement through the stochastic motion of the particles starts dominating. The measurement of the thermal conductivity with the given concentrations at the different temperatures in Fig. 7 indicates the necessity for a better theoretical model for the entire range of temperature.

3.1.3 Effects of particle size on thermal conductivity

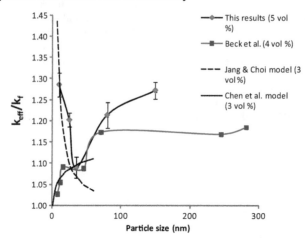

Fig. 8. Effect of diameter of nanoparticle on effective thermal conductivity of the Al_2O_3 water-based nanofluids.

As shown in Fig. 8, the experimental data in the study are compared with the predictions from the thermal conductivity model by Jang & Choi (2004), and a good agreement was found for 10 nm, 25 nm and 35 nm Al_2O_3 water-based nanofluids. Our experimental data indicate that the effective thermal conductivity decreases quickly with the decreasing size of nanoparticles from 10 nm to 35 nm, however as the nanoparticle size increases, the thermal conductivity deviates from the Jang & Choi model. As the nanoparticle diameter is reduced, the effective thermal conductivity of nanofluids becomes larger. Jang & Choi (2004) explained that this phenomenon is based on the Brownian motion, and that the smaller nanoparticles in average might produce a higher velocity of the Brownian motion in the fluid. As a result, the heat transfer by the convection would be enhanced, the effective thermal conductivity of nanofluids increases. However, if the particles approach to the micrometer size, they might not remain well suspended in the base fluid. Thus, large microparticles do not have the Brownian motion any more, and there would be no enhancement of the effective thermal conductivity. Our experimental results for the nanofluids with the 80 nm and 150 nm Al_2O_3 nanoparticles did not show a similar trend as described in the model of Jang & Choi (2004). Instead our experimental data shows that the thermal conductivity of the Al_2O_3 nanofluids increases as the particle size increases above 35 nm, similar to the data of Beck et al. (2009) above 50 nm. When the particles become larger, it can be better explained by the model of Chen (1996),

$$k_p = k_{bulk} \frac{0.75 d_p / l_p}{0.75 d_p / l_p + 1} \qquad (3)$$

where k_p, d_p, l_p and k_{bulk} are the thermal conductivity of nanoparticle, the characteristic length of nanoparticles, the mean free path of nanoparticle, and the thermal conductivity of bulk materials respectively. The correlation of Chen (1996) is built on solving the Boltzmann transport equation. The solution approaches the prediction of the Fourier law when the particle radius is much larger than the heat-carrier mean free path of the host medium, which implies that the diffusive heat transport is dominant. The model shows a trend of the thermal conductivity enhancement as the particle size increases. In sum, there may be a threshold in particle size where either the Brownian motion or the diffusive heat transport is more dominant.

3.2 Viscosity of Al$_2$O$_3$ nanofluids

3.2.1 Effect of volume concentrations on viscosity

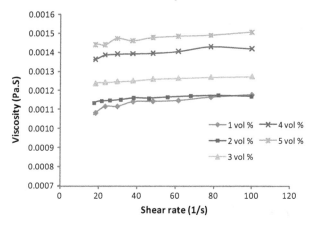

Fig. 9. Viscosity as a function of shear rate for the Al$_2$O$_3$ nanofluids at different volume concentrations.

The viscosity is illustrated in Fig. 9 as a function of the shear rate. The viscosity of the well-mixed Al$_2$O$_3$-water nanofluid is independent from the shear rate. The naofluids exhibit a Newtonian behavior. Figure 10 shows that the effective viscosity ratio increases as the volume concentrations increase. The results of Masoumi et al. (2009) (28 nm) and Nguyen et al. (2007)

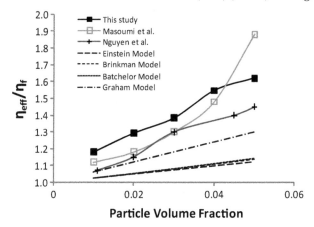

Fig. 10. Relative viscosity of the Al$_2$O$_3$ nanofluids as a function of volume concentration.

(36 nm) show a similar trend. From our experiments, the measured viscosity of the Al_2O_3 nanofluids is significantly higher than the base fluid by about 20% and 61% at 1 vol % and 5 vol % respectively. The results of Masoumi, Nguyen and ours are much higher than those of predicted values using the Einstein, Brinkman, Batchelor and Graham equations, as shown in Fig. 10. It is suggested that these equations have underestimated the nanofluid viscosities. The Einstein formula, and the others originating from it, were obtained based on the theoretical assumption of a linear fluid surrounded by the isolated particles. Such a model may worked under the situation of a liquid that contains a small number of dispersed particles. However, for higher particle concentrations the departure of these formulae from our experimental data is considerable, indicating that the linear fluid theory may be no longer appropriate to represent the nanofluids. Even the Batchelor formula, considering the Brownian effect, also performs poorly. A possible explanation is mentioned by Chandrasekar et al. (2010), the large difference may be a result of the hydrodynamic interactions between particles which become important at higher volume concentrations. Hence the conventional models cannot explain the high viscosity ratio. Noted that there are also discrepancies between our experimental results and the previous studies in Fig. 10. Although the nanofluids prepared have slight differences in the size of particles, it is inappropriate to account for such a large difference in the viscosity ratio. It is difficult to draw any conclusive remarks for such results, unless this intriguing behavior may be attributed to the various factors such as nanofluid preparation methods and how the experiment is conducted.

3.2.2 Effect of temperature on viscosity

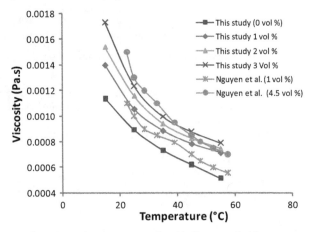

Fig. 11. Viscosity as a function of temperature for Al_2O_3 nanofluids.

The viscosities were measured for the nanofluids as a function of temperature. The viscosity under the particle volume fraction ranging at 1% , 2%, and 3% from 15 to 55 °C is shown in Fig. 11, the nanofluid viscosity decreases with an increase in temperature. The increasing temperature would weaken the inter-particle and inter-molecular adhesion forces. For all the nanofluids measured, the temperature gradient of viscosity is generally steeper at the temperature from 15 to 30 °C. Such the viscosity gradient is particularly more pronounced as the particle volume concentration increases. This observation is supported by Nguyen et al. (2007) results if we compare the gradient from 15 to 30 °C at 1 vol % and 4.5 vol %. The results suggest that the temperature effect on the particle suspension properties may be different for high particle fractions and for low ones. With an increase of temperature, the measured viscosity data have shown a gentle decrease with an increase of temperature. In our experiments, we have attempted to measure viscosity at the temperature higher than 55 °C, but a critical temperature has been observed, above the temperature, an 'erratic' increase of nanofluid viscosity was observed. The phenomenon may be resulted from the fast evaporation of nanofluids in the related small volume at a relative high temperature. Another possibility is that beyond the critical temperature, the surfactant might be broken down and accordingly the performance was considerably reduced or even destroyed, affecting the suspension capabilities. Thus, the particles have a tendency to form aggregation, resulting in the observed unpredictable increase of the nanofluid viscosity.

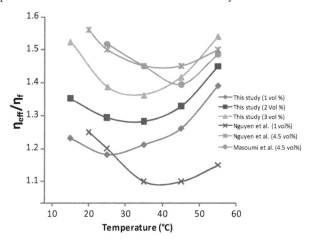

Fig. 12. Relative viscosity as a function of temperature for various concentrations of the Al₂O₃-water nanofluids.

As known, the water viscosity decreases with an increase of temperature. The viscosity values of the different concentration nanofluids measured from 15 to 55 °C are compared with a reference of the viscosity of water at these temperatures. As seen from Fig. 12, the effective viscosity under the different volume concentrations shows similar trends. For a given nanofluid and a particle fraction, the effective viscosity decreases at first and starts to increase at a certain temperature. This value implies that there should be an optimum temperature whereby when temperature increases, the decrease in viscosity is not effective. This observation can be substantiated by Nguyen et al. (2007) and Masoumi et al. (2009).

3.2.3 Effect of particle size on viscosity

From Fig. 13, our experimental results show that as the particle sizes increase, the effective viscosity decreases significantly and reaches an almost constant value at the end. This trend is similar to the results of other researchers shown in Fig. 13 except for particle size greater than 100 nm. Timofeeva et al. (2007) suggested the small particle size can form larger aggregates. The Krieger model (Krieger & Dougherty, 1959) can be used to estimate the relative viscosity between a nanofluid (nf) and its base fluid (f),

$$\frac{\eta_{nf}}{\eta_f} = (1 - \frac{\phi_a}{\phi_m})^{-2.5\phi_m} \qquad (4)$$

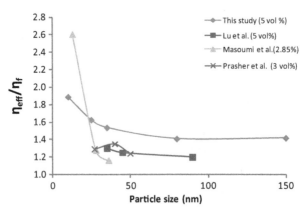

Fig. 13. Relative viscosity as a function of diameter of the Al_2O_3 nanoparticles in the base fluids.

where 2.5 is the intrinsic viscosity of spherical particles, ϕ_a is the volume fraction of aggregates, ϕ_m is the volume fraction of densely packed spheres and the volume fraction of aggregates is expressed as $\phi_a = \phi(\frac{d_a}{d})^{3-d_f}$, in which d_a is the diameter of aggregates, d is the nominal diameter of particle, d_f is the fractal dimension of the aggregates, ϕ is the volume fraction of the well-dispersed individual particles. For well-dispersed individual particles, ϕ_a is equal to ϕ, and the Krieger model reduces to the Einstein model. This is a very ideal case where there is zero aggregation. However, none of the researches is able to obey fully the Einstein model until now. The reason may be that it is unlikely to eliminate the aggregation completely (Duan et al., 2011). When nanoparticle size increases, the magnitude of $\frac{d_a}{d}$ decreases, thus the volume fraction of the aggregates decreases and relative viscosity ratio decreases. In addition, due to aggregation, the shape of the aggregate is no longer spherical. Theoretically, Einstein obtained the intrinsic viscosity at 2.5 for spherical particles, however the intrinsic viscosity would be greater than 2.5 for the other shapes (Rubio-hernandez et al., 2006) as the aggregate shape becomes disordered. This can also account for the increase of viscosity ratio as the particle diameter decreases.

Slight aggregation is likely to remain in our nanofluids measured just after preparation since the measurements are made for different particle sizes at a constant 5 % volume concentration, which is considered high. Based on Equation (4), the viscosity ratio would be higher after aggregation.

3.3 Surface tension measurements of nanofluid

Figure 14 shows the surface tension as a function of the volume concentration. The results demonstrate that the surface tensions of the Al_2O_3 water-based nanofluids are significantly lower than those of the base fluid, pure water. At each point, the error bars are too small to be observed. However, as the volume concentration increases, the surface tension remains almost unchanged in the Al_2O_3 nanofluids. Hence we can deduce that particle volume concentration does not have a major effect on the surface tension of the nanofluids. The experimental results of Golubovic et al. (2009) and Kim et al. (2007) have shown that the surface tensions of the Al_2O_3 nanofluids without surfactant is independent on concentration and has the same values as that of pure water. In our prepared nanofluids, the surfactant, CTAB was added to obtain a well-dispersed suspension. The addition of a small amount of surfactant into the liquid reduced the surface tension (Binks, 2002; Bresme & Faraudo, 2007).

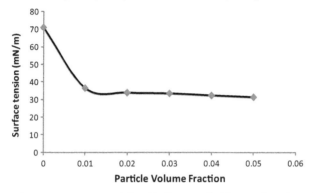

Fig. 14. Surface tension measurement of the Al_2O_3 nanofluids as a function of the volume concentration.

4. Conclusion

The thermal conductivity, viscosity, and surface tension of the Al_2O_3 water-based nanofluids were measured. It is found the thermal conductivity increases significantly with the nanoparticle volume fraction. With an increase of temperature, the thermal conductivity increases for a certain volume concentration of nanofluids, but the viscosity decreases. The size of nanoparticle also influences the thermal conductivity of nanofluids. It is indicated that existing classical models cannot explain the observed enhanced thermal conductivity in the nanofluids. Similarly, the viscosity increases as the concentration increases at room temperature. At the volume concentrations of 5%, the viscosity has an increment of 60%. The effect of particle sizes on the viscosity is limited. The addition of surfactant is believed to be the reason behind the decrease in surface tension in comparison with the base fluid. The significant deviation between the experimental results and the existing theoretical models is still unaccounted for. More comprehensive models therefore need to be developed. Particles sizes, particle dispersions, clustering, and temperature should be taken into account in the model development for nanofluids. Hence, to reach universal models for the thermal properties, more complete experiments involving a wide range of nanoparticle sizes would be conducted in future.

5. Acknowledgments

The research mainly depends on the experimental work of Mr. D. Kwok under the support of NTU-SUG and AcRF Tier 1 funding. The author would like to thank Profs. Kai Choong Leong and Charles Yang for their generosity in sharing their HWT and surface tension equipment. The author would also like thank to Dr. Liwen Jin for sharing his knowledge on the thermal conductivity measurement.

6. References

Batchelor, G.K. (1977). The effect of Brownian motion on the bulk stress in the suspension of spherical particles. *J. Fluid Mech.*, 83, 97-117.

Beck, M.P.; Yuan, Y.; Warrier, P. & Teja, A.S. (2009). The effect of particle size on the thermal conductivity of alumina nanofluids. *J. Nanopart. Res.*, 11, 1129-1136.

Bruggeman, D.A.G. (1935). Calculation of various physics constants in heterogenous substances. I. Dielectricity constants and conductivity of mixed bodies from isotropic substances. *Ann. Phys. (Paris)*, 24, 636-664.

Bentley, J.P. (1984). Temperature sensor characteristics and measurement system design. *J. Phys. E*, 17, 430-439.

Bhattacharya, P.; Saha, S.K.; Yadav, A.; Phelan, P.E. & Prasher, R.S. (2004). Brownian dynamics simulation to determine the effective thermal conductivity of nanofluids. *J. Appl. Phys.*, 95, 6492-6494.

Binks, B. (2002). Particles as surfactants - similarities and differences. *Curr. Opin. Colloid Interface Sci.*, 7, 21-41.

Bresme, F. & Faraudo, J. (2007). Particles as surfactants - similarities and differences. *J. Phys.: Condens. Matter*, 19, 375110.

Brinkman, H.C. (1952). The viscosity of concentrated suspensions and solution. *J. Chem. Phys.*, 20, 571.

Chandrasekar, M.; Suresh, S. & Chandra Bose, A. (2010). Experimental investigations and theoretical determination of thermal conductivity and viscosity of Al2O3/water nanofluid. *Exp. Therm. Fluid Sci.*, 34, 210-216.

Challoner, A.R. & Powell. R.W. (1956). Thermal conductivity of liquids: new determinations for seven liquids and appraisal of existing values. *Proc. R. Soc. Lond.*, 238, 90-106.

Chen, G. (1996). Nonlocal and nonequilibrium heat conduction in the vicinity of nanoparticles. *ASME J. Heat Transfer*, 118, 539-545.

Choi, S.U.S. (1995). Enhancing thermal conductivity of fluids with nanoparticles, In: *Developments and Applications of Non-Newtonian Flows*, Sinfiner, D.A. & Wang, H.P., PP 99-105, ASME, FED-vol.231 MD-vol.66, USA.

Chon, C.H.; Kihm, K.D.; Lee, S.P. & Choi, S.U.S. (2005). Empirical correlation finding the role of temperature and particle size for nanofluid (Al2O3) thermal conductivity enhancement. *Appl. Phys. Lett.*, 87, 1-3.

Czarnetzki, W. & Roetzel W. (1995). Temperature oscillation techniques for simultaneous measurement of thermal diffusivity and conductivity. *Int. J. Thermophys.*, 16, 413-422.

Das, S.K.; Putta, N.; Thiesen, P. & Roetzel W. (2003). Temperature dependence of thermal conductivity enhancement for nanofluids. *J. Heat Transfer*, 125, 567-574.

Duan, F; Kwek D & Crivoi A. (2011). Viscosity affected by nanoparticle aggregation in Al2O3-water nanofluids. *Nanoscale Res. Lett.*, 6, 248.

Eastman, J.A.; Choi, S.U.S.; Li, S.; Thompson, L.J. & Lee, S. (1997). Enhanced thermal conductivity through the development of nanofluids. *Mater. Res. Soc. Symp. Proc.*, 457, 3-11.

Eastman, J.A.; Choi, S.U.S.; Li, S.; Yu, W. & Thompson, L.J. (2001). Anomalously increased effective thermal conductivities of ethylene glycol-based nanofluids containing copper nanoparticles. *Appl. Phys. Lett.*, 78, 718-720.

Einstein, A. (1906). A new determination of the molecular dimensions. *Ann. Phys.*, 19, 289-306.

Frankel, N.A. & Acrivos, A. (1967). On the viscosity of a concentrated suspension of solid spheres. *Chem. Eng. Sci.*, 22, 847-853.

Golubovic, M.N.; Madhawa Hettiarachchi, H.D.; Worek, W.M. & Minkowycz, W.J. (2009). Nanofluids and critical heat flux, experimental and analytical study. *Appl. Therm. Eng.*, 29, 1281-1288.

Graham, A.L. (1981). On the viscosity of suspension of solid spheres. *Appl. Sci. Res.*, 37, 275-286.

Hamilton, R.L. & Crosser, O.K. (1962). Thermal conductivity of heterogeneous two component systems. *Ind. Eng. Chem. Fundam.*, 1, 187-191.

Jang, S.P. & Choi, S.U.S. (2004). Role of Brownian motion in the enhanced thermal conductivity of nanofluids. *Appl. Phys. Lett.*, 84, 4316-4318.

Keblinski, P.; Eastman, J.A. & Cahill, D.G. (2005). Nanofluids for thermal transport. *Mater Today*, 8, 36-44.

Keblinski, P.; Phillpot, S.R.; Choi, S.U.S. & Eastman, J.A. (2002). Mechanisms of heat flow in suspensions of nano-sized particles (nanofluids). *Int. J. Heat Mass Transfer*, 45, 855-863.

Kestin, J. & Wakeham W.A. (1978). A contribution to the theory of the transient hot-wire technique for thermal conductivity measurements. *Physica A*, 92, 102-116.

Kim, J.; Kang, Y.T. & Choi, C.K. (2004). Analysis of convective instability and heat transfer characteristics of nanofluids. *Phys. Fluids*, 16, 2395-2401.

Kim, S.J.; Bang, I.C.; Buongiorno, J. & Hu, L.W. (2007). Surface wettability change during pool boiling of nanofluids and its effect on critical heat flux. *Int. J. Heat Mass Transfer*, 50 (2007) 4105-4116.

Koo, J. & Kleinstreuer, C. (2004). A new thermal conductivity model for nanofluids. *J. Nanopart. Res.*, 6, 577-588.

Krieger, I.M. & Dougherty, T.J. (1959). A mechanism for non-newtonian flow in suspensions of rigid spheres. *Trans. Soc. Rheology*, 3, 137-152.

Kumar, D.H. Patel, H.E.; Kumar, V.R.R.; Sundararajan, T.; Pradeep, T. & Das, S.K. (2004). Model for heat conduction in nanofluids. *Phys. Rev. Lett.*, 93, 1-4.

Kumar, R. & Milanova, D.(2009). Effect of surface tension on nanotube nanofluids. *Appl. Phys. Lett.*, 94, 073107.

Kwek, D.; Crivoi, A. & Duan, F. (2010). Effects of temperature and particle size on the thermal property measurements of Al2O3-water nanofluids. *J. Chem. Eng. Data*, 55, 5690-5695.

Lee, S.; Choi, S.U.S.; Li, S. & Eastman, J.A. (1999). Measuring thermal conductivity of fluids containing oxide nanoparticles. *J. Heat Transfer*, 121, 280-289.

Li, C.H. & Peterson G.P.(2006). Experimental investigation of temperature and volume fraction variations on the effective thermal conductivity of nanoparticle suspensions (nanofluids). *J. Appl. Phys.*, 99, 1-8.

Li, X.; Zhu, D. & Wang X. (2007). Evaluation on dispersion behaviour of the aqueous copper nano-suspensions. *J. Colloid Interface Sci.*, 310, 456-463.

Li, Y.; Zhou J.; Tung, S.; Schneider, E. & Xi, S. (2009). A review on development of nanofluid preparation and characterization. *Powder Technol.*, 196, 89-101.

Lundgren, T.S. (1972). Slow flow through stationary random beds and suspensions of spheres. *J. Fluid Mech.*, 51, 273-299.

Maxwell, J.C. (1892). *A Treatise on Electricity and Magnetism, 3rd ed.*, Oxford University Press, London.

Mandelbrot; B.B. (1982). *The Fractal Geometry of Nature*, W.H. Freeman Press, San Francisco.

Masoumi, N.; Sohrabi, N. & Behzadmehr, A. (2009). A new model for calculating the effective viscosity of nanofluids. *J. Phys. D: Appl. Phys.*, 42, 055501.

Nagasaka, Y. & Nagashima A. (1981). Absolute measurement of the thermal conductivity of electrically conducting liquids by the transient hot-wire method. *J. Phys. E*, 14, 1435-1440.

Nguyen, C.T.; Desgranges, F.; Roy, G.; Galanis, N.; Mare, T.; Boucher, S. & Angue Mintsa, H. (2007). Temperature and particle-size dependent viscosity data for water-based nanofluids - hysteresis phenomenon. *Int. J. Heat Fluid Flow*, 28, 1492-1506.

Pantzali, M.N.; Kanaris, A.G.; Antoniadis, K.D.; Mouza, A.A. & Paras, S.V. (2009). Effect of nanofluids on the performance of a miniature plate heat exchanger with modulated surface. *Int. J. Heat Fluid Flow*, 30, 691-699.

Prasher, R.; Bhattacharya, P.; Phelan, P.E. & Das, S.K. (2005). Thermal conductivity of nanoscale colloidal solutions (nanofluids). *Phys. Rev. Lett.*, 94, 025901.

Roetzel, W.; Prinzen, S. & Xuan Y. (1990). Measurement of thermal diffusivity using temperature oscillations, In: *Measurement of conductivity 21*, Cremers C.J. & Fine, H.A., Ed., 201-207, Plenum Press, New York.

Rubio-hernandez, F.J.; Ayucar-Rubio, M.F.; Velazquez-Navarro J.F. & Galindo-Rosales, F.J. (2006). Intrinsic viscosity of SiO2, Al2O3 and TiO2 aqueous suspensions. *J. Colloid Interface Sci.*, 298, 967-972.

Sakamoto, M.; Kanda, Y.; Miyahara, M. & Higashitani, K. (2002). Origin of long-range attractive force between surfaces hydrophobized by surfactant adsorption. *Langmuir*, 18, 5713-5719.

Schwartz, L.; Garboczi, E. & Bentz, D. (1995). Interfacial transport in porous media: application to dc electrical conductivity of mortars. *J. Appl. Phys.*, 78, 5898-5908.

Timofeeva, E.V.; Gavrilov, A.N.; McCloskey, J.M. & Tolmachev, Y.V. (2007). Thermal conductivity and particle agglomeration in alumina nanofluids: Experiment and theory. *Phys. Rev. E*, 76, 061203.

Vafaei, S.; Purkayastha, A.; Jain, A.; Ramanath, G. & Borca-Tasciuc, T. (2009). The effect of nanoparticles on the liquid-gas surface tension of Bi2Te3 nanofluids. *Nanotechnology*, 20, 185702.

Vargaftik, N.B.; Volkov, B.N. & Voljak, L.D. (1983). International tables of the surface tension of water. *Phys. Chem. Ref. Data*, 12, 817.

Wang, B.X. Zhou, L.P. & Peng, X.F. (2003). A fractal model for predicting the eeffective thermal conductivity of liquid with suspension of nnanoparticles. *Int. J. Heat Mass Transfer*, 46, 2665-2672.

Wang, X.; Xu, X. & Choi, S.U.S. (1999). Thermal conductivity of nanoparticle - fluid mixture. *J. Thermophys. Heat Tr.*, 13, 474-480.

Wasan, D.T. & Nikolov, A.D. (2003). Spreading of nanofluids on solids. *Nature*, 423, 156-159.

Xie, H.; Fujii, M. & Zhang, X. (2005). Effect of interfacial nanolayer on the effective thermal conductivity of nanoparticle-fluid mixture. *Int. J. Heat Mass Transfer*, 48, 2926-2932.

Xuan Y. & Li, Q. (2000). Heat transfer enhancement of nanofluids. *Int. J. Heat Fluid Flow*, 21, 58-64.

Xuan, Y.; Li, Q. & Hu, W. (2003). Aggregation structure and thermal conductivity of nanofluids. *AICHE J.*, 49, 1038-1043.

Xue, Q. (2003). Model for effective thermal conductivity of nanofluids. *Phys. Lett. A*, 307, 313-317.

Xue, Q. & Xu, W.-M. (2005). A model of thermal conductivity of nanofluids with interfacial shells. *Mater. Chem. Phys.*, 90, 298-301.

Yang, B. & Han, Z.H. (2006). Temperature-dependent thermal conductivity of nanorod based nanofluids. *Appl. Phys. Lett.*, 89, 1-3.

Yu, W. & Choi, S.U.S. (2003). The role of interfacial layers in the enhanced thermal conductivity of nanofluids: a renovated Maxwell model. *J. Nanopart. Res.*, 5, 167-171.

Yu, W. & Choi, S.U.S. (2004). The role of interfacial layers in the enhanced thermal conductivity of nanofluids: a renovated Hamilton-Crosser model. *J. Nanopart. Res.*, 6, 355-361.

2

Thermal Effects on the Ferromagnetic Resonance in Polymer Composites with Magnetic Nanoparticles Fillers

Mirosław R. Dudek[1], Nikos Guskos[2,3] and Marcin Kośmider[1]
[1]*Institute of Physics, University of Zielona Góra, Zielona Góra*
[2]*Solid State Physics Section, Department of Physics,*
University of Athens, Panepistimiopolis
[3]*Institute of Physics,West Pomeranian University of Technology, Szczecin*
[1,3]*Poland*
[2]*Greece*

1. Introduction

Magnetic nanopowders placed in the nonmagnetic polymer matrices become a new type of smart materials which combine mechanical properties of temperature responsive polymer matrix and magnetic response of nanoparticles. These properties are used in some biotechnological and medical applications like hyperthermia treatment, nanocolloids, magnetic nanocapsules for drug targeting, magnetic resonance imaging (MRI), intracellular manipulation etc. (e.g. (Gao & Xu, 2009; Liu et al., 2009)), in the processes of mechanical and electrical micropower generation, in nanoelectromechanical systems as MEMS/NEMS devices (e.g. (Zahn, 2001)), electromagnetic interference suppression (Wilson et al., 2004). Recently, the unusual polymer/magnetic nanoparticles systems with a negative Poisson's ratio (e.g. ferrogels Dudek & Wojciechowski (2008); Wood & Camp (2011)) have begun to be studied. They belong to the so-called auxetic materials Evans et al. (1991); Lakes (1987); Smith & Wojciechowski (2008).

Ferromagnetic resonance experiment (FMR) (Vleck, 1950) is one of the basic tools to study the magnetic properties of magnetic agglomerates in viscoelastic nonmagnetic polymer matrix. As a particular example, we consider the FMR experiment with the γ-Fe_2O_3 (maghemite) ferrimagnetic nanoparticles embedded in a multiblock poly(ether-ester) copolymer nonmagnetic matrix which has been studied both experimentally (Guskos et al., 2006; 2008) and theoretically (Dudek et al., 2010). However, the obtained results are general and applicable to other nanoparticles and other viscous materials. Note that in medical applications magnetic iron oxides are used due to their low toxicity to human. Their saturation magnetization is practically equal to the bulk value at high temperatures, with negligible coercivity and no exchange bias below the blocking temperature. These properties of the iron oxide magnetic nanoparticles suggest nearly perfect nanocrystals without significant structural disorder (Dutta et al., 2004).

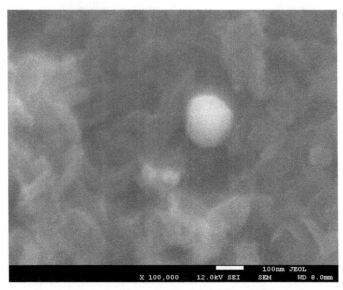

Fig. 1. SEM picture: an example of stucking a single magnetic nanoparticle in a pore - here, a carbon coated nickel nanoparticle in the porous sodium borosilicate glass.

The peculiar feature of the synthesized PEN-block-PTMO copolymer is that the magnetic fillers form agglomerates numbering from several to tens of nanoparticles. In the agglomerates the interparticle dipole - dipole magnetic interaction becomes important as well as the interaction of the magnetic nanoparticles with a non-magnetic matrix. Although the agglomerates are uniformly dispersed in the matrix their FMR spectra show additional peaks in low temperatures which originate from the orientational anisotropy of the frozen polymer blocks. The orientational dependence of the FMR spectra has been found earlier by Owens (Owens, 2003) for a colloidal suspension of γ-Fe$_2$O$_3$ nanoparticles which have been solidified in a static magnetic field (dc magnetic field). Similar observation has been found theoretically in a recent paper by Sukhov et al. (Sukhov et al., 2008). There is very instructive discussion on the shape of the ferromagnetic resonance spectra for the ensemble of the randomly distributed magnetic anisotropy axes as well as the discussion of the dependence of these spectra on temperature in terms of a stochastic model. The model is restricted to the case when the orientation of each anisotropy axis is frozen during computer simulation but it shares many features common with the experimental results, like the broadening of the FMR signal for the randomly distributed magnetic anisotropy axes as compared to the magnetic nanoparticles which all have the same orientation of the magnetic anisotropy. In paper (Dudek et al., 2008) it has been shown directly that blocking the rotational freedom of the magnetic nanoparticles, e.g. when the nanoparticles are stuck in the pores as it is suggested in Fig. 1, can produce additional resonance peaks in the FMR spectrum. In the latter case stochastic equations were used both for the magnetic nanoparticles magnetization and the rotational oscillations of the magnetic nanoparticles as a whole. The influence of the magnetic anisotropy orientation and temperature on the FMR spectra of magnetic agglomerates in polymer matrix was discussed in (Dudek et al., 2010). The most important property of the FMR spectrum depending on temperature will be discussed in the sections below.

2. Modeling ferromagnetic resonance experiment

Theoretical basis of ferromagnetic resonance can be found in the paper by van Vleck (Vleck, 1950) in which a magnetic resonance condition (Kittel's formula) for ferromagnetic materials has been derived with the help of a simple quantum model. It has been noted in the paper the importance of the effect of magnetic anisotropy on the resonance frequency. In our considerations we restrict to the case when an uniaxial anisotropy is the dominating magnetic anisotropy of the magnetic nanoparticles (Shliomis, 1975). Then, the term magnetic anisotropy axis is substituted for the easy axis of magnetization. The magnetization of magnetic nanoparticles changes after an external magnetic field is switched on and there are two mechanisms of this change: the reorientation process of magnetic nanoparticle as a whole (Brownian motion) and the Néel relaxation process of the magnetization itself. A magnetic nanoparticle, and by this the magnetic anisotropy axis, can rotate freely in a liquid carrier, but not in the case when the nanoparticle is part of a large agglomerate or its surrounding is a solid phase. The dominant interparticle interactions in the agglomerate are dipole interactions unless the nanoparticles do not form dense agglomerates where exchange interactions become important. So if we take into account the FMR experiment and we consider the agglomerate consisting of N single-domain magnetic nanoparticles, each of them experiences an effective magnetic field $\vec{H}_{\text{eff},i}$ of the form (Füzi, 2006):

$$\vec{H}_{\text{eff},i} = \vec{H}_{\text{dc}} + \vec{H}_{\text{ac}} + \frac{H_a}{|\vec{M}_i|}(\vec{M}_i \cdot \vec{n}_i)\vec{n}_i + \vec{H}_{i,\text{dipole}} \tag{1}$$

where $\vec{H}_{\text{dc}}$ is the external direct current (dc) magnetic field, $\vec{H}_{\text{ac}} = \vec{H}_{\text{ac}}^0 \cos(2\pi f t)$ is the external alternating current (ac) magnetic field of frequency f, $\vec{H}_{i,\text{dipole}}$ represents dipolar magnetic field produced by the magnetic nanoparticles of the agglomerate

$$\vec{H}_{i,\text{dipole}} = -\frac{1}{4\pi} \sum_{j=1,j\neq i}^{N} \left(\frac{\vec{M}_j}{r_{ij}^3} - 3\frac{(\vec{M}_j \cdot \vec{r}_{ji})\vec{r}_{ji}}{r_{ji}^5} \right), \tag{2}$$

$\vec{M}_i$ denotes magnetization of the nanoparticle i ($M = M_s V$ for the nanoparticle of volume V and saturation magnetization M_s), r_{ij} is the distance between nanoparticles i and j, H_a represents the magnetic anisotropy field which is defined as

$$H_a = \frac{2K_a}{\mu_0 M_s} \tag{3}$$

where K_a is magnetic anisotropy constant, and μ_0 is constant of permeability. The symbol $\vec{n}_i$ in Eq. (1) is a unit vector along the magnetic anisotropy direction of the nanoparticle i with the components

$$n_{x,i} = \sin(\varphi_i)\cos(\theta_i), \tag{4}$$
$$n_{y,i} = \sin(\varphi_i)\sin(\theta_i), \tag{5}$$
$$n_{z,i} = \cos(\varphi_i). \tag{6}$$

and φ_i and θ_i are the angles vector $\vec{n}_i$ makes with the z-axis and x-axis, respectively.

A rotating external magnetic field $\vec{H}_{ac}$ is transverse to $\vec{H}_{dc}$. We assume that the external dc magnetic field is oriented in the z-direction and the external ac magnetic field in the x-direction (Jung et al., 2002). Then, for the effective magnetic field defined in Eq. (1) the ferromagnetic resonance condition can be expressed as follows:

$$f = \frac{\gamma}{2\pi} H_{eff} \qquad (7)$$

where $\gamma = 2.21 \times 10^5 s^{-1}(A/m)^{-1}$ denotes the gyromagnetic ratio. In practice, the spectrometers EPR/FMR are built for one value of frequency f and then the dc magnetic field becomes a parameter to be changed to get the resonance condition. In our case the ac magnetic field frequency $f = 9.37$ GHz. Note that even in the case of a single magnetic nanoparticle (N=1) its resonance frequency strongly depends on the orientation of the magnetic anisotropy axis with respect to the dc magnetic field direction.

It turns out that the magnetic nanoparticle's magnetization dynamics can be modeled with the help of the classical spin model which represents stochastic version of the Landau-Lifshitz equation ((Gilbert, 1955; Landau & Lifshitz, 1953)) :

$$\frac{d\vec{M}_i}{dt} = -\gamma \vec{M}_i \times [\vec{H}_{eff,i} + \vec{B}_i] - \alpha \frac{\gamma}{M_s V} \vec{M}_i \times (\vec{M}_i \times [\vec{H}_{eff,i} + \vec{B}_i]), \qquad (8)$$

where $i = 1, 2, \ldots, N$, and α denotes the damping constant. The symbol $\vec{B}_i$ represents the white-noise field fluctuations (e.g. (Jönsson, 2003), (Usadel, 2006)). Then the thermal averages of $\vec{B}_i = (B_{x,i}, B_{y,i}, B_{z,i})$ fulfill the relations:

$$\langle B_{q,i}(t) \rangle = 0, \quad q = x, y, z, \qquad (9)$$

$$\langle B_{q,i}(t) B_{p,i}(t') \rangle = \frac{2\alpha k_B T}{\gamma M_s V} \delta_{q,p} \delta(t - t'), \quad p = x, y, z. \qquad (10)$$

The magnetic properties of the magnetic nanoparticles can be described with the help of the solutions $\vec{M}_i(t)$ of this set of equations. They strongly depend on the magnetic anisotropy axis orientation. In particular, the shape of the magnetic hysteresis loop can change from the almost square like to the case when it vanishes depending on the orientation of the external dc magnetic field with respect to the orientation of the magnetic anisotropy axis. This can be seen in Fig. 2 and Fig. 3 in which the magnetic hysteresis loops are presented for the z-component of the nanoparticle's magnetization in the case when the anisotropy axis orientation oscillations are close to the z-direction (parallel to dc magnetic field) and close to the x-direction (transverse to dc magnetic field), respectively. The hysteresis loops, which are shown in the figures, result from the computer simulation of a simplified model of a carbon coated magnetic nanoparticle where the coating is represented by C_{60} molecule. In the model, the magnetic nanoparticle is represented by magnetic anisotropy axis and its magnetization follows the Landau-Lifshitz equation (Eq. (8)). The carbon atoms in C_{60} molecule vibrate according to the molecular dynamics method. The rotational oscillations of a fullerene (and by this the magnetic anisotropy axis rotations) are harmonically bonded to the z-direction and x-direction, respectively, with a given spring constant. The latter means that the magnetic nanoparticle cannot rotate freely. The larger temperature is the larger the rotational oscillations are. Besides the spring force the anisotropy axis experiences the

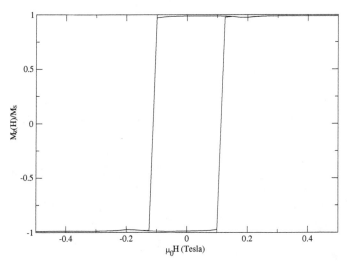

Fig. 2. Magnetic hysteresis loop of the z-component of the magnetic nanoparticle's magnetization in the case when the orientation of the magnetic anisotropy axis undergoes the small oscillations around the z-direction (parallel to the external dc magnetic field). Some parameters of the computer simulation: demagnetizing factor in shape anisotropy $D = 0.15$ $(K_a = \mu_0(1 - 3D)M_s^2/4)$, $\alpha = 0.066$, $R = 2$nm, $M_s = 450kA/m$, T=10K.

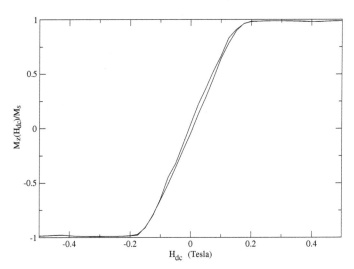

Fig. 3. The same as in Fig. 2 but the magnetic anisotropy axis orientation oscillates around the x-direction (transverse to the external dc magnetic field).

magnetic torque which is represented by two opposite point forces (Dudek et al., 2010) applied to the anisotropy axis with the strength

$$| \vec{F} | = \frac{1}{R} | K_a V \sin(2\Psi) |$$ (11)

where R is the fullerene's radius, and the greater the angle Ψ between the easy axis, represented by vector $\vec{n}$, and magnetization $\vec{M}$, the greater the magnetic torque. The model of the magnetic nanoparticles used in the computer simulations has been shown in Fig. 4 in the case when the magnetic anisotropy axis is, respectively, parallel and perpendicular to the external dc magnetic field.

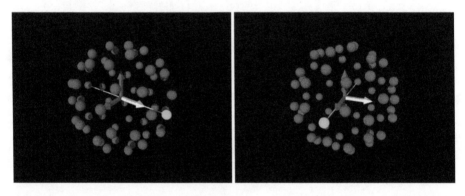

Fig. 4. Model of carbon coated magnetic nanoparticle in the case when the coating is represented by C_{60} molecule. The magnetic nanoparticle is represented by the anisotropy axis which, in the model, passes through the center of the fulleren and the carbon atom painted green, in the figure. The magnetic nanoparticle was not drawn with clarity reasons. The cartesian coordinate system has been plotted and the blue axis represents the z-direction which is the direction of the external dc magnetic field and the white axis represents the x-direction.

The same features of the magnetic hysteresis loops as those presented in Fig. 2 and Fig. 3 can be observed in magnetic nanowires, e.g. (Sorop et al., 2004), where the similar relationship between the shape anisotropy of a nanowire and the external dc magnetic is observed. In low temperatures, much below 100K, the large agglomerates of the γ-Fe_2O_3 (maghemite) ferrimagnetic nanoparticles embedded in a multiblock poly(ether-ester) copolymer nonmagnetic matrix Guskos et al. (2006; 2008) are practically frozen into the matrix with a random orientation of the magnetic anisotropy axes. Then, the observed magnetic hysteresis loop represents the averaged one which is approximately a mixture of the cases discussed in Fig. 2 and Fig. 3. It is worth to add that in high temperatures, where the block copolymer is dissolved, the magnetic properties resemble the properties of ferrofluids and there is no observed magnetic hysteresis loop.

Much more information on the magnetic properties of the polymers filled with nanoparticles can be get from the analyses of the absorption lines in FMR experiment. They can be represented by the imaginary part of the dynamic magnetic susceptibility

$$\chi = \chi' - i\chi'',$$ (12)

where χ'' denotes total hysteresis losses per volume of magnetic nanoparticle through a cycle of the magnetization. For the chosen magnetic fields $H_z = H_{dc}$ and $H_x = H_{ac}$ the components of the complex ac susceptibility (Eq. (12)) can be calculated by performing the Fourier transform on the time averaged x-component of the magnetization, i.e.,

$$\chi = \frac{1}{\tau H_{ac}^0} \int_0^\tau dt M_x(t) e^{-i2\pi ft}, \tag{13}$$

where $\tau = 1/f$. In the case of theoretical modeling, the values $M_x(t)$ can be obtained from the Landau-Lifshitz equation (Eq. 8). In real FMR experiments, the absorption lines derivatives, $d\chi''/dH_{dc}$ (the derivative of the out-of-phase susceptibility) are measured instead of direct measuring χ''. In Fig. 5 there are presented the absorption lines derivatives obtained for the

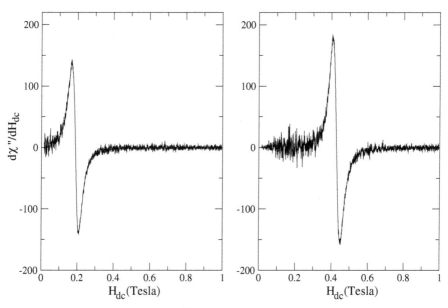

Fig. 5. Absorption lines derivatives, $d\chi''/dH_{dc}$ resulting from the computer simulations in the case when the magnetic anisotropy axis orientation oscillates around the direction parallel and transverse to the external dc magnetic field. The parameters of the computer simulation are the same as in Fig. 2.

model of the carbon coated nanoparticles shown in Fig. 4 in the case when their magnetic anisotropy axis oscillations are controlled by the harmonic forces applied to the ends of the axis and the forces are coupled with the z-direcion and x-direction, respectively. Note that if the magnetic anisotropy axis is linked to the direction which is tranverse to the direction of H_{dc} then the corresponding resonance magnetic field H_r becomes shifted to higher values of H_{dc} compared with the case when it is linked to the direction which is parallel. At the value of $H_{dc} = H_r$ the dynamic susceptibility χ'' takes its maximum value. It is worth emphasizing that the energy absorbed by the magnetic nanoparticles from the external AC magnetic field is proportional to χ''. After the energy is converted into heat there is observed an increase ΔT of temperature which can be estimated at each cycle of the applied AC magnetic field with a given frequency $\omega = 2\pi f$ as follows (Sellmyer & Skomski, 2006):

$$\Delta T = \frac{\mu_0 V H_0^2}{c \, m_{\text{ferro}}} f \chi'' \tag{14}$$

where c is the average specific heat of carbon and magnetic nanoparticle $c = (c_{\text{carbon}} + c_{\text{ferro}})/2$, m_{ferro} represents mass of magnetic nanoparticle. The remote heating of the magnetic nanoparticles can be important in viscous materials in low temperatures for reorientation processes among the magnetic nanoparticles. This particular feature of the magnetic nanoparticles to posses the different values of H_r for different orientations of their magnetic anisotropy axis (Fig. 5) becomes another interesting property of materials filled with magnetic nanoparticles where the static magnetic field H_{dc} can be used as a remote switcher for the local heating different groups of nanoparticles.

3. Temperature dependence of the spectral lines in viscous magnetic materials

In the case when the magnetic nanoparticles are placed in a viscous material for which the viscosity varies significantly depending on temperature, magnetic properties of such materials are also beginning to significantly depend on temperature. This special property of viscous magnetic materials has been studied experimentally for maghemite nanoparticles embedded in a multiblock poly(ether-ester) copolymer nonmagnetic matrix Guskos et al. (2006; 2008). The experiments were performed in a wide range of temperatures, 3.5-288 K. In addition to the experimental results were also carried out theoretical studies Dudek et al. (2010). Several examples of spectral lines discussed in Dudek et al. (2010) are presented in Fig. 6 for concetration of 0.1% of γ-Fe_2O_3 nanoparticles dispersed in the polymer matrix. The figure shows the completely different FMR spectra in the range of low temperatures and high temperatures. These two ranges of temperatures are also evident in the resonance field H_r (Fig. 7) as a function of temperature, where at low temperatures a marked decrement of H_r is observed. In the latter case the experimental results for two different concentrations of magnetic nanoparticles are presented and up to 50 K there is no signicant difference between them. This could mean that the thermal properties of non-magnetic matrix, in this case of a multiblock poly(ether-ester) copolymer, are of decisive importance for the magnetic properties of the magnetic agglomerates and not the reorientation processes between magnetic nanoparticles.

In low temperature region we have solid-like nonmagnetic matrix where magnetic relaxation takes place through the process of magnetization relaxation (Neél relaxation). Once the magnetic anisotropy axes are oriented randomly some additional peaks appear on the spectral lines at the higher values of H_{dc}. This property of spectral lines is shown in the previous section on the example of a single magnetic nanoparticle. The presence of many magnetic agglomerates consisting of different numbers of magnetic nanoparticles is one of the mechanisms for observing the broadening of the spectral lines. At higher temperatures the magnetic relaxation takes place both through the magnetization relaxation and rotation of the whole nanoparticle in a nonmagnetic surrounding and the additional peaks observed on spectral lines at low temepratures do vanish.

A simplified theoretical model was constructed in (Dudek et al., 2010) where a cluster consisting of N magnetic nanograins is placed randomly into a non-magnetic polymer matrix.

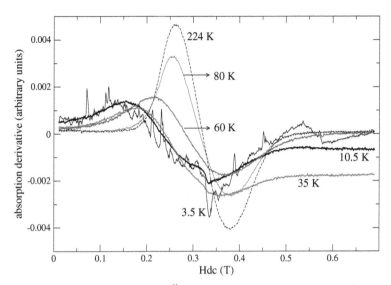

Fig. 6. The temperature dependence of $d\chi''/dH_{dc}$ for 0.1% γ-Fe$_2$O$_3$ dispersed in PEN-block-PTMO matrix.

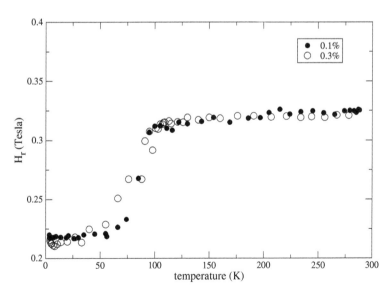

Fig. 7. An example of the dependence of the resonance field H_r on temperature for magnetic nanoparticles γ-Fe$_2$O$_3$ in PEN-block-PTMO matrix. The plots correspond to two concentrations of nanoparticles of 0.1% and 0.3%.

They occupy a permanent position but may rotate. Each of the magnetic nanograins $i = 1, 2, \ldots, N$ has magnetization M_i which dynamics is described with the help of the stochastic version of the Landau-Lifshitz equation in Eq. (8). The rotational dynamics of magnetic nanoparticles is described with the help of the Langevin equations for the magnetic anisotropy axis orientation. These equations take the following form (Dudek et al., 2010; 2008):

$$\frac{d\varphi_i}{dt} = -\frac{2}{R\zeta}|K_a V \sin(2\psi_i)| \sin(\varphi_i - \varphi_i') - \frac{K_{el}}{\zeta}\sin(\varphi_i - \varphi_{0,i}) + \frac{1}{\zeta}\lambda_{\varphi_i}, \tag{15}$$

$$\frac{d\theta_i}{dt} = -\frac{2}{R\zeta}|K_a V \sin(2\psi_i)| \sin(\theta_i - \theta_i') - \frac{K_{el}}{\zeta}sin(\theta_i - \theta_{0,i}) + \frac{1}{\zeta}\lambda_{\theta_i}. \tag{16}$$

in the diffusion limit, where ζ represents the friction of the i-th nanoparticle in the elastic non-magnetic polymer matrix and $\lambda_{\varphi,i}$ and $\lambda_{\theta,i}$ represent the white-noise driving torque (Coffey et al., 1984; Gardiner, 1983) for i-th nanoparticle, and K_{el} represents the spring constant which controls the rotational oscillations of the magnetic anisotropy axis. In the above stochastic equations the thermal rotational fluctuations of the i-th magnetic nanoparticle are characterized by temperature T and λ_{φ_i} and λ_{θ_i} and they fulfill the relations:

$$\langle \lambda_q(t) \rangle = 0 \tag{17}$$

$$\langle \lambda_q(t)\lambda_{q'}(t') \rangle = 2k_B T\zeta\delta(t - t'), \tag{18}$$

where $q = \varphi_i, \theta_i$. The angles φ' and θ' represent the angles which the magnetization $\overrightarrow{M}$ makes with z-axis and x-axis, and the angles φ_0 and θ_0 are the initial angles of the easy axis after the magnetic nanoparticle has been built into polymer matrix. The numerical scheme applied to the stochastic equations in (Dudek et al., 2010) is the Euler-Maruyama method. The theoretical model introduced in (Dudek et al., 2010) reproduces qualitatively the results of the experiment (Guskos et al., 2006; 2008). In particular, the FMR spectrum and the dependence of H_r on temperature have qualitatively the same properties. It is evident from Fig. 8 and Fig. 9 if we compare them with Fig. 6 and Fig. 7.

The results of the theoretical model have been obtained by two assumptions. The first one is assuming an empirical model for the viscosity parameter ν, the Arrhenius law,

$$\nu(T) = \nu_0 e^{E/k_B T} \tag{19}$$

where E is the activation energy. In the model, the viscosity parameter $\nu(T)$ is related to the rotational friction parameter ζ of magnetic nanoparticles in a polymer surounding (used in Eqs. (15) and (16)), as follows:

$$\zeta = 8\pi\nu(T)r^3, \tag{20}$$

where $r = R/2$ denotes the radius of a sphere representing magnetic nanoparticle and its polymer coating. Hence the simple assumption in Eq. (19) and not the presence of a phase transition is responsible for the qualitatively different behavior of H_r in the low and high temperatures in the theoretical model (Fig. 9).

The second assumption is introducing the Bloch law approximation

$$M_s(T) = M_0(0)\left(1 - \left(\frac{T}{T_0}\right)^\delta\right) \tag{21}$$

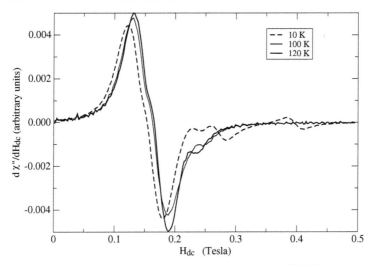

Fig. 8. Computer simulations of the temperature dependence of $d\chi''/dH_{dc}$ calculated for $N=30$ magnetic nanoparticles in the case when their magnetic anisotropy axes are randomly oriented (Dudek et al., 2010). The parameters of the computer simulation are the same as in Fig. 2.

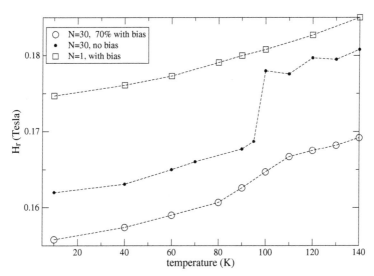

Fig. 9. Computer simulations of the dependence of the resonance field H_r on temperature for agglomerates of magnetic nanoparticles (Dudek et al., 2010). The agglomerates consisting of $N = 30$ magnetic nanoparticles represent two cases: when all magnetic nanoparticles are randomly oriented and when 70% of them is aligned with the dc magnetic field. In the case of a single magnetic nanoparticle ($N=1$) its magnetic anisotropy axis is aligned with the dc magnetic field.

for the magnetization of the magnetic nanoparticles where T is temperature, $\delta = 1/3$ and T_0 is some constant. The value of α is a parameter of the model under consideration. Another value of α can be also found in publications on magnetic materials.

The complexity of the FMR spectral lines can be seen in the example in Fig. 10 where the absorption lines derivatives $d\chi''/dH_{dc}$ have been plotted for a single magnetic nanoparticle in the case when its easy magnetic axis oscillates around the direction perpendicular to the external dc magnetic field in a surrounding with temperature-dependent viscosity $\nu(T)$. In low temperatures the magnetic resonance field H_r moves towards the lower values of H_{dc} with increasing temperature instead to move towards the higher values of H_{dc} as it is in the case of magnetic nanoparticles oscillating around the direction of the dc magnetic field. Only above a certain temperature there is no qualitative difference in the FMR spectrum for magnetic nanoparticles with magnetic easy axis oriented parallel or perpendicular to the external dc magnetic field.

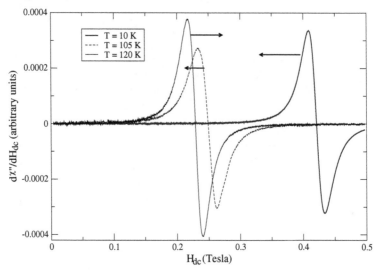

Fig. 10. Absorption lines derivatives, $d\chi''/dH_{dc}$ in theoretical model (Dudek et al., 2010) for a single ($N = 1$) magnetic nanograin in the case when its magnetic anisotropy axis oscillates around the direction transverse to the direction of the external dc magnetic field. The plotted curves correspond to temperatures $T = 10, 105, 120K$, respectively. In low temperatures the magnetic resonance field H_r moves towards the lower values of H_{dc} with increasing temperature and only above a certain temperature it begins to move toward the higher values of H_{dc}.

In the case of magnetic agglomerates dispersed in a viscous medium and which consist of a large number of magnetic nanoparticles with randomly oriented axes relative to the field H_{dc} the mechanism shown in Fig. 10 can be important in low temperatures.

4. Conclusions

Both the discussion in section 2 and FMR spectrum in Fig. 5 show that the static magnetic field H_{dc} can be used as a remote switcher for the local heating different groups of nanoparticles corresponding to different orientations of their magnetic easy axis. This property of the dependence of a maximum of χ'' on the orientation of the magnetic anisotropy axis with respect to the external dc magnetic field may be useful in designing new materials such as multi-functional magnetic nanocapsules. The thermal effects on the FMR spectrum in polymer composites filled with magnetic nanoparticles provide additional information about the magnetic structure of the material.

5. Acknowledgments

Some of the computer simulations have been performed in Wroclaw Centre for Networking and Supercomputing, Poland.

6. References

Coffey, W., Evans, M. & Grigolini, P. (1984). *Molecular Diffusion and Spectra*, Wiley, New York.

Dudek, M., N.Guskos, Senderek, E. & Rosłaniec, Z. (2010). Temperature dependence of the fmr absorption lines in viscoelastic magnetic materials, *J. Alloy Compd.* Vol.504: 289–295.

Dudek, M. R., Guskos, N., Grabiec, B. & Maryniak, M. (2008). Magnetization dynamics in landau-lifshitz-gilbert formulation. fmr experiment modeling, *J. Non-Cryst. Solids* Vol.354: 4146–4150.

Dudek, M. & Wojciechowski, K. (2008). Magnetic films of negative poisson's ratio in rotating magnetic fields, *J. Non-Cryst. Solids* Vol.354: 4304–4308.

Dutta, P., Manivannan, A., Seehra, M. S., Shah, N. & Huffman, G. P. (2004). Magnetic properties of nearly defect-free maghemite nanocrystals, *Phys. Rev. B* Vol.70: 174428.

Evans, K., Nikansah, M., Hutchinson, I. & Rogers, S. (1991). Molecular network design, *Nature* Vol.353: 124.

Füzi, J. (2006). Magnetic characteristics of dipole clusters, *Physica B* Vol.372: 239–242.

Gao, J. & Xu, B. (2009). Applications of nanomaterials inside cells, *Nano Today* Vol.4: 37–51.

Gardiner, C. (1983). *Handbook of Stochastic Methods for Physics, Chemistry and the Natural Sciences*, Springer-Verlag.

Gilbert, T. (1955). A lagrangian formulation of the gyromagnetic equation of the magnetization field, *Phys. Rev.* Vol.100: 1243.

Guskos, N., Glenis, S., Likodimos, V., Typek, J., Maryniak, M., Rosłaniec, Z., Kwiatkowska, M., Baran, M., Szymczak, R. & Petridis, D. (2006). Matrix effects on the magnetic properties of γ-fe$_2$o$_3$ nanoparticles dispersed in a multiblock copolymer, *J. Appl. Phys.* Vol.99: 084307.

Guskos, N., Likodimos, V., Glenis, S., Maryniak, M., M.Baran, Szymczak, R., Roslaniec, Z., Kwiatkowska, M. & Petridis, D. (2008). Magnetic properties of γ-fe$_2$o$_3$/poly(esther-ester) copolymer nanocomposites, *Nanosci. Nanotech.* (No.8): 2127.

Jönsson, P. E. (2003). Superparamagnetism and spin glass dynamics of interacting magnetic nanoparticle systems, *arXiv:cond-mat/0310684v2* .

Jung, S., Ketterson, J. & Chandrasekhar, V. (2002). Micromagnetic calculations of ferromagnetic resonance in submicron ferromagnetic particles, *Phys. Rev. B* Vol.66: 132405.

Lakes, R. (1987). Foam structures with a negative poisson's ratio, *Science* Vol.235: 1038–1040.

Landau, L. & Lifshitz, E. (1953). On the theory of the dispersion of magnetic permeability in ferromagnetic bodies, *Phys. Z. Sowjetunion* Vol.8: 153.

Liu, T.-Y., Hu, S.-H., Liu, D.-M., Chen, S.-Y. & Chen, I.-W. (2009). Biomedical nanoparticle carriers with combined thermal and magnetic responses, *Nano Today* Vol.4: 52–65.

Owens, F. J. (2003). Ferromagnetic resonance of magnetic field oriented fe3o4 nanoparticles in frozen ferrofluids, *J. Phys. Chem. Solids* Vol.64: 2289–2292.

Sellmyer, D. & Skomski, R. (2006). *Advanced Magnetic Materials*, Springer Science+Business Media, Inc., Place of publication.

Shliomis, M. I. (1975). Magnetic fluids, *Sov. Phys. Usp.* Vol.17: 153–169.

Smith, C. & Wojciechowski, K. (2008). Preface: phys. stat. sol. (b), *Phys. Status Solidi b* Vol.245: 486–488.

Sorop, T., Nielsch, K., Göring, P., Kröll, M., Blau, W., Werspohn, R., Gösele, U. & de Jongh, L. (2004). Study of the magnetic hysteresis in arrays of ferromagnetic fe nanowires as a function of the template filling fraction, *J. Magn. Magn. Mater.* Vol.272-276: 1656–1657.

Sukhov, A., Usadel, K. & Nowak, U. (2008). Ferromagnetic resonance in an ensemble of nanoparticles with randomly distributed anisotropy axes, *J. Magn. Magn. Mater.* Vol.320: 31–35.

Usadel, K. D. (2006). Temperature-dependent dynamical behavior of nanoparticles as probed by ferromagnetic resonance using landau-lifshitz-gilbert dynamics in a classical spin model, *Phys. Rev. B* Vol.73: 212405.

Vleck, J. V. (1950). Concerning the theory of ferromagnetic resonance absorption, *Phys. Rev.* Vol.78: 266–274.

Wilson, J., Poddar, P., Frey, N., Srikanth, H., Mohomed, K., Harmon, J., Kotha, S. & Wachsmuth, J. (2004). Synthesis and magnetic properties of polymer nanocomposites with embedded iron nanoparticles, *J. Appl. Phys.* Vol.95: 1439–1443.

Wood, D. & Camp, P. (2011). Modeling the properties of ferrogels in uniform magnetic fields, *Phys. Rev. E* Vol.83: 011402.

Zahn, M. (2001). Magnetic fluid and nanoparticle applications to nanotechnology, *J. Nanopart. Res.* Vol.3: 73–78.

3

Magnetic Nanoparticles: Its Effect on Cellular Behaviour and Potential Applications

Hon-Man Liu[1] and Jong-Kai Hsiao[2]
[1]National Taiwan University, Department of Radiology
[2]Tzu-Chi University, Department of Radiology
Taiwan

1. Introduction

With the advancement of nanotechnology and the development of molecular medicine, molecular and imaging becomes one of the most popular researches in the latest medicine. Molecular imaging can be defined as the imaging of targeted molecules non-invasively and repetitively in living organisms and cellular imaging can be defined similarly as the imaging of cells or cellular process non-invasively and repetitively in living organisms. At present, the common imaging tools for clinical study include ultrasound, computed tomography (CT), and magnetic resonance imaging (MRI). However, MRI is superior to CT for its better in soft tissue contrast, more sensitive in pathology detection, and lack of ionization irradiation.

In clinical practice, gadolinium-contained compound is the commonest contrast medium used in MRI study. Molecular imaging differs from traditional imaging in that contrast agents are typically used to help identify particular biomarkers or pathways with high sensitivity and selectivity (Achilefu, 2010). However, gadolinium (Gd) is not proper for the molecular imaging and cellular imaging due to its low relaxivity, that further decrease upon cellular internalization; not biocompatible and potential toxicity following cellular dechelation over time. Iron oxide (IO) nanoparticle contrast medium is another contrast medium used in clinical study. They provide the most significant signal change per unit of metal atom, especially on T2* MR imaging. Iron oxide nanoparticle are made of thousands of iron atoms in Fe_3O_4 or γ-Fe_2O_3 form so that they can increase the contrast-to-noise ratio and make its sensitivity superior to Gd contrast agent on MR examination. Their main component, oxidized iron, can be metabolized in liver and recycled as important component of red blood cells. Iron oxide nanoparticle have a relatively long circulation time and low toxicity (Bradbury and Hricak 2005; Funovics et al., 2004; Harisinghani et al., 2003; Jain et al., 2005; Montet et al., 2006). Their surfaces coating may strategically contain chemical linkage of functional groups and ligands for multimodal imaging purpose (Rogers & Basu, 2005). They can be easily detected by light and electron microscopy. Iron oxide nanoparticle posses some novel properties not seen with the other macromolecules. They can be manipulated by conjugating both targeting ligands or peptide and therapeutic components such as photosensitizer to help in diagnosis and treatment. Iron oxide nanoparticles can be used to monitor cellular migration, molecular events, and signal pathway associated with different pathological status. Owing to its magnetic character, iron oxide nanoparticle can be manipulated magnetically and altered their magnetic character according to size of core and the condition of the coating. In this assay, we are going to review the characteristic and types of magnetic nanoparticles (MNP), especially the IO nanoparticles, the mechanism of internalization of MNP into the cell, the impact to cellular and other behaviour of macrophage and stem cell after labelling with MNP, and the future of application of MNP in nanomedicine.

2. Magnetic resonance imaging and magnetic nanoparticles

When we put a body into a strong magnetic field and then apply an external radiofrequency (RF) for a period. The RF may causes disturbance of the thermal equilibrium in the body system. After the RF stopped, MR imaging detects the different signals due to the different proton relaxation times (T) of hydrogen atom of the tissue in different body part. This makes MR offers great contrast between different soft tissues in the body. There are two types of MR imaging mechanisms: T1-weighted and T2-weighted.

T1, the "longitudinal" (spin-lattice) relaxation time is defined as the time required for a substance to regain the 63% of its original longitudinal magnetization after an RF pulse. It represents the correlation of frequency between molecular motions and the Larmor frequency. The frequencies of small molecule (e.g. water) and large molecule (e.g. protein) are significantly different from the Larmor frequency and thereby have long T1 and present as low signal intensity (dark) on T1-weighted images. The motion frequency of medium-sized molecule such as cholesterol close to those used for MR imaging, thereby it has a short T1 relaxation time and thus appear high signal intensity (bright) on T1-weighted images. T1 relaxation time can be shortened from the interaction between the unpaired electrons in the paramagnetic iron such as Gd ions in contrast medium and the protons in water. This makes those pathology with pooling of Gd contrast agent appear bright on T1-weighted images.

T2 is the "transverse" (spin-spin) relaxation time. Following a 90 degree RF pulse, the protons lose their coherence and transverse magnetization. The tissue inhomogeneity causes fluctuations of the magnetic field randomly, leading to variations in the precession frequency of different spins on x-y plane. Consequently, the net x-y magnetization is lost since the initial phase coherence is lost. This results in T2 relaxation. Thus T2 relaxation is a measure of how long the resonating protons of a substance can be changed from coherent to de-coherent and then back to coherent stage following 90 degree RF pulse in x-y plane. T2 relaxation time is defined as the time needed for the transverse magnetization decreases to 37% of its original magnitude after a 90 degree RF pulse. Generally, T2 relaxation is much less dependent on the magnetic field strength than T1 relaxation time. However, the magnetic field is not homogenous, and the process is depending on the exact location of the molecules in the magnet. In such circumstances, a special transverse relaxation time constant is defined as T2*, which is usually much smaller than T2 and highly sensitive to magnetic field strength.

The MR contrast medium can be divided into positive and negative contrast media according to their characteristic appearance on T1- weighted or T2-weighted images.

Positive contrast media appear brighter on MR images owing to a reduction in T1 relaxation time. They include those containing Gd, manganese or iron ions. Negative contrast agents appear dark on MR imaging due to shortening T1, T2, and T2* relaxation times. Iron oxide is the most common negative contrast medium used clinically.

As mentioned before, gadolinium agent is not suitable for molecular or cellular imaging. In the last 10 years, most research of molecular imaging using MRI is focused on the application of IO nanoparticle.

Compared to larger particles of the same chemical composition, nanoparticles can pass some biological barriers such as capillaries. Human albumin, a circulatory macromolecule, is similar to nanoparticles with a diameter of 5-10 nm (Wiwanitkit, 2006). Enzymes and receptors are also ranged in the similar size (Rawat, 2006). A nanoparticle of such size can have in excess of 1500 potential sites for chemical modification (Debbage et al., 2008; Harris et al., 2003) without loss of biological functionality. It is 150 times more than an antibody has. The high capacity for nanoparticle modification has led to their use as amplifiers for in vivo imaging. Both the surface properties and size of nanoparticles are important for their interaction with biological systems and therefore for their distribution in the circulation.

In considering the use in *in vivo* imaging, the ideal IO nanoparticles is with small size (5–150 nm) (table 1), high mass magnetization value, and great surface functionality. If the diameter of the MNPs is larger than 200 nm, they are usually taken up by the liver, spleen, and reticuloendothelial system and resulting in decreased blood circulation times. If their diameters are less than 5 nm, they are rapidly removed through the kidney (Gupta & Gupta, 2005). Different sizes of IO nanoparticles including SPIO (superparamagnetic IO, 60–150 nm), USPIO (ultrasmall SPIO, 10–50 nm), and MION (monocrystalline IO, 5–10 nm) can lead to different magnetic properties and function differently in various applications (Choi et al., 2006; Corot et al., 2006; de Vries et al., 2005; Thorek et al., 2006; Wang et al. 2001;).

The magnetism of MNP and its effect on MR imaging can depend significantly on their morphology, crystal structure, size and uniformity. The crystal structure of SPIO nanoparticle has the general formula of $Fe^{3+}O_3M^{2+}O$, where M^{2+} represents a divalent metal ion (i.e., iron, manganese, nickel, cobalt or magnesium) (Kateb et al., 2011). The ferric iron (Fe^{3+}) makes the complex magnetic (Daldrup-Link et al., 2003; Wang et al., 2001) and large, unpaired, thermodynamically independent spines (single domain particles) makes the complex superparamagnetic. Single domain particles or magnetic domains have a net magnetic dipole. External magnetic fields can cause the magnetic domain to re-orient. The signal intensity of these MNP is related to the size of the particle, its position, its concentration within a given voxel, data acquisition parameters, the magnetic field, and dosage of the SPIO (Wang et al., 2001). In order to achieve higher relaxivity, types of MNPs have also been designed and included those doped with alternative metals such as $CoFe_2O_4$, $NiFe_2O_4$, $MnFe_2O_4$, Gd_2O_3 and gold-coated cobalt nanoparticles (Bouchard, et al., 2009; Bridot et al., 2007; Lee et al., 2007). Magnetism in MNPs is highly sensitive to its size because it arises from the collective interaction of atomic magnetic dipoles. At a critical size, MNPs will change from a state that has multiple magnetic domains to only a single domain. Below this critical size, the thermal energy becomes comparable to what is needed for spins to flip, and the magnetic dipoles are in status of rapid randomization. Such MNPs do not have permanent magnetic moments in the absence of an external field but can quickly respond to an external magnetic field and are referred to as superparamagnetic.

Name	Coating	Size (nm)	Relaxivity (mM^{-1}sec^{-1}) r_1	Relaxivity (mM^{-1}sec^{-1}) r_2
Feridex/Endorem, Ferumoxides AMI-25	Dextran T10	120-128	10.1	120
Resovist, Ferucarbotran SHU-555 A	Carboxydextran	60	9.7	189
Combidex/ Sinerem Ferumoxtran-10 AMI-227	Dextran T10, T1	15-30	9.9	65
Supravist SHU- 555 C	Carboxydextran	21	10.7	38
Clariscan, Feruglose NC-100150	Pegylated starch	20	n.a.	n.a.

Table 1. Examples of available SPIO and USPIO agents. Modified from Corot et al., 2006.

MION has a magnetically labeled cell probe MR imaging agent with size about 5-10nm. It has monocrystallinity and can be used for receptor-directed MR imaging. Its small size make MION can easily pass through capillary endothelium without changing its supermagnetism. It has been stated that it is possible to be detected by MR imaging at concentration as low as 1 ug Fe/g tissue. Though it is still in the experimental state, the preliminary targeted MR imaging with MION prove to be a powerful tool for cellular and molecular MR imaging in the future.

Many different chemical methods can be used for synthesizing magnetic nanoparticles. The most commonly used are precipitation-based approaches, either by co-precipitation or reverse micelle synthesis (Nitin et al., 2004; Shen et al., 1993). MNP without any surface coating are not stable in aqueous media, readily aggregate, and precipitate. For *in vivo* applications via intravenous route, these particles aggregates in blood frequently and are recognized and phagocytosed by macrophages (Lee et al., 2006). Therefore, the surface of MNP should be coated with a variety of different moieties that can eliminate or minimize their aggregation under physiological conditions. Usually, two main approaches are used for coating MNP, including in situ coatings and post-synthesis coatings (Berry et al., 2004; Horak et al., 2007; Jodin et al., 2006). With in situ coating, the MNP are coated during the synthesis process. This coating approach involves a co-precipitation process in the presence of the polysaccharide dextran and a cross linked chemically to increase its stability. This particular coating approach has been very successful in producing dextran SPIOs which are biocompatible and water – soluble. Other coatings in this class include carboxydextran coating, starch-based coating, and dendrimer-based coatings. The post-synthesis coatings can be used for coating MNP with a variety of materials, including, monolayer ligands, polymers, combinations of polymers and biomolecules such as phospholipids and carbohydrates, and silica.

Multiple MNP can also be encapsulated in liposomes to create magnetoliposomes (De Cuyper & Joniau, 1988). Polyethylene glycol (PEG)-modified, phospholipid micelles coating is favourable since this can results in satisfactory solubility and stability in aqueous solutions, well biocompatibility, and also with prolonged blood circulation time when they are delivered intravenously. The PEG can be modified for bioconjugation of various moieties such as antibody, oligonucleotides, and peptides and may allow for molecular specific intracellular targeting of specific proteins and nucleic acid (Gupta & Gupta, 2005; Kohler et al., 2004; Kumagai et al., 2007; Lee et al., 2006, 2007a; Mikhaylova et al., 2004; Nitin et al., 2004; Veiseh et al., 2005). PEG-coated MNP has the disadvantage such as limited binding sites available for further ligand binding (Gupta & Gupta, 2005), and the coating thickness can significantly affect their relaxivity (Laconte et al., 2007). In addition to PEG coating, other materials such as antibiofouling poly(TMSMA-r-PEGMA) (Lee et al., 2006), hyaluronic acid layers (Kumar et al., 2007) and carboxylfunctionalized poly(amidoamine) dendrimers of generation 3 (Shi et al., 2007) have also been used to coat the surface of IO nanoparticles for either increasing circulation time in the blood or delivering peptides at high efficiency.

3. Impact of magnetic nanoparticles in immunologic cell

For most of the clinical imaging application on magnetic nanoparticles, the delivery route is intravenous injection. The human immune system, mostly reticuloendothelial system, recognizes these magnetic nanoparticles and ingests them. The size and surface charge of the magnetic nanoparticles determine which kind of cells that interact with magnetic nanoparticles (Moghimi & Bonnemain 1999) For particles larger than 20 nm, macrophage and Kuppfer cell is the corresponding cells that deal with MNP (Moghimi & Hunter 2001, 2005). If the MNPs are less than 20 nm, these MNPs have greater opportunity to reside in lymph nodes, after they extravasate into interstitial spaces (Moghimi & Bonnemain 1999). Currently clinical approved iron oxide nanoparticles for MR images ranged mostly from 50-100 nm, in which macrophages play important roles in ingestion of these MNPs. Macrophages are cells that prevent invading bacteria, viruses by phagocytosis of these microbes. It initiates inflammatory change by secreting cytokines such as tumor necrosis factor-alpha and interleukin 2-beta which recruits more circulating cells for repairing damaged tissue. Recent studies reveal macrophages also play important roles in tumor invasion. Consequently, alteration of macrophage behaviour has potential influence on human immunity, inflammatory process and cancer invasion. Understanding of impacts of macrophages toward ingested magnetic nanoparticles is herein clinically important.

Two different MNPs are now under clinical use. Ferucarbotran is composed of both Fe_3O_4 (magnetite) and g-Fe_2O_3 (maghemite) and coated with carboxydextran that is negatively charged. Ferumoxides is also composed of iron oxide that coated with dextran. Protamine sulfate is usually added in cell culture for more efficient ferumoxides labelling (Arbab et al., 2006).

Studies on clinically used MNPs, ferucarbotran, toward murine macrophage cell line revealed MNPs ingestion stimulates TNF-alpha and IL-2 Beta secretion. The migratory ability of MNPs laden macrophage increased but the phagocytotic activity of macrophages decreased (Hsiao et al., 2008) However, these findings are based on 100 ug Fe/mL MNP concentration that is 11 times higher than plasma concentration (Metz et al., 2004). Similar findings could be observed on murine peritoneal macrophage cultured with 100 ug Fe/mL MNPs. The secretion of TNF-alpha, IL-2 Beta and Nitric oxide, a bactericidal chemical, are all increased in conjunction with the promotion of macrophage migration ability (Yeh et al., 2010).

Long term exposure to MNPs has significant influence on macrophages. Research on human macrophages treated with ferucarbotran show increased apoptosis after 120 hours of incubation even at the concentration of 1 ug Fe/mL. Human macrophage also shows apoptotic change when facing smaller MNPs, supravist, a smaller particle of 20.8 nm in diameter, for 120 hours at the concentration of 0.1 ug Fe/mL (Lunov et al., 2010a). The apoptotic event is inducted by N-terminal kinase (JNK) pathway that is activated by reactive oxygen species (Lunov et al., 2010a; 2010b). There is evidence that elevated TNF-alpha induce human macrophage apoptosis after these macrophages expose to ferucarbotran for 3- 5 days. However, there is no evidence that support ferucarbotran stimulate TNF-alpha secretion on human macrophage. All of studies performed above are in vitro experiments that intravenous injection of MNPs and collecting of circulating macrophage are still pending. Moreover, under intravenous injection condition, all of clinical MNPs are eliminated by reticuloendothelial system within 30 minutes in which no toxic event are observed.

Human monocyte cell line, THP-1, is a precursor of macrophage and it has been evaluated for its interaction with ferumoxides. The ferumoxides has been mixed with 1mg/mL of protamine sulfate for higher labeling efficiency. Under incubation concentration of 4.5 ug Fe/mL of ferumoxides-protamine complex for 2 hours, there is no significant TNF-alpha secretion level change upon lipopolysaccharide stimulation (Janic et al., 2008). The CD-54 and CD-83 is not upregulated in response to lipopolysaccharide.

Lymphocytes are important immune cells that regulate both cellular and humoral immunity against invading organism and cancer cells. Although lymphocytes are not easily labeled with MNPs, it is still possible by modifying surface of MNPs with tat peptide, a HIV membrane translocating peptide that is specific to CD4+ lymphocytes (Garden et al., 2006). The synthesized Tat linked MNPs are 31.3± 8.5nm which is slightly larger than original MNPs that is 25.7± 6.1 nm. Under TEM, these particles located at both cytoplasm and nucleus, which is different from other MNPs that only located at lysosomes. There is neither proliferation ability nor IL-2 secretion capability change of CD4+ CD25+ lymphocytes after labelling with tat-linked MNPs (Garden et al., 2006). Dendritic cells are antigen presenting cells that express antigens to other immune cells, mostly lymphocytes, to continue immune response. Labelling of dendritic cells allows monitoring migration of these cells in vivo (Tavaré et al., 2011; Noh et al., 2011) The mouse dendritic cells were labelled with endorem, a clinically proved MNPs in Europe with corresponding product named ferumoxide in USA. There is no drastic effect of labelled dendritic cells such as T lymphocyte proliferation, in vivo growth rate of lymph nodes after labelled or unlabeled dendritic cells labelling. Under B16 melanoma lung metastatic model, both labelled and unlabeled dendritic cells show protective effect upon pulmonary metastasis (Tavaré et al., 2011).

In conclusion, the effects of MNPs toward immune cells are diverse, the cell type, particle size, charge and labelling amount all contribute to cell behaviour change. Although some reports show immunological response change after MNPs labelling, most of the MNPs exceeds the daily clinical practice. However, systemically analysis of MNPs and immune cells interaction is important and this study may have potential impact on immune therapy.

4. Impact of magnetic nanoparticles in stem cell

Stem cells play promising roles in tissue regeneration and engineering. They could be used for tissue transplantation and it is now understand that stem cells also interact with cancer cells. Some of the stem cell promotes the growth of cancer cells whereas some animal model shows stem cell suppresses the tumor growth.

There are different types of stem cells. Embryonic stem cells are pluripotent, which means the cells could differentiate into almost all cells. However, the ethics concern and current stem cell technology progress makes it less interesting for cell labelling. Mesenchymal stem cells are multi-potent cells that could differentiate into different kinds of cells of medical interest such as bone tissue and cartilage.

Bone marrow derived mesenchymal stem cells are capable of differentiating into many tissue that is essential for tissue repair. However, when these cells delivered into damaged tissue, it is hard to differentiate where these cells are. Labelling cells with MNPs are then important to monitor the location, migration in vivo. It has been proved that MNPs labelled mesenchymal stem cells can be visualized for implantation into damaged cardiac tissue in porcine model (Kraitchman et al., 2002). In the study, ferumoxides incorporated with poly-L-lysine were incubated with swine mesenchymal stem cells and injected into myocardiocytes under X-ray guidance. Post-mortem histology shows injected cells resides in designated myocardial tissue. The labelled mesenchymal stem cells are also applied for monitoring the repair of lipopolysaccarides induced damaged brain tissue in rat model by using MNPs-tat peptide conjugate. The result shows cell migratory behaviour into the damaged brain (Jackson et al., 2010). For understanding of interaction between mesenchymal stem cells and tumor, labelled mesenchymal stem cells is also monitored for its interaction with glioma in mouse model by using ferucarbontran in conjunction with protamine sulfate and proved that mesenchymal stem cells reduce glioma growth and mesenchymal stem cells is capable of migration into glioma tissue (Chien et al., 2010).

The mechanism of MNPs uptake by different kinds of stem cells are not fully investigated but recent study shows endocytosis by clathrin receptor is one of the mechanisms (Huang et al., 2005; Lu et al., 2007). These study shows inhibitor of clathrin receptor, phenylarsine oxide, can block the ingestion of mesoporous iron oxide nanoparticles into human mesenchymal stem cells. Macropinocytosis also play significant role once if protamine sulfate is used. It is also proved that tat peptide linked MNPs enter cell by macropinocytosis (Arbab et al., 2006).

Most of stem cell labelling for MR imaging is based on T2 weighted contrast. However, some efforts aiming on T1 contrast agent such as gadolinium based chelates conjugating into mesoporous silica nanoparticles has been proved for its imaging capability in animals injected with human mesenchymal stem cells. The viability and differentiating capacity of these mesenchymal stem cells are preserved (Hsiao et al., 2008; Tsai et al., 2008). The mesoporous nanoparticles has also been labelled with fluorescent dyes that monitoring the cells with fluorescent imaging modality is also possible.

In addition to cell viability, labelled mesenchymal stem cells has been verified for its mitochondrial potential and reactive oxygen species, both of which represents intracellular stress. Neither mitochondrial potential nor reactive oxygen species change under ferucarbotran incubation at the concentration of 100 ug Fe/mL for 24 hours (Hsiao et al., 2007). Long term incubation up to 72 hours has also been investigated and shows no adverse effect upon mesenchymal stem cells (Yang et al., 2011). Similar results are found on ferumoxide-polylysine and ferumoxides-protamine sulfate complex toward mesenchymal stem cells (Arbab et al., 2003; Pawelczyk et al., 2006).

Stem cells are valuable for its differentiation capacity. Concerns for preserving its differentiating capacity are essential. For clinical used MNPs such as ferucarbotran for directly labelling, it has been showed that labelled mesenchymal stem cells is still capable of differentiating to adipose tissue, and bone tissue at the labelling period of 24 hours (Hsiao et al., 2007). The long term effect has also been evaluated for its cartilage differentiation capacity (Yang et al., 2011). The activity of chondrogenesis of ferucarbontran labelled mesenchymal stem cells decreased as iron content increases (Hinning et al., 2009). Similar finding upon osteogenesis is also found. Dose dependent osteogenesis inhibition is observed on human mesenchymal stem cells (Chen et al., 2010). The labelling dose is consequently very important.

Labelling of mesenchymal stem cells with ferumoxides in conjunction with transfection agent is also popular. The differentiation capacity has also been studied. The adipogenesis and osteogenesis capacity is preserved but there is debate upon chondrogenesis (Kostura et al., 2004; Arbab et al., 2005). The model of ferumoxides-polylysine shows inhibition of chondrogenesis whereas ferumoxide-protamine sulfate shows no inhibitory effect. Although there is no study comparing these two labelling method, the ferumoxide-protamine sulfate and ferumoxide-polylysine complex, labelling mesenchymal stem cells with ferumoxide-protamine sulfate might be better for further investigation. Besides, labelling dose of MNPs should be suitable for preserving imaging capability and differentiating capacity.

In conclusion, labelling of stem cells for imaging is medically important that could be used for cell trafficking and potentially tumor inhibition. Although imaging capability of these labelled mesenchymal stem cells is concerned, the differentiation capacity of these cells should be preserved. Meanwhile, no satisfactory methods or consensus about labelling stem cell with MNP established though direct labelling using ferucarbotran or labelling ferumoxides with protamine sulfate are popular. Efforts on designing novel MNPs for cell labelling is still demanding.

5. Bifunctional, multi-functional, and theragnostic magnetic nanoparticles

Nanoparticles have advantages for their multi-conjugating capability that makes it possible to exhibit imaging and therapeutic character in one particle. The capability of imaging is mostly rely on the core that is magnetic. Either the shell or the core itself exhibit therapeutic effect. The therapeutic effects include gene delivery, hyperthermia, chemotherapy and photodynamic therapy. The benefits of theragnostic design are based on the following advantages. First, the magnetic core plays both imaging and magnetic guidance character. Targeting to specific organ or tissue is theoretically possible once if a guiding magnetic is applied. Secondly, the location where MNPs resides and acts as therapeutic agent can be visualized. Unlike most drugs that are small molecules, MNPs has different, specific organ and cell distribution that makes it possible for different treatment strategy. Lastly, MNPs can traverse vasculature barrier and go into intercellular space or even cell surface once if recognizing molecules has been conjugated at the MNPs surface.

Hyperthermia with MNPs is based on the fact that tumor cells are more liable toward temperature change. It have been investigated that temperature between 41°C and 42°C can induce tumor cell death by destruction of cell membrane (Sellins & Cohen, 1991). The enzymatic system is also influenced. The hyperthermia is achieved by alternating current magnetic field system around the frequency of 100 KHz at the magnetic field intensity in 30.6 kA/m (Silva et al., 2011). Limited clinical trial was done and showed controversial effect (Maier-Hauff et al., 2007; 2011). In one study, 66 patients of glioblastoma, a high grade brain tumor, were enrolled and MNPs were injected into tumor of these patient. Hyperthermia associated with radiotherapy was done and there is statistical difference between hyperthermia group and traditional radiotherapy group. The survival after first diagnosis is 8.6 months longer in hyperthermia group compared with conventional treatment group. In addition, the adverse effect of hyperthermia is not significant according to the observation of the study (Maier-Hauff et al., 2011).

Photodynamic therapy is one of the cancer treatment methods that have also been used for theragnostic purpose. The mechanism of photodynamic therapy is based on synthesis of singlet oxygen at the expense of photon activation of photosensitizer. The produced singlet oxygen is capable of destruct adjacent cells by oxidation. Some of the photosensitizers are clinically available. Efforts trying to conjugate MNPs with photosensitizers have potential benefits such as understanding the location of drugs accumulation and MNPs can also be guided by magnetic fields. The model of multi-functional MNPs has been proved possible in vitro. Hela cells can be imaged and killed by iridium complexes conjugated iron oxide nanoparticles (Lai et al., 2008). The iridium complexes have been also conjugated to MnO based mesoporous silicate nanoparticles that exhibit T1 weighted contrast enhancement. The photodynamic therapy effect is proved efficacious at in vitro HeLa cell model (Peng et al., 2011).

Gene therapy is at the edge of new strategy for cancer therapy. MNPs is capable of serving gene delivery carrier and also used for magnetic guidance. Studies focused on cancer related gene such as E1A has been successfully delivered into HeLa cells after E1A gene incoporated with iron oxide nanoparticles. Intratumoral injection of the plasmid-MNPs complex results in tumor size reduction compared with control group, whereas only radiation therapy was done (Shen et al., 2010).

In conclusion, multifunctional MNPs are at the initial stage of development. The benefits of biodistribution and magnetic character make theragnostic strategy different from other treatment. However, more efforts upon toxicity and therapeutic range should be done before it has been used widely in the clinical medical fields.

6. Future

Cellular Imaging can be an application of MNP as cellular marker for imaging of macrophage activity and as cellular marker for imaging of cell migration and cell trafficking. With the advancement of modern molecule design, we can also have the capability of design a MNP with the role of both diagnostic and treatment.

A major limitation of IO MNP is the loss of signal on T2-weighted MRI and creating 'black holes' on images; that (1) prevent direct anatomical MR evaluation of the tissue in question (requiring comparison of pre- and post-contrast images), and (2) make it difficult to discriminate between targeted cells and image artefacts (i.e. as caused by susceptibility artifacts or imperfect pulse sequences).One such approach could be the use of a 'white marker' MR T1-weighted sequence, that creates positive MNP contrast. For cellular imaging, as labelling is not permanent and self-replicable like reporter genes, with dilution of label upon cell division, iron oxide detection may rapidly become impossible. Finally, careful iron oxide titration and cellular differentiation studies need to be performed. Short- and long-term toxicity studies are warranted. It needs a comprehensive study on the fate of the particles in vivo following biodegradation; quantify the number of iron oxide labelled molecules or cells per voxel and to increase the specificity of detection of iron oxides.

Perhaps the least studied limitation is the potential acute and chronic systemic toxicity of the particles themselves. Toxicity can result from the MNP themselves or the individual components of the MNP that can be released during degradation in vivo. Nanomaterials may influence a living organism through different biological pathways (Nel et al., 2006). From previous limited report IO MNP and gold colloids seem to be less of a concern in terms of toxicity and IO can be cleared from the body via various routes with minimal toxicity (Briley-Saebo et al., 2004; Corot et al., 2006; Jain et al., 2008). Different types of nanoparticles have been shown to be cytotoxic to human cells (Lewinski et al., 2008), induce oxidative stress (Long et al., 2006), or elicit an immune response (Dobrovolskaia et al., 2007). After administration, nanoparticles must traverse a complex and often hostile environments that have evolved to seek out and exclude foreign material (Minchin et al., 1999). The first few steps of this dangerous journey include the interacting with plasma proteins and accumulating in macrophages or the reticuloendothelial system of the liver, spleen, or lymph nodes. The types of proteins that absorb to the surface are affected by size, shape, and surface characteristics. Importantly, there is now strong evidence that the proteins that surround the nanoparticles play a critical role in determining their fate in vivo (Kreuter et al., 2002; Owens et al.; 2006, Lynch et al., 2006). Dextran is clinically approved for modifying IO MNP but liver accumulation is still evident. Silica nanoparticles have been evaluated for potential hepatoxicity because of their propensity to be taken up by the liver (Nishimori et al., 2009).Whereas large particles (>300nm) showed little adverse effects, particles less than 100 nm induced acute liver damage and cytokine release.

7. Conclusion

The nanotechnology offer great opportunities for molecular imaging and future medicine. However, they are difficulty in designing and administration. The possible acute or chronic toxicity associated with the nanoparticle is still under investigated. The implementation of nanotechnology in medicine will depend on more understand and depth knowledge about them.

8. References

Arbab, A.S., Bashaw, L.A., Miller, B.R., Jordan, E.K., Lewis, B.K., Kalish, H., & Frank, J.A. (2003). Characterization of biophysical and metabolic properties of cells labeled with superparamagnetic iron oxide nanoparticles and transfection agent for cellular MR imaging. *Radiology*, 229(3),838-846.

Arbab, A.S., Yocum, G.T., Rad, A.M., Khakoo, A.Y., Fellowes, V., Read, E.J., & Frank J.A. (2005). Labeling of cells with ferumoxides-protamine sulfate complexes does not inhibit function or differentiation capacity of hematopoietic or mesenchymal stem cells. *NMR in Biomed*, 18(8), 553-559.

Arbab, A.S., Liu, W., & Frank, J.A. (2006).Cellular magnetic resonance imaging: current status and future prospects. *Expert Rev Med Devices*, 3(4), 427-439.

Achilefu, S. (2010) Introduction to concept and strategies for molecular imaging. *Chem. Rev.* 110, 2575–2578

Berry, C.C., Wells, S., Charles, S., Aitchison, G., & Curtis, A.S. (2004). Cell response to dextranderivatised iron oxide nanoparticles post internalisation. *Biomaterials*, 25, 5405–5413.

Bouchard, L.S., Anwar, M.S., Liu, G.L., Hann, B., Xie, Z.H., Gray, J.W., Wang, X., Pines, A., & Chen, F.F. (2009). Picomolar sensitivity MRI and photoacoustic imaging of cobalt nanoparticles. *Proc. Natl. Acad. Sci.* 106, 4085–4089

Bradbury, M. & Hricak, H. (2005). Molecular MR imaging in oncology. *Magn Reson Imaging Clin N Am*, 13,225–240.

Bridot, J.L., Faure, A.C., Laurent, S., Rivière, C., Billotey, C., Hiba, B., Janier, M., Josserand, V., Coll, J.L., Elst, L.V., Muller, R., Roux, S., Perriat, P., & Tillement, O. (2007). Hybrid gadolinium oxide nanoparticles: multimodal contrast agents for in vivo imaging. *J. Am. Chem. Soc.* 129, 5076–5084

Briley-Saebo, K., Bjørnerud, A., Grant, D., Ahlstrom, H., Berg, T., & Kindberg, G.M. (2004). Hepatic cellular distribution and degradation of iron oxide nanoparticles following single intravenous injection in rats: implications for magnetic resonance imaging. *Cell Tissue Res.* 316, 315–323.

Chen, Y.C., Hsiao, J.K., Liu, H.M., Lai, I.Y., Yao, M., Hsu, S.C., Ko, B.S., Chen, Y.C., Yang, C.S., & Huang, D.M. (2010). The inhibitory effect of superparamagnetic iron oxide nanoparticle (Ferucarbotran) on osteogenic differentiation and its signaling mechanism in human mesenchymal stem cells. *Toxicology and Applied Pharmacology*, 245(2), 272-279.

Chien, L.Y., Hsiao, J.K., Hsu, S.C., Yao, M., Lu, C.W., Liu, H.M., Chen, Y.C., Yang, C.S., & Huang, D.M. (2011). In vivo magnetic resonance imaging of cell tropism, trafficking mechanism, and therapeutic impact of human mesenchymal stem cells in a murine glioma model. *Biomaterials*, 32(12), 3275-3284.

Choi, S.H., Han, M.H., Moon, W.K., Son, K.R., Won, J.K., Kim, J.H., Kwon, B.J., Na, D.G., Weinmann, H.J., & Chang, K.H. (2006). Cervical lymph node metastases: MR imaging of gadofluorine M and monocrystalline iron oxide nanoparticle-47 in a rabbit model of head and neck cancer. *Radiology*, 241 (3), 753-762.

Corot, C. Robert, P., Idée, J.M., & Port, M. (2006) Recent advances in iron oxide nanocrystal technology for medical imaging. Adv. Drug Deliv. Rev. 58, 1471–1504.

Daldrup-Link, H.E., Rudelius, M., Oostendorp, R.A., Settles, M., Piontek, G., Metz, S., Rosenbrock, H., Keller, U., Heinzmann, U., Rummeny, E.J., Schlegel, J., & Link, T.M., (2003). Targeting of hematopoietic progenitor cells with MR contrast agents. *Radiology*, 228, 760–767.

Debbage, P., & Jaschke,W. (2008). Molecular imaging with nanoparticles: giant roles for dwarf actors. *Histochem Cell Biol,* 130,845–875.

De Cuyper, M., & Joniau, M. (1988). Magnetoliposomes. Formation and structural characterization. *Eur Biophys J,* 15,311–319.

de Vries, I.J., Lesterhuis, W.J., Barentsz, J.O., Verdijk, P., van Krieken, J.H., Boerman, O.C., Oyen, W.J., Bonenkamp, J.J., Boezeman, J.B., Adema, G.J., Bulte, J.W., Scheenen, T.W., Punt, C.J., Heerschap, A., & Figdor, C.G., (2005). Magnetic resonance tracking of dendritic cells in melanoma patients for monitoring cellular therapy. *Nat. Biotechnol,* 23 (11), 1407–1413.

Dobrovolskaia, M.A., & McNeil, S.E. (2007). Immunological properties of engineered nanomaterials. *Nat Nanotechnol,* 2,469–478.

Funovics, M.A., Kapeller, B., Hoeller, C., Su, H.S., Kunstfeld, R., Puig, S., & Macfelda, K. (2004). MR imaging of the her2/neu and 9.2.27 tumor antigens using immunospecific contrast agents. *Magn Reson Imaging,* 22,843–850.

Garden, O.A., Reynolds, P.R., Yates, J., Larkman, D.J., Marelli-Berg, F.M., Haskard, D.O., Edwards, A.D., & George, A.J. (2006). A rapid method for labelling CD4+ T cells with ultrasmall paramagnetic iron oxide nanoparticles for magnetic resonance imaging that preserves proliferative, regulatory and migratory behaviour in vitro. *Journal of Immunological Methods,* 314(1-2), 123-133.

Gupta, A.K., & Gupta, M. (2005). Synthesis and surface engineering of iron oxide nanoparticles for biomedical applications. *Biomaterials,* 26, 3995–4021.

Harisinghani, M.G., Barentsz, J., Hahn, P.F., Deserno, W.M., Tabatabaei, S., van de Kaa, C.H., de la Rosette, J., & Weissleder, R. (2003). Noninvasive detection of clinically occult lymph-node metastases in prostate cancer. *N Engl J Med,* 348,2491–2499.

Harris, L.A., Goff, J.D., Carmichael, A.Y., Riffle, J.S., Harburn, J.J., St. Pierre, T.G., & Saunders, M. (2003) Magnetite nanoparticle dispersions stabilized with triblock copolymers. *Chem Mater,* 15,1367–1377.

Henning, T.D., Sutton, E.J., Kim, A., Golovko, D., Horvai, A., Ackerman, L., Sennino, B., McDonald, D., Lotz, J., & Daldrup-Link, H.E. (2009). The influence of ferucarbotran on the chondrogenesis of human mesenchymal stem cells. *Contrast media & molecular imaging,* 4(4),165-173.

Horak, D., Babic, M., Jendelova, P., Herynek, V., Trchová, M., Pientka, Z., Pollert, E., Hájek, M., & Syková, E. (2007). D-mannose-modified iron oxide nanoparticles for stem cell labeling. *Bioconjug Chem,* 18,635–644.

Hsiao, J.K., Tai, M.F., Chu, H.H., Chen, S.T., Li, H., Lai, D.M., Hsieh, S.T., Wang, J.L., & Liu, H.M. (2007). Magnetic nanoparticle labeling of mesenchymal stem cells without transfection agent: cellular behavior and capability of detection with clinical 1.5 T magnetic resonance at the single cell level. *Magn Reson Med,* 58(4),717-724.

Hsiao, J.K., Chu, H.H., Wang, Y.H., Lai, C.W., Chou, P.T., Hsieh, S.T., Wang, J.L., & Liu, H.M. (2008). Macrophage physiological function after superparamagnetic iron oxide labeling. *NMR Biomed,* 21(8),820-829.

Hsiao, J.K., Tsai, C.P., Chung, T.H., Hung, Y., Yao, M., Liu, H.M., Mou, C.Y., Yang, C.S., Chen, Y.C., & Huang, D.M. (2008). Mesoporous silica nanoparticles as a delivery system of gadolinium for effective human stem cell tracking. *Small,* 4(9),1445-1452.

Huang, D.M., Hung, Y., Ko, B.S., Hsu, S.C., Chen, W.H., Chien, C.L., Tsai, C.P., Kuo, C.T., Kang, J.C., Yang, C.S., Mou, C.Y., & Chen, Y.C. (2005). Highly efficient cellular labeling of mesoporous nanoparticles in human mesenchymal stem cells: implication for stem cell tracking. *FASEB J.,* 19(14),2014-2016.

Jackson, J.S., Golding , J.P., Chapon, C., Jones, W.A., & Bhakoo, K.K. (2010) Homing of stem cells to sites of inflammatory brain injury after intracerebral and intravenous administration: a longitudinal imaging study. *Stem Cell Res Ther.* 1(2),17.

Jain, K.K. (2005). Role of nanobiotechnology in developing personalized medicine for cancer. *Technol Cancer Res Treat,* 4,645–650.

Jain, T.K. Reddy, M.K., Morales, M.A., Leslie-Pelecky, D.L., & Labhasetwar, V. (2008). Biodistribution, clearance, and biocompatibility of iron oxide magnetic nanoparticles in rats. *Mol. Pharm.,* 5, 316–327.

Janic, B., Iskander,,A.S., Rad, A.M., Soltanian-Zadeh, H., & Arbab, A.S. (2008). Effects of Ferumoxides – Protamine Sulfate Labeling on Immunomodulatory Characteristics of Macrophage-like THP-1 Cells. *PLoS ONE,* 3(6), e2499.

Jodin, L., Dupuis, A.C., Rouviere, E., & Reiss, P. (2006). Infl uence of the catalyst type on the growth of carbon nanotubes via methane chemical vapor deposition. *J Phys Chem B,* 110,7328–7333.

Kateb, B., Chiu, K., Black, K.L., Yamamoto, V., Khalsa, B., Ljubimova, J.Y., Ding, H. , Patil, R., Portilla-Arias, J.A. , Modo, M. , Moore, D.F., Farahani, K., Okun, M.S., Prakash, N., Neman, J., Ahdoot, D., Grundfest, W., Nikzad, S., & Heiss, J.D. (2011). Nanoplatforms for constructing new approaches to cancer treatment, imaging, and drug delivery: What should be the policy? *NeuroImage,* 54, S106–S124

Kohler, N., Fryxell, G.E., & Zhang, M. (2004). A bifunctional poly(ethylene glycol) silane immobilized on metallic oxide-based nanoparticles for conjugation with cell targeting agents. *J Am Chem Soc,* 126,7206–7211.

Kostura, L., Kraitchman, D.L., Mackay, A.M., Pittenger, M.F., & Bulte, J.W. (2004). Feridex labeling of mesenchymal stem cells inhibits chondrogenesis but not adipogenesis or osteogenesis. *NMR in Biomed,* 17(7),513-517.

Kraitchman, D.L., Heldman, A,W,, Atalar, E,, Amado, L.C., Martin, B.J., Pittenger, M.F., Hare, J.M.,& Bulte, J.W. (2003). In Vivo Magnetic Resonance Imaging of Mesenchymal Stem Cells in Myocardial Infarction. *Circulation,* 107(18), 2290-2293.

Kreuter, J., Shamenkov, D., Petrov, V., Ramge, P., Cychutek, K., Koch-Brandt, C., & Alyautdin, R. (2002). Apolipoprotein-mediated transport of nanoparticle-bound drugs across the blood-brain barrier. *J Drug Target,* 10,317–325.

Kumagai, M., Imai, Y., Nakamura, T., Yamasaki, Y., Sekino, M., Ueno, S., Hanaoka, K., Kikuchi, K., Nagano, T., Kaneko, E., Shimokado, K., & Kataoka, K. (2007). Iron hydroxide nanoparticles coated with poly(ethylene glycol)-poly(aspartic acid) block copolymer as novel magnetic resonance contrast agents for in vivo cancer imaging. *Colloids Surf B Biointerfaces,* 56,174–81.

Kumar, A., Sahoo, B., Montpetit, A., Behera, S., Lockey, R.F., & Mohapatra, S.S. (2007). Development of hyaluronic acid-Fe2O3 hybrid magnetic nanoparticles for targeted delivery of peptides. *Nanomedicine,* 3,132–137.

Laconte, L.E., Nitin, N., Zurkiya, O., Caruntu, D., O'Connor, C.J., Hu, X., & Bao, G. (2007). Coating thickness of magnetic iron oxide nanoparticles affects R(2) relaxivity. *J Magn Reson Imaging,* 26,1634–41.

Lai, C.W., Wang, Y.H., Lai, CH., Yang, M.J., Chen, C.Y., Chou, P.T., Chan, C.S., Chi, Y., Chen, Y.C., & Hsiao, J.K. (2008). Iridium-complex-functionalized Fe3O4/SiO2 core/shell nanoparticles: a facile three-in-one system in magnetic resonance imaging, luminescence imaging, and photodynamic therapy. *Small,* 4(2), 218-224.

Lee, H., Lee, E., Kim, do K., Jang, N.K., Jeong, Y.Y., & Jon, S. (2006). Antibiofouling polymer-coated superparamagnetic iron oxide nanoparticles as potential magnetic resonance contrast agents for in vivo cancer imaging. *J Am Chem Soc,* 128,7383–7389.

Lee, J.H., Huh, Y.M., Jun, Y.W., Seo, J.W., Jang, J.T., Song, H.T., Kim, S., Cho, E.J., Yoon, H.G., Suh, J.S., & Cheon, J. (2007) Artificially engineered magnetic nanoparticles for ultra-sensitive molecular imaging. *Nat. Med.* 13, 95–99.

Lee, H., Yu, M.K., Park, S., Moon, S., Min, J.J., Jeong, Y.Y., Kang, H.W., & Jon, S. (2007). Thermally cross-linked superparamagnetic iron oxide nanoparticles: synthesis and application as a dual imaging probe for cancer in vivo. *J Am Chem Soc,* 129,12739–12745.

Lewinski, N., Colvin, V., & Drezek, R. (2008). Cytotoxicity of nanoparticles. *Small,* 4,26–49.

Long, T.C., Saleh, N., Tilton, R.D., Lowry, G.V., & Veronesi, B. (2006). Titanium dioxide (P25) produces reactive oxygen species in immortalized brain microglia (BV2): implications for nanoparticle neurotoxicity. *Environ Sci Technol,* 40,4346–4352.

Lu, C.W., Hung, Y., Hsiao, J.K., Yao, M., Chung, T.H., Lin, Y.S., Wu, S.H., Hsu, S.C., Liu, H.M., Mou, C.Y., Yang, C.S., Huang, D.M., & Chen, Y.C. (2007). Bifunctional magnetic silica nanoparticles for highly efficient human stem cell labeling. *Nano Lett,* 7(1), 149-154.

Lunov, O., Syrovets, T., Büchele, B., Jiang, X., Röcker, C., Tron, K., Nienhaus, G.U., Walther, P., Mailänder, V., Landfester, K., & Simmet, T. (2010a). The effect of carboxydextran-coated superparamagnetic iron oxide nanoparticles on c-Jun N-terminal kinase-mediated apoptosis in human macrophages. *Biomaterials,* 31(19),5063-5071.

Lunov, O., Syrovets, T., Röcker, C., Tron, K., Nienhaus, G.U., Rasche, V., Mailänder, V., Landfester, K., & Simmet, T. (2010b). Lysosomal degradation of the carboxydextran shell of coated superparamagnetic iron oxide nanoparticles and the fate of professional phagocytes. *Biomaterials,* 31(34), 9015-9022.

Lynch, I., Dawson, K.A., & Linse, S. (2006). Detecting cryptic epitopes created by nanoparticles. *Sci STKE,* 2006, pe14

Maier-Hauff, K., Rothe, R., Scholz, R., Gneveckow, U., Wust, P., Thiesen, B., Feussner, A., von Deimling, A., Waldoefner, N., Felix, R., & Jordan, A. (2007). Intracranial thermotherapy using magnetic nanoparticles combined with external beam radiotherapy: results of a feasibility study on patients with glioblastoma multiforme. *Journal of Neuro-Oncology,* 81(1), 53-60.

Maier-Hauff, K., Ulrich, F., Nestler, D., Niehoff, H., Wust, P., Thiesen, B., Orawa, H., Budach, V., & Jordan, A. (2011). Efficacy and safety of intratumoral thermotherapy using magnetic iron-oxide nanoparticles combined with external beam radiotherapy on patients with recurrent glioblastoma multiforme. *Journal of Neuro-Oncology,* 103(2), 317-24.

Metz, S., Bonaterra, G., Rudelius, M., Settles, M., Rummeny, E.J., & Daldrup-Link, H.E. (2004). Capacity of human monocytes to phagocytose approved iron oxide MR contrast agents in vitro. *Eur Radiol,* 14(10), 1851-1858.

Mikhaylova, M., Kim, D.K., Bobrysheva, N., Osmolowsky, M., Semenov, V., Tsakalakos, T., & Muhammed, M. (2004). Superparamagnetism of magnetite nanoparticles: dependence on surface modification. *Langmuir*, 20,2472–2477.

Minchin, R.F., Orr, R.J., Cronin, A.S., & Puls, R.L. (1999) The pharmacology of gene therapy. *Croat Med J* 40,381–391.

Moghimi, S.M. & Bonnemain, B. (1999). Subcutaneous and intravenous delivery of diagnostic agents to the lymphatic system: applications in lymphoscintigraphy and indirect lymphography. *Advanced Drug Delivery Reviews*, 37(1-3), 295-312.

Moghimi, S.M. & Hunter, A.C. (2001). Recognition by Macrophages and Liver Cells of Opsonized Phospholipid Vesicles and Phospholipid Headgroups. *Pharmaceutical Research*, 18(1), 1-8.

Moghimi, S.M., Hunter, A.C., & Murray, J.C. (2005). Nanomedicine: current status and future prospects. *Faseb J.*, 19(3),311-330.

Montet, X., Montet-Abou, K., Reynolds, F., Weissleder, R., & Josephson, L. (2006). Nanoparticle imaging of integrins on tumor cells. *Neoplasia*, 8,214–222.

Nel, A., Xia, T., Mädler, L., & Li, N. (2006). Toxic potential of materials at the nanolevel. Science 311, 622–627.

Nishimori, H., Kondoh, M., Isoda, K., Tsunoda, S., Tsutsumi, Y., Yagi, K. (2009) Silica nanoparticles as hepatotoxicants. *Eur J Pharmaceut Biopharmaceut*, 72,496–501.

Nitin, N., LaConte, L.E., Zurkiya, O., Hu, X., Bao, G. (2004). Functionalization and peptidebased delivery of magnetic nanoparticles as an intracellular MRI contrast agent. *J Biol Inorg Chem*, 9,706–712.

Noh, Y.-W., Jang, Y.S., Ahn, K.J., Lim, Y.T., & Chung, B.H. (2011). Simultaneous in vivo tracking of dendritic cells and priming of an antigen-specific immune response. *Biomaterials*, 32(26), 6254-6263.

Owens, 3rd D.E., & Peppas, N.A. (2006). Opsonization, biodistribution, and pharmacokinetics of polymeric nanoparticles. *Int J Pharmaceut*, 307,93–102.

Pawelczyk, E., Arbab, A.S., Pandit, S., Hu, E., & Frank, J.A. (2006). Expression of transferrin receptor and ferritin following ferumoxides-protamine sulfate labeling of cells: implications for cellular magnetic resonance imaging. *NMR in Biomed*, 19(5), 581-592.

Peng, Y.K., Lai, C.W., Liu, C.L., Chen, H.C., Hsiao, Y.H., Liu, W.L., Tang, K.C., Chi, Y., Hsiao, J.K., Lim, K.E., Liao, H.E., Shyue, J.J., & Chou, P.T. (2011). A new and facile method to prepare uniform hollow MnO/functionalized mSiO core/shell nanocomposites. *ACS nano*, 5(5), 4177-4187.

Rawat, M., Singh, D., Saraf, S., & Saraf, S. (2006). Nanocarriers: promising vehicle for bioactive drugs. *Biol. Pharm. Bull.*, 29, 1790–1798.

Rogers, W.J. & Basu, P. (2005). Factors regulating macrophage endocytosis of nanoparticles: implications for targeted magnetic resonance plaque imaging. *Atherosclerosis*, 178, 67–73.

Sellins, K.S. & Cohen, J.J. (1991). Hyperthermia induces apoptosis in thymocytes. *Radiation Research*, 126(1), 88-95.

Shen, L.-F., Chen, J., Zeng, S., Zhou, R.R., Zhu, H., Zhong, M.Z., Yao, R.J., & Shen, H. (2010). The Superparamagnetic Nanoparticles Carrying the E1A Gene Enhance the Radiosensitivity of Human Cervical Carcinoma in Nude Mice. *Molecular Cancer Therapeutics*, 9(7), 2123-2130.

Shen, T., Weissleder, R., Papisov, M., Bogdanov, A. Jr, & Brady, T.J. (1993). Monocrystalline iron oxide nanocompounds (MION): physicochemical properties. *Magn Reson Med,* 29,599–604.

Shi, X., Thomas, T.P., Myc, L.A., Kotlyar, A., & Baker, J.R. Jr. (2007). Synthesis, characterization, and intracellular uptake of carboxyl-terminated poly(amidoamine) dendrimer-stabilized iron oxide nanoparticles. *Phys Chem Chem Phys,* 9, 5712–5720.

Silva, A.C., Oliveira, T.R., Mamani, J.B., Malheiros, S.M., Malavolta, L., Pavon, L.F., Sibov, T.T., Amaro, E. Jr, Tannús, A., Vidoto, E.L., Martins, M.J., Santos, R.S., & Gamarra, L.F. (2011). Application of hyperthermia induced by superparamagnetic iron oxide nanoparticles in glioma treatment. *International journal of nanomedicine,* 6, 591-603.

Tavaré, R., Sagoo, P., Varama, G., Tanriver, Y., Warely, A., Diebold, S.S., Southworth, R., Schaeffter, T., Lechler, R.I., Razavi, R., Lombardi, G., & Mullen, G.E. (2011). Monitoring of in vivo function of superparamagnetic iron oxide labelled murine dendritic cells during anti-tumour vaccination. *PLoS ONE,* 6(5), e19662.

Thorek, D.L., Chen, A.K., Czupryna, J., & Tsourkas, A. (2006). Superparamagnetic iron oxide nanoparticle probes for molecular imaging. *Ann Biomed Eng,* 34,23–38.

Tsai, C.P., Hung, Y., Chou, Y.H., Huang, D.M., Hsiao, J.K., Chang, C., Chen, Y.C., & Mou, C.Y. (2008). High-contrast paramagnetic fluorescent mesoporous silica nanorods as a multifunctional cell-imaging probe. *Small,* 4(2),186-191.

Veiseh, O., Sun, C., Gunn, J., Kohler, N., Gabikian, P., Lee, D., Bhattarai, N., Ellenbogen, R., Sze, R., Hallahan, A., Olson, J., & Zhang, M. (2005). Optical and MRI multifunctional nanoprobe for targeting gliomas. *Nano Lett,* 5,1003–1008.

Wang, Y.X., Hussain, S.M., & Krestin, G.P. (2001). Superparamagnetic iron oxide contrast agents: physicochemical characteristics and applications in MR imaging. *Eur. Radiol.,* 11, 2319–2331.

Wiwanitkit, V. (2006). Glomerular pore size corresponding to albumin molecular size, an explanation for underlying structural pathology leading to albuminuria at nanolevel. *Renal Fail,* 28,101.

Yang, C.Y., Hsiao, J.K., Tai, M.F., Chen, S.T., Cheng, H.Y., Wang, J.L., & Liu, H.M. (2011). Direct labeling of hMSC with SPIO: the long-term influence on toxicity, chondrogenic differentiation capacity, and intracellular distribution. *Mol Imaging Biol.* 13(3), 443-451.

Yeh, C.-H., Hsiao, J.K., Wang, J.L., & Fuu Sheu, F. (2010). Immunological impact of magnetic nanoparticles (Ferucarbotran) on murine peritoneal macrophages. *Journal of Nanoparticle Research,* 12(1), 151-160.

4

Self-Assembly of Nanoparticles at Solid and Liquid Surfaces

Peter Siffalovic, Eva Majkova, Matej Jergel,
Karol Vegso, Martin Weis and Stefan Luby
Institute of Physics, Slovak Academy of Sciences
Slovakia

1. Introduction

The research field of nanoparticle synthesis and related nanoparticle applied sciences have been steadily growing in the past two decades. The chemical synthesis of nanoparticles was improved up to the point that the organic and inorganic nanoparticle colloids are produced with a low size dispersion and with a well defined nanoparticle shape in large quantities. A stunning feature of a drying nanoparticle colloidal solution is the ability to create self-assembled arrays of nanoparticles. The self-assembled nanoparticle arrays mimic the natural crystals. The size of perfectly ordered domains is limited by the size dispersion of nanoparticles. Consequently the defects in the self-assembled structure are obvious and unavoidable. Despite these defects, the self-assembled nanoparticle arrays represent a new class of nanostructures built on "bottom-up" technological approach to fabrication. The traditional way of "top-down" fabrication technology primarily based on nano-lithography is complex, including many technological steps, time consuming and expensive. The main advantage is the tight control of all parameters governing the final nanostructures. On the other hand, the emerging fabrication technologies based on the self-assembled nanoparticles are fast, less complex and more price competitive. An extensive research is now focused on a deeper understanding of processes that control the self-assembly. New routes for directed or stimulated self-assembly are studied to achieve a tighter control than readily available in the spontaneous self-assembly. In this chapter we will discuss the spontaneous nanoparticle self-assembly with emphasis on characterization of nanoparticle arrays at various stages of the self-assembly process. The main diagnostic technique used throughout this chapter will be the grazing-incidence small-angle X-ray scattering (GISAXS) that represents a reliable and simple monitor of nanoparticle arrangement. The theoretical background of GISAXS and required instrumentation are described in Section 2. The most flexible surface to study the nanoparticle self-assembly processes is the liquid surface. The Section 3. reviews the latest results of studies combing the GISAXS technique with Langmuir nanoparticle layers on the water subphase. Almost all relevant nanoparticle applications rely on self-assembled arrays on solid surfaces. The Section 4 describes in detail the possibilities of nanoparticle transfer from liquid onto solid surfaces. The post-processing of self-assembled nanoparticle arrays and their applications are reviewed in the last Section 5.

2. SAXS/GISAXS techniques and their employment for nanoparticle research

The transmission (TEM) and scanning (SEM) electron microscopy provide information on the nanoparticle shape, average size and size distribution. However, this information is usually obtained after numerical evaluation of real space micrographs from limited data sets. Alternative approach is based on the angle-resolved analysis of scattered X-rays or neutrons from the nanoparticles and their assemblies. In this chapter we will employ the small-angle X-ray scattering (SAXS) (Guinier and Fournet 1955) for the nanoparticle colloidal solutions. For nanoparticles immobilized at interfaces, a related technique so-called grazing-incidence small-angle X-ray scattering (GISAXS) is used that has been recently reviewed (Renaud, Lazzari et al. 2009). A general scheme of the GISAXS experiment is shown in Fig. 1.

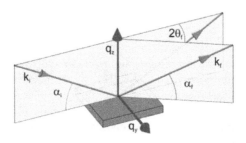

Fig. 1. The GISAXS measurement geometry

The collimated X-ray beam defined by $\vec{k}_i$ is incident under a small grazing angle on the sample surface. The scattered radiation is recorded by a two dimensional X-ray detector. Each point at the detector plane receives the scattered radiation given by a set of two angles ($2\theta_f$, α_f) that corresponds to a unique scattering vector $\vec{q}$ in the reciprocal space. The relationship between the scattering vector in reciprocal space and the scattering angles in the real space is given by the following equations (Müller-Buschbaum 2009)

$$q_x = \frac{2\pi}{\lambda}\Big[cos(2\theta_f)cos(\alpha_f) - cos(\alpha_i)\Big]$$

$$q_y = \frac{2\pi}{\lambda}\Big[sin(2\theta_f)cos(\alpha_f)\Big]$$ \hfill (1)

$$q_z = \frac{2\pi}{\lambda}\Big[sin(\alpha_f) + sin(\alpha_i)\Big]$$

The SAXS/GISAXS signal is given by constructive interferences of X-ray waves partially scattered on individual nanoparticles. The total scattered intensity (also called the scattering cross-section) at specific $\vec{q}$ vector in the reciprocal space is given as (Feigin, Svergun et al. 1987)

$$I(\vec{q}) = \sum_{i=1}^{N}\sum_{j=1}^{N} F^i(\vec{q}).F^{j,*}(\vec{q}).exp\Big[i\vec{q}.(\vec{r}_i - \vec{r}_j)\Big]$$ \hfill (2)

where N is the total number of nanoparticles, $F^i(\vec{q})$ is the form-factor of the ith nanoparticle and $\vec{r}_i$ defines the position of the ith nanoparticle. Within the simple Born (kinematic) approximation (BA) the nanoparticle form-factor is simply given by the Fourier transform of the nanoparticle density function $\rho_i(\vec{r})$ as follows (Glatter and Kratky 1982)

$$F^i(\vec{q}) = \int \rho_i(\vec{r}).exp(i\vec{q}.\vec{r})d\vec{r} \qquad (3)$$

For the nanoparticles immobilized at interfaces we have to include the refraction/reflection phenomena at the interfaces and the associated multiple scattering events. This is treated in detail within the framework of the distorted-wave Born approximation (DWBA) which introduces a modified form-factor for each nanoparticle confined near the interface (Holý, Pietsch et al. 1999). A detailed survey of the DWBA theory can be found in the following reference (Renaud, Lazzari et al. 2009). A typical DWBA effect is the presence of the Yoneda enhancement at the critical exit angle in the GISAXS patterns (Yoneda 1963). In many cases we can avoid the DWBA multiple scattering terms by recording the GISAXS pattern at the incident angle several times larger than the critical angle for the total X-ray reflection of the supporting substrate (Daillant and Gibaud 2009). If we assume that the nanoparticles can be described by an average form-factor $\left\langle |F(\vec{q})|^2 \right\rangle$ than the eq. (2) in BA can be rearranged as follows

$$I(\vec{q}) = N\left\langle |F(\vec{q})|^2 \right\rangle S(\vec{q}) \qquad (4)$$

Here the $S(\vec{q})$ represent the nanoparticle interference function. The nanoparticle interference function is the reciprocal space equivalent of the nanoparticle pair correlation function $P(\vec{r})$ defined in real space (Lazzari 2009). The pair correlation function is proportional to the probability of finding a nanoparticle at the position vector $\vec{r}$ centered at an arbitrarily selected nanoparticle. This function is directly accessible from the TEM/SEM micrographs.

The GISAXS experimental technique was confined for a long time to synchrotron facilities as the scattering cross-section is generally very low. Each synchrotron ring has a dedicated SAXS beamline that can support conventional GISAXS setup. The Fig. 2 shows the typical GISAXS scheme of the BW4 beamline at the DORIS III ring at HASYLAB, Hamburg (Stribeck 2007). The front-end of the experimental setup is a wiggler that generates the X-ray radiation. The crystal monochromator is used to select a single wavelength typically at 0.139 nm. The radiation is further conditioned with slits and two cylindrical mirrors to focus the radiation in both directions at the detector plane. The additional beryllium X-ray lenses can be attached to focus the radiation at the sample position (Roth, Dohrmann et al. 2006).

The distance between the sample and detector can vary between 3 m and 13 m that allows flexibility in the accessible range of the reciprocal space. The two-dimensional (2D) X-ray CCD detector is used to record the X-ray radiation scattered by the sample. The primary and specularly reflected beams are suppressed by the beamstops.

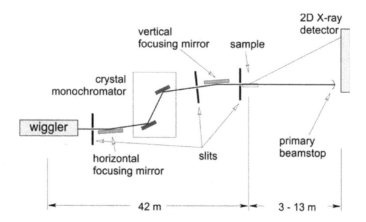

Fig. 2. The sketch of the experimental GISAXS geometry at BW4 beamline, HASYLAB

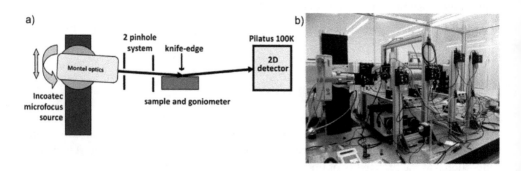

Fig. 3. a) The scheme of the laboratory GISAXS setup and (b) the photograph of its realization at Institute of Physics SAS.

The latest advances in the low-power X-ray generators and the efficient X-ray optics opened a new era of laboratory equipments suitable for GISAXS measurements (Michaelsen, Wiesmann et al. 2002). Nowadays already several companies (Bruker AXS, Anton Paar, Hecus XRS, Rigaku) supply complete X-ray solutions supporting GISAXS measurement modes for solid-state samples. The Fig. 3a and Fig. 3b show the laboratory setup scheme and the photograph of a home-built GISAXS instrumentation developed at the Institute of Physics SAS, respectively (Siffalovic, Vegso et al. 2010). This setup supports GISAXS measurements on solid as well as liquid surfaces. The core of the experimental apparatus is a compact low-power (30 W) X-ray source (Cu-K$_\alpha$) equipped with a loosely focusing X-ray Montel optics (Wiesmann, Graf et al. 2009). The source can be freely rotated and translated in the vertical direction. This is important for the precise adjustment of the incident angle in the GISAXS measurements at liquid surfaces. The unwanted scattered radiation is eliminated by laser-beam precisely cut tungsten pinholes. The sample is fixed on a goniometer that allows precise height and tilt adjustments.

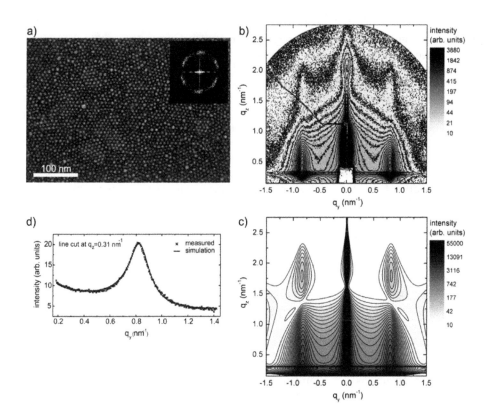

Fig. 4. a) The SEM micrograph of Fe-O self-assembled nanoparticles. Measured (b) and simulated (c) GISAXS pattern of self-assembled nanoparticles. d) The extracted line-cut from the measured GISAXS pattern along with the simulation.

The auxiliary knife-edge blade is used to reduce the parasitic air-scattering. The additional vacuum flight-tube can be inserted between the sample and the X-ray detector to reduce the air scattering and absorption. The detector used is a fast acquisition CMOS based 2D X-ray detector of PILATUS detector family (Kraft, Bergamaschi et al. 2009).

To illustrate the capability of the GISAXS technique to characterize the self-assembled nanoparticle monolayers we use an example of iron oxide nanoparticles (Siffalovic, Majkova et al. 2007). The Fig. 4a shows the SEM image of a self-assembled array of iron oxide nanoparticles. The inset of Fig. 4a shows the Fourier transform of SEM micrograph with partially smeared-out spots corresponding to the hexagonal arrangement. The smearing-out is due to mutually misaligned nanoparticle domains originating from finite nanoparticle size dispersion which is in sharp contrast to natural atomic crystals. The Fig. 4a and 4b show the measured and simulated GISAXS pattern, respectively. The characteristic side maxima located at the $q_y \approx \pm 0.82 nm^{-1}$ are the "finger prints" of the self-assembly in the nanoparticle array. In the first approximation, the mean interparticle separation can be estimated from the side maximum position in the reciprocal space as $\Delta \approx 2\pi/q_y \approx 7.7nm$. This simple estimation is valid only for a slowly varying nanoparticle form-factors within the kinematic BA. A precise fitting of the measured GISAXS data using the full DWBA theory can provide further information on the nanoparticle size and size dispersion as well as their correlation length (Lazzari 2002). The Fig. 4d shows a line cut extracted from the measured GISAXS pattern with the corresponding fit. The fitted nanoparticle diameter was 6.1±0.6 nm and the lateral correlation length in the nanoparticle array was 87 nm. It has to be noted that colloidal nanoparticles are covered by a surfactant shell to avoid their spontaneous agglomeration in colloidal suspensions. In the case of Fe-O nanoparticles discussed above, oleic acid and oleylamine were used. A GISAXS pattern fitting provides basic information on the metallic-like nanoparticle core size while the organic shell is rather invisible for X-rays. On the other hand, the positions of the side maxima in the GISAXS pattern are always connected with the interparticle distance which is affected by the surfactant shell. This example clearly demonstrates the ability of GISAXS technique to extract main nanoparticle parameters in the self-assembled arrays. The main advantage is that the GISAXS technique does not require any specific sample environment conditions such as vacuum nor special sample preparation. On the other hand it can be applied even in very aggressive environments such as UV/ozone reactor (Siffalovic, Chitu et al. 2010). Moreover, a rapid GISAXS data acquisition in millisecond range can be used for a real-time in-situ probing of nanoparticle reactions and self-assembly processes (Siffalovic, Majkova et al. 2008).

3. Nanoparticle self-assembly at liquid/air interfaces

In the last ten years we have seen a tremendous progress in the colloidal nanoparticle chemistry (Feldheim 2002; Nagarajan 2008; Niederberger and Pinna 2009). The refined chemical synthesis routes can produce large quantities of highly monodisperse nanoparticles in colloidal solutions with the size dispersion below 10 % (Park, An et al. 2004). The low nanoparticle dispersion is the "holy grail" of the large-scale nanoparticle self-assembly (Pileni 2005). Being able to prepare nanoparticles with zero size dispersion, we could fabricate genuine artificial nanoparticle crystals competing with natural ones in terms of the structure perfection and long-range order. However the finite nanoparticle size dispersion permits only a limited extent of ordering in nanoparticle self-assembled arrays. A typical model for description of the real nanoparticle assemblies is the paracrystal model (Hosemann and Bagchi 1962; Guinier 1963). Here a paracrystal order parameter summed up with the mean interparticle distance defines degree of the array perfection.

The colloidal nanoparticle solutions can be applied on a solid substrate directly or in two steps, utilizing liquid surface for self-assembly with a subsequent transfer onto a solid substrate. Drop casting followed by solvent evaporation is an example of the former method (Chushkin, Ulmeanu et al. 2003) that proved to be successful e.g. for preparation of large-area self-assembled arrays of noble metal nanoparticles with the diameter of a few tens nm. In addition to the nanoparticle size, surfactant type affects the self-assembly as well. For smaller nanoparticles, such as those presented in this chapter with the diameter below 10 nm, a direct application of the colloidal nanoparticle solutions on solid substrate produces only locally well assembled regions but is not suitable for large-area nanoparticle depositions. Here, the latter above mentioned method is promising as it will be shown later. The GISAXS technique can be employed to track the nanoparticle assemblies in rapidly drying colloidal solution at solid surfaces (Siffalovic, Majkova et al. 2007). We used the focused X-ray beam to map the nanoparticle self-assembly at arbitrary selected position within the colloidal drop. The Fig. 5 shows the three typical GISAXS patterns.

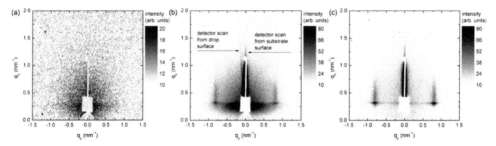

Fig. 5. The GISAXS pattern recorded from a drying colloidal Fe-O nanoparticle drop at three different stages: a) directly after drop casting, b) intermediate phase. c) dried colloidal drop.

The Fig. 5a shows the GISAXS pattern directly after application of a colloidal Fe-O nanoparticle solution onto silicon substrate. The GISAXS pattern does not show any maxima typical for self-assembled nanoparticle layers. The visible scattering in the GISAXS pattern is characteristic for a diluted nanoparticle solution and can be described by the nanoparticle form-factor. The Fig. 5b shows the intermediate state when the X-ray beam partially passes through the colloidal drop surface. The scattering streaks originating from interfaces also called "detector scans" are visible. The first one can be attributed to the scattering from the substrate surface and the second one originates from the colloidal drop surface. The angle between the two detector streaks directly maps the angle between the normal of substrate surface and the normal of the probed colloidal drop surface. The side maxima belong to the already dried self-assembled areas. The Fig. 5c shows the final GISAXS pattern after the colloidal solution is completely evaporated. The interparticle distance of final nanoparticle assembly are clearly manifested in the GISAXS pattern by the side maxima.

The spatially static GISAXS technique can track the nanoparticle assembly only in one selected probing volume within the evaporating colloidal drop. In order to monitor various probe volumes inside the colloidal nanoparticle drop during the self-assembly process we introduced a scanning GISAXS technique. The scanning GISAXS method is based on the fast vertical or horizontal scanning of the evaporating colloidal drop by the probing X-ray beam (Siffalovic, Majkova et al. 2008). The sketch of the scanning GISAXS technique is shown in Fig. 6a. The colloidal drop composed of iron oxide nanoparticles dispersed in toluene was applied onto silicon substrate located on a vertically scanning goniometer. As the evaporating drop was gradually scanned across the incoming X-ray beam we continuously recorded X-ray scattering from three different drop zones. In the zone Z0 the X-ray beam passed above the evaporating drop. These data were used for the background correction. In the zone Z1 we recorded exclusively the X-ray scattering originating from the drying drop surface and drop interior. In the zone Z2 we additionally detected the X-ray scattering coming from the substrate surface. The Fig. 6b shows the line cuts extracted from the GISAXS frames taken in zone Z1 corresponding to the three different stages of the colloidal drop evaporation process: 1.) directly after drop casting, 2.) intermediate state, and 3.) final state characterized by the complete solvent evaporation. It is important to notice that the experimental data for all three evaporation stages can be fitted solely using the nanoparticle form-factor function. According to the eq. (4) the interference function is constant in this case, i.e. $S(\vec{q}) = 1$.

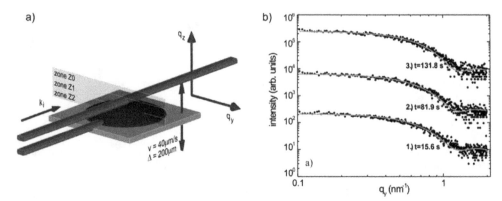

Fig. 6. a) The scheme of the GISAXS scanning technique. b) The GISAXS pattern line cuts at the critical exit angle for the three different stages of the colloidal Fe-O nanoparticle drop evaporation.

This means that the nanoparticles do not create self-assembled domains at the evaporating drop surface or in its volume at any time that suggests the origin of the nanoparticle self-assembly to be located at the three-phase boundary as predicted for a drying drop of dispersed particles (Deegan, Bakajin et al. 1997). The scanning GISAXS technique clearly demonstrates the ability to track the nanoparticle self-assembly process in real-time with millisecond time resolution.

As mentioned above, colloidal nanoparticles are usually terminated by surfactant molecules to avoid spontaneous agglomeration in colloidal suspensions. The nanoparticles with hydrophobic termination allow self-assembly at liquid/air interfaces and formation of Langmuir films in the form of simple 2D systems (Ulman 1991). Controlling the surface pressure by changing the nanoparticle layer area and the temperature of the subphase, we can produce large-area and homogenous self-assembled nanoparticle layers. The electron microscopy techniques including SEM, TEM or scanning probe techniques (AFM, STM) cannot be utilized to monitor the nanoparticle self-assembly at liquid/air interface. The visible/UV optical microscopy and Brewster angle microscopy are limited in resolution due to diffraction limit (Born and Wolf 1999). For a certain kind of metal and metal oxide nanoparticles exhibiting plasmonic properties (Au, Ag, Al, Cu) the interparticle distance can be indirectly monitored by the energy shift in localized surface plasmon resonance due to the dipole-dipole coupling of excited plasmons in the self-assembled nanoparticle arrays (Rycenga, Cobley et al. 2011). On the other hand the GISAXS technique can be employed to directly monitor the interparticle distance in self-assembled arrays directly in the Langmuir trough. The laboratory GISAXS setup shown in Fig. 3 was used to record the GISAXS patterns of Ag nanoparticles (6.2±0.7 nm) directly in the Langmuir trough. The GISAXS patterns of self-assembled Ag nanoparticles with oleic acid as surfactant at the surface pressures of 16 mN/m and 26 mN/m are shown in Fig. 7a and Fig. 7b, respectively.

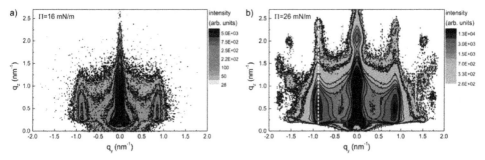

Fig. 7. The GISAXS patterns of self-assembled Ag nanoparticle Lagmuir films at surface pressure a) 16 mN/m and b) 26 mN/m.

The surface pressure of 16 mN/m corresponds to a closed nanoparticle monolayer on the water surface. The interference function produces two symmetrical side maxima at $q_y \approx \pm 0.87 nm^{-1}$ (truncation rods) corresponding to the average interparticle distance of 7.2 nm. The higher order side maxima are absent due to the short exposure time. The two-dimensional nanoparticle monolayer has a constant interference function in the q_z direction where the modulation visible on the truncation rods is produced solely by the nanoparticle form-factor (Holý, Pietsch et al. 1999). At the surface pressure of 26 mN/m, the second nanoparticle layer forms and changes the observed GISAXS pattern (Vegso, Siffalovic et al. 2011). The newly formed nanoparticle vertical correlation perpendicular to the Langmuir film plane results in the modulation of the observed truncation rod depicted by the dashed white line in Fig. 7b. It can be shown that the modulation along the truncation rod is associated with the second nanoparticle layer laterally shifted in analogy with the "AB stacking" in solid state crystals (Kittel 2005). The presence of the second layer can be verified also by distinct second order maxima in Fig. 7b. The presented GISAXS results show the possibility to study not only the lateral but also the vertical nanoparticle correlations in 3D nanoparticle assemblies that is due to the ability of GISAXS to inspect non-destructively buried layers and interfaces. This useful feature of the GISAXS technique to study the buried vertical correlations of interfaces was already applied in studies of multilayered thin films (Salditt, Metzger et al. 1994; Siffalovic, Jergel et al. 2011).

Recently we have performed in-situ real-time studies of compression and decompression of Ag nanoparticle Langmuir films. We were interested in the correlation between the macroscopic elastic properties of nanoparticle layers and microscopic layer parameters like the interparticle distance. As a convenient measure of macroscopic elastic properties we use the surface elastic modulus defined as (Barnes, Gentle et al. 2005)

$$E = -A\left(\frac{\partial \Pi}{\partial A}\right)_T \qquad (1)$$

Here Π is the measured surface pressure of the nanoparticle layer with the area A at a constant subphase temperature T. The Fig. 8 shows the evaluated side maximum position along the q_y direction in the GISAXS reciprocal space map similar to the one shown in Fig. 7a.

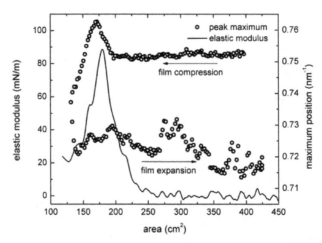

Fig. 8. The evaluated GISAXS peak maximum position and the surface elastic modulus of the Ag nanoparticle layer at water/air interface as a function of the layer area.

After spreading the nanoparticle solution onto the water subphase, the nanoparticles assemble into small clusters with hexagonal ordering that has been identified by independent ex situ experiments (to be published). Increasing the surface pressure by reducing the layer area results in the formation of a continuous monolayer without a change of the interparticle distance. This compression stage is characterized by a constant elastic modulus as the isolated nanoparticle clusters are joining into larger entities. At surface area of approximately 250 cm² we observe an increase in the elastic modulus peaking at the area of 180 cm². This stage can be associated with the densification of the nanoparticle layer accompanied by the nanoparticle rearrangements along the individual cluster boundaries and cluster coalescence. At the maximum of surface elastic modulus we observe also a slight compaction of the nanoparticle layer at nanoscale indicated by the change of the interparticle distance. This phase ends up with a compact nanoparticle layer. A further compression of the nanoparticle layer results in the formation of a second nanoparticle layer that induces a sudden drop in the elastic modulus and significant release of the mean interparticle distance. The nanoparticles forming the second layer create vacancies in the first one that is accompanied by deterioration of the order in the first nanoparticle layer. In this case the paracrystal model of the nanoparticle layer predicts a shift of the maximum to lower q_y values in the reciprocal space (Lazzari 2009) that was confirmed by this experimental observation. After the decompression the interparticle distance in the nanoparticle layer does not relax to the initial value. It has to be noted that the second layer formation and tendency to form 3D ordered nanoparticle assemblies was demonstrated here for Ag nanoparticles with oleic acid as surfactant, however, other types of metallic nanoparticles with other type of surfactant may behave differently. This example shows the benefit of GISAXS technique to precisely monitor microscopic parameters of the nanoparticle assemblies prior to the deposition onto solid substrates that will be discussed in the following section.

4. Transfer of self-assembled layers from liquid onto solid surfaces

In the previous section we discussed the formation of nanoparticle monolayers at water/air interface. The Langmuir film represented by self-assembled nanoparticle monolayer seems to be the most promising candidate for the homogenous deposition of large-area nanoparticle arrays. The two important questions are remaining. The first one is: "What is the suitable surface pressure for deposition and how to monitor it?" The second one is: "How to transfer the Langmuir film onto solid substrate with a minimum damage of the self-assembled layer?" In this section we try to give answers to them.

The first question was partially addressed in the previous section. We have shown the GISAXS technique gives a precise tool to monitor the monolayer formation at nanoscale. In Fig. 8 we showed the evolution of the interparticle distance with increasing surface pressure and we related formation of the second nanoparticle layer to a sudden drop in the observed surface elastic modulus. Additionally, we can track the evolution of the interference function in the q_z direction. We showed that the interference along the q_z axis is a constant function for the nanoparticle monolayer. A new vertical correlation between the two layers may appear with the monolayer collapse accompanying the formation of the second nanoparticle layer as discussed in the previous section. This transition is manifested in the modulation of the X-ray scattered intensity along the truncation rod. The Fig. 7b shows the GISAXS pattern of the nanoparticle multilayer with a new peak formed along the first truncation rod (marked with dashed white line). For the nanoparticle monolayer, the intensity is at maximum at the critical exit angle, i.e. at the Yoneda peak. The formation of the second layer shifts the maximum intensity upward in the q_z direction.

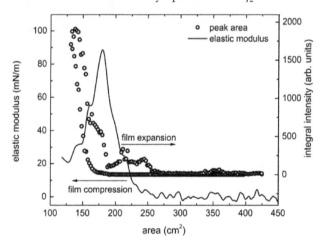

Fig. 9. The integral intensity of the first Bragg peak along the first truncation rod corresponding to the formation of a vertically correlated Ag nanoparticle multilayer as a function of the layer area.

The Fig. 9 shows the integral intensity of the newly formed Bragg peak along the first truncation rod corresponding to the vertically correlated nanoparticles as a function of the surface area. The GISAXS measurement clearly shows that the decrease in the elastic modulus is associated with the formation of the second nanoparticle layer. Moreover we observe a hysteretic behavior during the Langmuir film decompression associated with the irreversibility of the expanded nanoparticle layer that is also documented by the interparticle distance behavior shown in Fig. 8. After opening the barriers the nanoparticle layer does not relax into a monolayer but fragments into small islands still exhibiting a certain amount of nanoparticles in the second layer (see also further). The GISAXS measurements confirmed the assumption that the fully closed nanoparticle monolayer forms short before the monolayer collapse evidenced by a maximum in surface elastic modulus.

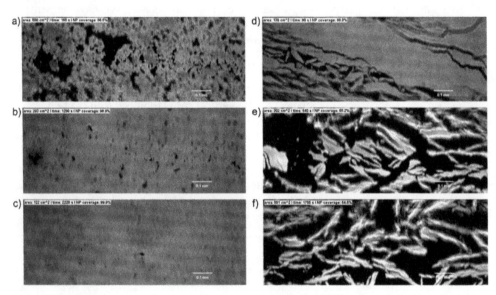

Fig. 10. The BAM images taken at surface areas a) 500 cm², b) 293 cm² and c) 122 cm² taken during the Ag nanoparticle layer compression and BAM images at surface areas d) 139 cm², e) 302 cm² and f) 501 cm² taken during the nanoparticle layer expansion.

The Brewster angle microscopy (BAM) provides further evidence of the nanoparticle monolayer formation at microscale (Henon and Meunier 1991). The laser based BAM provides much better contrast between the nanoparticle monolayer and water subphase than the conventional normal incident microscopy. The Fig. 10a)-10c) show three images taken during the nanoparticle layer compression and Fig. 10d)-10f show three images taken during the nanoparticle layer decompression. The nanoparticle layer was composed of surfactant terminated Ag nanoparticles with a core size of 6.2±0.7 nm. The nanoparticle surfactant was oleic acid. The nanoparticle layer shows vacant areas in Fig. 10a). Decreasing the film area, we close the vacancies and a compact nanoparticle monolayer forms as shown in Fig. 10c). The subsequent expansion of the nanoparticle layer is accompanied by the generation of millimeters long cracks across the nanoparticle layer as shown in Fig. 10d). A further increase of the area available for the nanoparticle expansion leads to the disruption of nanoparticle layer into micrometer large needle-like clusters as shown in Fig. 10e) and Fig. 10f). The hysteretic behavior of the nanoparticle layer at microscale during the compression and decompression cycle is obvious and supports the interpretation of the GISAXS measurements. The Fig. 10 shows selected BAM images during the compression and expansion cycles. However we have recorded a full series of BAM images at 15 second time intervals during the compression cycle. Based on the BAM images we can calculate the average nanoparticle surface coverage based on the ratio between the bright areas that can be attributed to the nanoparticle layer and the black areas corresponding to the water subphase.

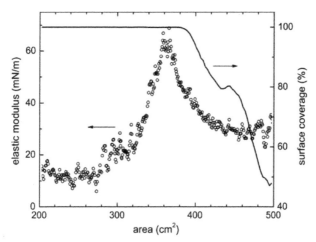

Fig. 11. The nanoparticle surface coverage based on BAM measurement along with the surface elastic modulus as a function of the Ag nanoparticle layer area during compression.

We have to keep in mind that the calculation is correct only at microscale as the nanoscale vacancies are invisible due to the BAM diffraction limit. The Fig. 11 shows the calculated nanoparticle surface coverage as a function of the film area. The graph shows also the calculated elastic modulus based on the measured nanoparticle layer surface pressure. The nanoparticle surface coverage reaches its maximum value of 100% short before the maximum in the film elastic modulus appears during the compression cycle. This is in a very good correlation with the GISAXS measurement that relates the nanoparticle monolayer collapse to the maximum in elastic modulus. The BAM measurements underestimate the nanometer-sized vacancies in the forming monolayer. This is the reason that the BAM indicate formation of nanoparticle monolayer already before the monolayer collapse. An alternative would be the imaging ellipsometry being able to track the nanoparticle layer formation at microscale more quantitatively than the BAM technique (Roth and et al. 2011).

In order to understand the formation of nanoparticle monolayer at nanoscale we deposited the nanoparticle layers on silicon substrates. The probes were deposited at different surface pressures by simply immersing the substrate into the nanoparticle covered water subphase. The selected areas of nanoparticle layers were studied by the non-contact atomic force microscopy (AFM) rather than the scanning electron microscopy as the latter one cannot provide the information on the layer height.

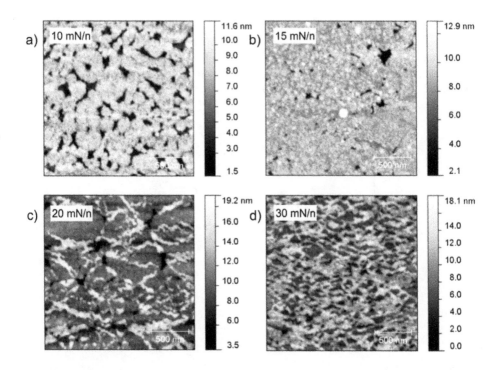

Fig. 12. The AFM images of Ag nanoparticle layers taken at the following surface pressures: a) 10 mN/m, b) 15 mN/m, c) 20 mN/m and d) 30 mN/m.

The Fig. 12 shows the AFM images of Ag nanoparticle layers deposited at different surface pressures. The nanoparticle monolayer deposited at the 10 mN/m shown in Fig. 12a displays vacancies in the nanoparticle coverage. At this stage the isolated nanoparticle clusters are coalescing into a single nanoparticle layer. The Fig. 12b shows a nanoparticle layer deposited at 15 mN/m. This AFM image shows the nanoparticle clusters forming almost a closed nanoparticle monolayer. The maximum of the surface elastic modulus was reached shortly after 15 mN/m. The AFM image shown in Fig. 12c deposited at the 20 mN/m clearly demonstrates the formation of the second nanoparticle layer after the monolayer collapse. The preferential sites for the formation of the second layer are located at the boundaries of the nanoparticle clusters. The final AFM image shown in Fig. 12d deposited at the surface pressure of 30 mN/m exhibits already a significant number of nanoparticles forming the second layer. The Fig. 13 shows calculated AFM height histograms of the nanoparticle layers deposited at different surface pressures. Only a single peak located at 6 nm corresponding to the height of monolayer is present up to the surface pressure of 15 mN/m. For the sample deposited at 20 mN/m shown in Fig. 12c, appearance of a shoulder suggests onset of formation of a second nanoparticle layer. For higher surface pressures, the newly formed peak at 12 nm in the height histogram distribution gives clear evidence of the second nanoparticle layer.

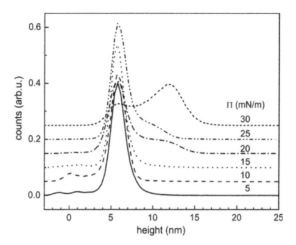

Fig. 13. The height histograms of the Ag nanoparticle layers deposited at different surface pressures obtained by analysis of the AFM images.

The number of nanoparticles occupying the second layer is steadily growing with the increasing surface pressure. At the surface pressure of 30 mN/m already more than 50% of the second nanoparticle layer was formed. The ex-situ AFM measurements provide important additional information to the in-situ GISAXS and BAM measurements. However we cannot rule out possible relaxations in the nanoparticle assemblies due to their transfer from the liquid to solid surface.

Based on the previous analyses we can conclude that the optimum deposition conditions for the nanoparticle monolayer deposition occur at the surface pressure slightly below the threshold pressure for the monolayer collapse. To achieve homogenous nanoparticle deposition over large areas of solid substrates, we modified the conventional Langmuir-Schaefer deposition (Chitu, Siffalovic et al. 2010). The scheme of the deposition trough is shown in Fig. 14.

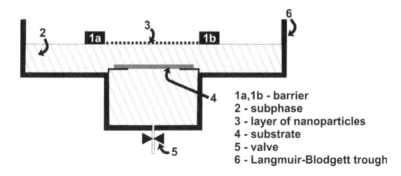

1a,1b - barrier
2 - subphase
3 - layer of nanoparticles
4 - substrate
5 - valve
6 - Langmuir-Blodgett trough

Fig. 14. The scheme of the modified Langmuir-Blodgett trough.

Differently to the conventional Langmuir-Schaefer deposition, the deposited substrate is immersed into the subphase. After spreading the nanoparticles at the water subphase and adjusting the deposition surface pressure, the water is slowly removed by opening an outlet valve. The moving water/air interface will slowly cross the inclined substrate, depositing the nanoparticle array onto it. This deposition technique produces highly homogenous nanoparticle layers on large substrates. The Fig. 15a shows a silicon wafer with the total area of some $18 \, cm^2$ homogenously covered with an iron oxide nanoparticle monolayer $(6.1\pm0.6 \, nm)$.

Fig. 15. a) Photograph of the homogenous Fe-O nanoparticle monolayer deposited onto silicon substrate. b) The SEM micrographs of a selected spot at the different magnifications.

To check the monolayer homogeneity we arbitrarily selected one spot at the deposited substrate and analyzed it with the SEM. The Fig. 15b shows four SEM micrographs of the selected spot at different magnification levels.

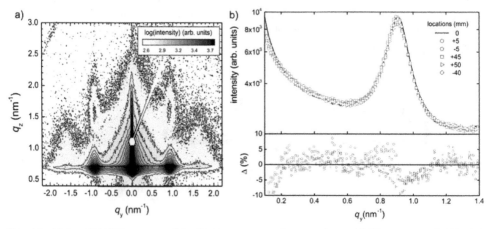

Fig. 16. a) The GISAXS pattern of the Fe-O nanoparticle monolayer. b) The extracted GISAXS line-cuts at the critical exit angle from six different locations at the substrate.

At the lowest magnification we notice the absence of any cracks in the deposited monolayer. On the contrary the traditional vertical Langmuir-Blodgett deposition is forming a series of long cracks and is not suitable for large-scale deposition. At the highest magnification we can observe a dense hexagonally ordered layer of the iron oxide nanoparticles. The SEM is suitable for detailed analysis of the selected areas of the nanoparticle monolayer but is not convenient for a rapid screening across the large areas. We have already shown that the scanning GISAXS technique provides a fast probe of the nanoparticle order at nanoscale over macroscopic areas. The Fig. 16a shows the GISAXS reciprocal space map of an arbitrarily selected location at the substrate. The integral intensity and the position of the side maxima are the measure of the nanoparticle order in the X-ray probed area. Comparing the GISAXS patterns from the different locations at the substrate we obtain the information on the homogeneity of the deposited nanoparticle monolayer. The Fig. 16b shows six line cuts extracted from the GISAXS patterns measured at different locations. The differences between the measured curves are less than ±5% that indicates a relatively high homogeneity of the deposited monolayer.

5. Processing and application of the self-assembled nanoparticle layers

In this section we focus on the issues connected with applications of deposited self-assembled nanoparticle layers. We discuss possibilities of removing the nanoparticle surfactant to increase the electrical conductivity of the nanoparticle layer as required for many applications. We address deposition of the nanoparticle layers onto thin membranes for sensor applications. We present also embedded self-assembled nanoparticle layers for organic solar cells and spintronic devices.

The surfactant molecules terminating the nanoparticles are inevitable for the synthesis and deposition of nanoparticles. However for many applications the electrical conductivity is required (Schmid 2010) while non-conductive organics is mostly used as surfactant. The surfactant molecules can be eliminated by the vacuum annealing, plasma etching, UV/ozone cleaning and many other techniques. In this section we analyze the impact of the UV/ozone cleaning on the Fe-O nanoparticle arrangement in self-assembled arrays. The UV/ozone cleaning is based on the reaction of UV light (λ=6.7 eV) with the oxygen molecules producing the highly reactive ozone. The UV light initiates photo-dissociation of the surfactant molecules that further react with the ozone molecules and are removed from the nanoparticle surface. Also a direct reaction of the surfactant molecules with the ozone molecules also called ozonolysis removes the surfactant molecules from the nanoparticle surface. In our experiment we removed the surfactant molecules from the self-assembled monolayer of iron oxide nanoparticle with the core diameter of 6.1±0.6 nm. The SEM micrographs along with the calculated nanoparticle pair correlation functions for the as-deposited sample and the sample processed in UV/ozone reactor are shown in Fig. 17a and Fig. 17b, respectively. For the as deposited nanoparticle monolayer the mean interparticle distance is given by the position of the first maximum in the pair correlation function that is 7.4 nm. After removal of the surfactant molecules terminating the nanoparticles the mean interparticle distance decreased to 6.4 nm. Moreover the nanoparticle array re-assembled into a labyrinth-like structure as shown by the SEM micrograph in Fig. 17b. This is very important for the electrical conductivity as the new nanoparticle assembly contains percolated conductive paths across the nanoparticle array.

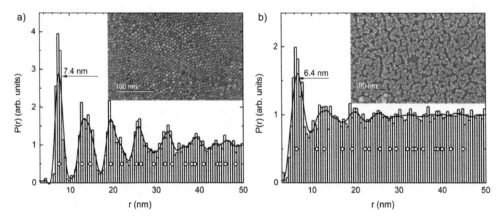

Fig. 17. The SEM micrograph and the corresponding pair correlation function for a) as deposited monolayer and b) monolayer treated in UV/ozone reactor.

We have demonstrated that the GISAXS technique is very suitable as an in-situ probe of the processes at nanoscale. We performed a time-resolved measurement of the nanoparticle re-assembly directly in the UV/ozone reactor. The above described changes in the nanoparticle pair correlation function in the direct space are manifested here as changes of the interference function in the reciprocal space. The best way of extracting the shape of the nanoparticle interference function from the GISAXS pattern is its lateral line cut along the q_y direction at the critical exit angle. The Fig. 18a shows the temporal evolution of such a line cut constructed from a series of time-resolved GISAXS frames.

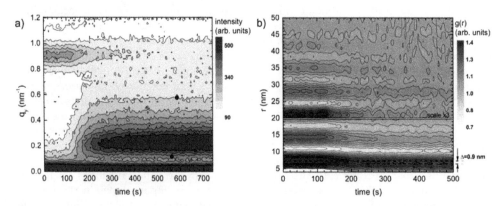

Fig. 18. a) The temporal evolution of the GISXAS line cut along the q_y direction at the critical exit angle. b) The corresponding temporal evolution of the nanoparticle pair correlation function.

The initial as-deposited self-assembled state is characterized by a maximum located at $q_y \approx 0.9 nm^{-1}$. After switching on the UV/ozone reactor the maximum corresponding to the initial self-assembled state moves slightly to higher q_y-values and its integral intensity significantly drops. Simultaneously a new peak located at $q_y \approx 0.2 nm^{-1}$ develops. The new peak corresponds to the cluster formation that can be seen in the SEM micrograph in Fig. 17b. The measured GISAXS data can be recalculated into a time-resolved nanoparticle pair correlation function shown in Fig. 18b. This function reflects in detail the nanoparticle re-assembly due to the removal of the surfactant molecules. The first maximum of the pair correlation function is shifted by some 0.9 nm to lower values within the first 200 seconds. This is in full agreement with the change of the interparticle distance calculated from the SEM micrographs in Fig. 17. This example demonstrates the possibilities of GISAXS to track fast temporal changes in the nanoparticle assemblies even in the strongly reducing environments.

Application of the conductive layers composed of metal oxide nanoparticles can be exemplified on the latest generation of the Fe-O nanoparticle-based gas sensors like SO_2, NO_x, CO, O_3 and CH_4. The NO_2 sensors are of primary importance for public security as they detect trace amounts of the explosives like EGDN, TNT, PETN, RDX, etc. A large nanoparticle-covered active surface for the gas adsorption is the main advantage when compared to the conventional thin films sensors. The Fig. 19a show a complete sensor based on the metal oxide nanoparticle multilayers (Luby, Chitu et al. 2011).

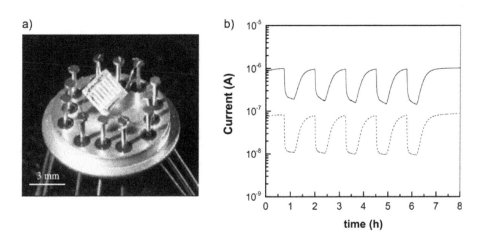

a)

b)

Current (A)

time (h)

Fig. 19. a) The photograph of a nanoparticle gas sensor. b) The electrical response of the sensors fabricated with iron oxide (full line) or cobalt iron oxide (dashed line) nanoparticles.

Visible is the heating meander as the sensor working temperature is 350°C. The active area of the sensor is composed of seven monolayers of Fe_2O_3 or $CoFe_2O_4$ nanoparticles. The Fig. 19b shows the dynamic electrical response of the sensors to 5 ppm of NO_2 gas.

The nanoparticle layers exhibiting plasmonic properties in the visible and near-infrared parts of the solar spectra are potential candidates for the next generation of plasmonic solar cells (Catchpole and Polman 2008; Atwater and Polman 2010). The enhanced scattering cross-section of the plasmonic nanoparticles can efficiently trap the light into the active layer of the solar cells and to increase their external quantum efficiency.

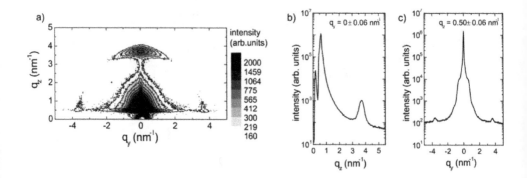

Fig. 20. a) The GISAXS reciprocal space map of the active layer deposited on Ag nanoparticle monolayer. The vertical b) and horizontal c) line-cuts across the GISAXS reciprocal space map.

The Ag nanoparticles fulfill both requirements for application in solar cells. In particular, they exhibit plasmon resonance in visible region and are highly electrically conductive. We deposited a monolayer of Ag nanoparticles (6.2±0.7 nm) at the ITO (indium tin oxide) transparent conductive layer supported on a glass substrate. Subsequently an organic active layer composed of polymer blend of P3HT (poly(3-hexylthiophene)) and PCBM (phenyl-C61-butyric acid methyl ester) of a 100 nm thickness was spin-coated on the nanoparticle monolayer. The Fig. 20a shows the GISAXS pattern of the final structure.

A prominent Bragg peak at $q_z = 3.65 nm^{-1}$ originates from the molecular P3HT stacking with the inter-molecular distance of 1.7 nm and is clearly visible also in the vertical line cut in Fig. 20b The nanoparticle correlation is visible as a small peak at $q_y = 0.66 nm^{-1}$ in the Fig. 20c that corresponds to the mean interparticle distance of some 9.5 nm. Here the GISAXS method provides the information on the correlations in the nanoparticle monolayer located at the buried interface hardly accessible by other analytical techniques.

Another example is the embedded nanoparticle monolayer in the hybrid tunnel junction of novel spintronic devices (Siffalovic, Majkova et al. 2009). Here the surfactant shell is inevitable to provide the tunnelling effect. The Fig. 21a shows schematically the multilayer structure containing iron oxide nanoparticle monolayer. The first fabrication step is the vacuum deposition of a metallic layer forming the bottom electrode. The second step is the deposition of the nanoparticle monolayer that is overcoated by another vacuum deposited metallic layer in the final step. The Fig. 21b shows the evolution of a line cut in the GISAXS pattern with the growing thickness of the metallic overlayer. The peak at $q_y = 0.83 nm^{-1}$ marked with the dashed line corresponding to the nanoparticle layer can be seen throughout the entire deposition process. These examples demonstrate that the buried nanoparticle monolayer confined to the interface with a thin metallic film can be monitored using the GISAXS technique.

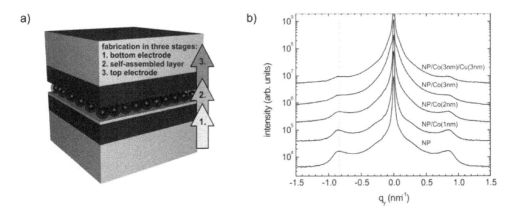

Fig. 21. a) A sketch of the spintronic structure that contains a Fe-O nanoparticle monolayer. b) Extracted line-cuts from the GISAXS reciprocal space maps at the critical exit angle in the different fabrication stages of spintronic structure.

The nanoparticle monolayers and multilayers can be deposited also on flexible membranes to be employed for monitoring mechanical properties like strain (Herrmann, Müller et al. 2007). The principle of a strain sensor is based on a change of electrical current across the nanoparticle layer as a function of the applied mechanical stress that modifies the interparticle distance in the film and consequently the electrical resistivity. The sensitivity of the nanoparticle-based strain sensors is roughly by two orders of magnitude better than that of the conventional thin metallic film ones. We investigated the nanoscale response of the nanoparticle monolayer to the applied external stress (Siffalovic, Chitu et al. 2010). We deposited a monolayer composed of iron oxide nanoparticles (6.2±0.7 nm) onto a mylar foil (1 μm thickness). The mylar foil was fixed in a stretching device for in-situ SAXS tensile stress measurements as shown in Fig. 22a.

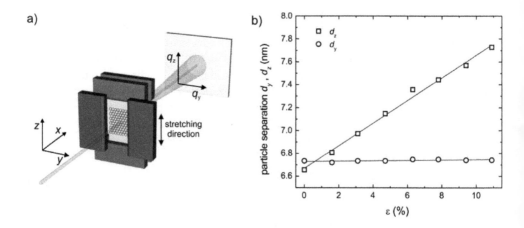

Fig. 22. a) Scheme of the experimental setup with an in-situ SAXS tensile stage. b) The evaluated interparticle separation as a function of the strain in two perpendicular directions.

The mylar foil was strained up to 11% in the z-direction and the SAXS patterns were recorded. Relying on them, the mean interparticle distance was evaluated in the applied stress direction and in the direction perpendicular to it. The results are shown in Fig. 22b. In the direction perpendicular to the applied stress the nanoparticle separation remained constant. However in the direction of the applied stress the interparticle distance followed linearly the measured foil strain. These measurements provide the test basis for the future strain sensors based on the nanoparticle layers.

In this section we included only a few of a large variety of practical applications of the nanoparticle monolayers. The nanoparticle deposition, eventual post-deposition processing of the nanoparticle layer and the test measurements of the macroscopic properties of interest are common for all these applications. The presented SAXS/GISAXS techniques offer an efficient and direct access to the nanoparticle arrangement within the final device.

6. Conclusion

The chapter provides an introductory guide to X-ray scattering studies of nanoparticle self-assembly processes at liquid/air and solid/air interfaces. It is primarily intended for graduate and post-graduate students but it is aimed also at other scientific community in the field addressing the issues of general interest. In particular, it shows the latest advances in the rapidly growing field of self-assembled nanoparticle layers. The X-ray scattering diagnostic technique was reviewed that provides an easy access even to buried nanoparticle assemblies. The main advantage of the X-ray scattering analysis is the possibility to track technologically important processes connected with the nanoparticle self-assembly or re-assembly in real time. The self-assembly process after colloidal drop casting and evaporation was described shortly while a detailed study of the self-assembly process at the liquid/air interface was the core of the chapter. This interface represents an ideal system for the nanoparticle assembling as the nanoparticles are confined to the interface but still keep translational mobility along it. The processes accompanying the formation of a nanoparticle monolayer and its transition to a multilayer were described in detail. Ideal deposition conditions for the nanoparticle monolayer formation were derived relying on the surface pressure and surface elastic modulus measurements. A modified Langmuir-Schaefer technique suitable for large-area deposition of nanoparticle arrays was presented. Selected applications of the deposited self-assembled layers were reviewed.

It has to be stressed that the colloidal nanoparticle self-assembly is a complex process resulting from an interplay between many factors where the nanoparticle type and size as well as the chemical composition of surfactant play a crucial role. Therefore none of the self-assembly techniques described in the chapter is generally applicable to any colloidal nanoparticle solution. It is also the reason why different techniques were presented with different types of nanoparticles.

It has to be also noted that in addition to the spontaneous nanoparticle self-assembly treated in this chapter of limited length, other approaches to assembling based on recent developments are of growing interest in the nanoparticle community. These include e.g. directed self-assembly of nanoparticles on pre-patterned substrates, chemically driven self-assembly, nanoparticle self-assembly stimulated by a magnetic or electro-magnetic field.

7. Acknowledgment

This publication is the result of the project implementation Center of Applied Nanoparticle research, ITMS code 26240220011, supported by the Research & Development Operational Program funded by the ERDF. The support of Grant Agency VEGA Bratislava, project No. 2/0041/11, is also acknowledged.

8. References

Atwater, H. A. and Polman, A. (2010). Plasmonics for improved photovoltaic devices. *Nature Materials* 9(3): 205-213.

Barnes, G., Gentle, I., et al. (2005). *Interfacial science: an introduction*. Oxford [u.a.], Oxford Univ. Press.

Born, M. and Wolf, E. (1999). *Principles of optics: electromagnetic theory of propagation, interference and diffraction of light*. Cambridge; New York, Cambridge University Press.

Catchpole, K. R. and Polman, A. (2008). Design principles for particle plasmon enhanced solar cells. *Applied Physics Letters* 93(19): 191113.

Chitu, L., Siffalovic, P., et al. (2010). Modified Langmuir-Blodgett deposition of nanoparticles - measurement of 2D to 3D ordered arrays. *Measurement Science Review* 10(5): 162-165.

Chushkin, Y., Ulmeanu, M., et al. (2003). Structural study of self-assembled Co nanoparticles. *Journal of Applied Physics* 94(12): 7743-7748.

Daillant, J. and Gibaud, A. (2009). *X-ray and neutron reflectivity : principles and applications.* Berlin Heidelberg, Springer.

Deegan, R. D., Bakajin, O., et al. (1997). Capillary flow as the cause of ring stains from dried liquid drops. *Nature* 389(6653): 827-829.

Feigin, L. A., Svergun, D. I., et al. (1987). *Structure analysis by small-angle X-ray and neutron scattering.* New York [etc.], Plenum Press.

Feldheim, D. L. (2002). *Metal nanoparticles: synthesis, characterization, and applications.* New York [u.a.], Dekker.

Glatter, O. and Kratky, O. (1982). *Small angle x-ray scattering.* London; New York, Academic Press.

Guinier, A. (1963). *X-ray diffraction in crystals, imperfect crystals, and amorphous bodies.* San Francisco,, W.H. Freeman.

Guinier, A. and Fournet, G. (1955). *Small-angle scattering of X-rays.* New York,, Wiley.

Henon, S. and Meunier, J. (1991). Microscope at the Brewster-Angle - Direct Observation of 1st-Order Phase-Transitions in Monolayers. *Review of Scientific Instruments* 62(4): 936-939.

Herrmann, J., Müller, K. H., et al. (2007). Nanoparticle films as sensitive strain gauges. *Applied Physics Letters* 91(18): 183105.

Holý, V., Pietsch, U., et al. (1999). *High-resolution X-ray scattering from thin films and multilayers.* Berlin; New York, Springer.

Hosemann, R. and Bagchi, S. N. (1962). *Direct analysis of diffraction by matter.* Amsterdam, North-Holland Publ. Comp.

Kittel, C. (2005). *Introduction to solid state physics.* Hoboken, NJ, Wiley.

Kraft, P., Bergamaschi, A., et al. (2009). Performance of single-photon-counting PILATUS detector modules. *Journal of Synchrotron Radiation* 16(3): 368-375.

Lazzari, R. (2002). IsGISAXS: a program for grazing-incidence small-angle X-ray scattering analysis of supported islands. *Journal of Applied Crystallography* 35: 406-421.

Lazzari, R. (2009). Grazing Incidence Small-Angle X-Ray Scattering from Nanostructures. In: *X-ray and Neutron Reflectivity.* J. Daillant and A. Gibaud, Springer Berlin / Heidelberg. 770: 283-342.

Luby, S., Chitu, L., et al. (2011). Oxide nanoparticle arrays for sensors of CO and NO2 gases. *Vacuum* In Press, Corrected Proof.

Michaelsen, C., Wiesmann, J., et al. (2002). Recent developments of multilayer mirror optics for laboratory x-ray instrumentation. *X-Ray Mirrors, Crystals, and Multilayers Ii* 4782: 143-151.

Müller-Buschbaum, P. (2009). A Basic Introduction to Grazing Incidence Small-Angle X-Ray Scattering. In: *Applications of Synchrotron Light to Scattering and Diffraction in Materials and Life Sciences.* M. Gomez, A. Nogales, M. C. Garcia-Gutierrez and T. A. Ezquerra, Springer Berlin / Heidelberg. 776: 61-89.

Nagarajan, R. (2008). *Nanoparticles: synthesis, stabilization, passivation, and functionalization.* Washington, DC, American Chemical Soc.

Niederberger, M. and Pinna, N. (2009). Metal Oxide Nanoparticles in Organic Solvents Synthesis, Formation, Assembly and Application. *Engineering Materials and Processes*. London, Springer London.

Park, J., An, K., et al. (2004). Ultra-large-scale syntheses of monodisperse nanocrystals. *Nature Materials* 3(12): 891-895.

Pileni, M.-P. (2005). *Nanocrystals forming mesoscopic structures*. Weinheim, Wiley-VCH.

Renaud, G., Lazzari, R., et al. (2009). Probing surface and interface morphology with Grazing Incidence Small Angle X-Ray Scattering. *Surface Science Reports* 64(8): 255-380.

Roth, S. V., Dòhrmann, R., et al. (2006). Small-angle options of the upgraded ultrasmall-angle x-ray scattering beamline BW4 at HASYLAB. *Review of Scientific Instruments* 77(8): 085106.

Roth, S. V. and et al. (2011). In situ observation of cluster formation during nanoparticle solution casting on a colloidal film. *Journal of Physics: Condensed Matter* 23(25): 254208.

Rycenga, M., Cobley, C. M., et al. (2011). Controlling the Synthesis and Assembly of Silver Nanostructures for Plasmonic Applications. *Chemical Reviews* 111(6): 3669-3712.

Salditt, T., Metzger, T. H., et al. (1994). Kinetic Roughness of Amorphous Multilayers Studied by Diffuse-X-Ray Scattering. *Physical Review Letters* 73(16): 2228-2231.

Schmid, G. (2010). *Nanoparticles from theory to application*. Weinheim, Wiley-VCH.

Siffalovic, P., Chitu, L., et al. (2010). Kinetics of Nanoparticle Reassembly Mediated by UV-Photolysis of Surfactant. *Langmuir* 26(8): 5451-5455.

Siffalovic, P., Chitu, L., et al. (2010). Towards strain gauges based on a self-assembled nanoparticle monolayer-SAXS study. *Nanotechnology* 21(38).

Siffalovic, P., Jergel, M., et al. (2011). GISAXS - probe of buried interfaces in multilayered thin films. In: *X-Ray Scattering*. C. M. Bauwens. New York, Nova Science Publishers.

Siffalovic, P., Majkova, E., et al. (2009). Fabrication and Characterization of Hybrid Tunnel Magnetoresistance Structures with Embedded Self-Assembled Nanoparticle Templates. *Acta Physica Polonica a* 115(1): 332-335.

Siffalovic, P., Majkova, E., et al. (2008). Real-Time Tracking of Superparamagnetic Nanoparticle Self-Assembly. *Small* 4(12): 2222-2228.

Siffalovic, P., Majkova, E., et al. (2007). Self-assembly of iron oxide nanoparticles studied by time-resolved grazing-incidence small-angle x-ray scattering. *Physical Review B* 76(19).

Siffalovic, P., Vegso, K., et al. (2010). Measurement of nanopatterned surfaces by real and reciprocal space techniques. *Measurement Science Review* 10(5): 153-156.

Stribeck, N. (2007). *X-ray scattering of soft matter*. Berlin, Springer.

Ulman, A. (1991). *An introduction to ultrathin organic films : from Langmuir-Blodgett to self-assembly*. Boston [u.a.], Acad. Press.

Vegso, K., Siffalovic, P., et al. (2011). In situ GISAXS monitoring of Langmuir nanoparticle multilayer degradation processes induced by UV photolysis. *physica status solidi (a)*: (accepted, in press).

Wiesmann, J., Graf, J., et al. (2009). X-Ray Diffractometry with Low Power Microfocus Sources - New Possibilities in the Lab. *Particle & Particle Systems Characterization* 26(3): 112-116.

Yoneda, Y. (1963). Anomalous Surface Reflection of X Rays. *Physical Review* 131(5): 2010.

5

Nanofluids

Wei Yu, Huaqing Xie and Lifei Chen
Shanghai Second Polytechnic University
P. R. China

1. Introduction

Nanofluids are a new class of fluids engineered by dispersing nanometer-sized materials (nanoparticles, nanofibers, nanotubes, nanorods, nanosheet, or droplets) in base fluids. In other words, nanofluids are nanoscale colloidal suspensions containing solid nanomaterials. They are two-phase systems with one phase (solid phase) in another (liquid phase). For a two-phase system, there are some important issues we have to face. One of the most important issues is the stability of nanofluids and it remains a big challenge to achieve desired stability of nanofluids. In this paper we will review the new progress in the methods for preparing stable nanofluids and summarize the stability mechanisms. In recent years, nanofluids have attracted more and more attention. The main driving force for nanofluids research lies in a wide range of applications. Although some review articles involving the progress of nanofluid investigation were published in the past several years [1-6], most of the reviews are concerned on the experimental and theoretical studies of the thermophysical properties or the convective heat transfer of nanofluids. The purpose of this paper will focuses on the new preparation methods and stability mechanisms, especially the new application trends for nanofluids in addition to the heat transfer properties of nanofluids. We will try to find some challenging issues that need to be solved for future research based on the review on these aspects of nanofluids.

2. Preparation methods for nanofluids

2.1 Two-step method

Two-step method is the most widely used method for preparing nanofluids. Nanoparticles, nanofibers, nanotubes or other nanomaterials used in this method are first produced as dry powders by chemical or physical methods. Then the nanosized powder will be dispersed into a fluid in the second processing step with the help of intensive magnetic force agitation, ultrasonic agitation, high-shear mixing, homogenizing and ball milling. Two-step method is the most economic method to produce nanofluids in large scale, because nanopowder synthesis techniques have already been scaled up to industrial production levels. Due to the high surface area and surface activity, nanoparticles have the tendency to aggregate. The important technique to enhance the stability of nanoparticles in fluids is the use of surfactants. However the functionality of the surfactants under high temperature is also a big concern, especially for high temperature applications.

Due to the difficulty in preparing stable nanofluids by two-step method, several advanced techniques are developed to produce nanofluids, including one-step method. In the following part, we will introduce one-step method in detail.

2.2 One-step method

To reduce the agglomeration of nanoparticles, Choi et al. developed a one-step physical vapor condensation method to prepare Cu/ethylene glycol nanofluids [7]. The one-step process consists of simultaneously making and dispersing the particles in the fluid. In this method the processes of drying, storage, transportation, and dispersion of nanoparticles are avoided, so the agglomeration of nanoparticles is minimized and the stability of fluids is increased [5]. The one-step processes can prepare uniformly dispersed nanoparticles and the particles can be stably suspended in the base fluid. The vacuum-SANSS (submerged arc nanoparticle synthesis system) is another efficient method to prepare nanofluids using different dielectric liquids [8, 9]. The different morphologies are mainly influenced and determined by various thermal conductivity properties of the dielectric liquids. The nanoparticles prepared exhibit needle-like, polygonal, square and circular morphological shapes. The method avoids the undesired particle aggregation fair well.

One-step physical method cannot synthesize nanofluids in large scale and the cost is also high, so the one-step chemical method is developing rapidly. Zhu et al. presented a novel one-step chemical method for preparing copper nanofluids by reducing $CuSO_4 \cdot 5H_2O$ with $NaH_2PO_2 \cdot H_2O$ in ethylene glycol under microwave irradiation [10]. Well-dispersed and stably suspended copper nanofluids were obtained. Mineral oil-based nanofluids containing silver nanoparticles with a narrow size distribution were also prepared by this method [11]. The particles could be stabilized by Korantin, which coordinated to the silver particle surfaces via two oxygen atoms forming a dense layer around the particles. The silver nanoparticle suspensions were stable for about 1 month. Stable ethanol based nanofluids containing silver nanoparticles could be prepared by microwave-assisted one-step method [12]. In the method, polyvinylpyrrolidone (PVP) was employed as the stabilizer of colloidal silver and reducing agent for silver in solution. The cationic surfactant octadecylamine (ODA) is also an efficient phase-transfer agent to synthesize silver colloids [13]. The phase transfer of the silver nanoparticles arises due to coupling of the silver nanoparticles with the ODA molecules present in organic phase via either coordination bond formation or weak covalent interaction.

However there are some disadvantages for one-step method. The most important one is that the residual reactants are left in the nanofluids due to incomplete reaction or stabilization. It is difficult to elucidate the nanoparticle effect without eliminating this impurity effect.

2.3 Other novel methods

Wei et al. developed a continuous-flow microfluidic microreactor to synthesize copper nanofluids. By this method, copper nanofluids can be continuously synthesized, and their microstructure and properties can be varied by adjusting parameters such as reactant concentration, flow rate and additive. CuO nanofluids with high solid volume fraction (up to 10 vol%) can be synthesized through a novel precursor transformation method with the help of ultrasonic and microwave irradiation [14]. The precursor $Cu(OH)_2$ is completely transformed to CuO nanoparticle in water under microwave irradiation. The ammonium citrate prevents the growth and aggregation of nanoparticles, resulting in a stable CuO aqueous nanofluid with higher thermal conductivity than those prepared by other dispersing methods. Phase-transfer method is also a facile way to obtain monodisperse noble metal colloids [15]. In a water-cyclohexane two-phase system, aqueous formaldehyde is transferred to cyclohexane phase via reaction with dodecylamine to form reductive intermediates in cyclohexane. The intermediates are capable of reducing silver or gold ions in aqueous solution to form dodecylamine protected silver and gold nanoparticles in cyclohexane solution at room temperature. Feng et al. used the aqueous-organic phase-transfer method for preparing gold, silver and platinum nanoparticles on the basis of the decrease of the PVP's solubility in water with the temperature increase [16]. Phase-transfer method is also applied for preparing stable kerosene based Fe_3O_4 nanofluids. Oleic acid is successfully grafted onto the surface of Fe_3O_4 nanoparticles by chemisorbed mode, which lets Fe_3O_4 nanoparticles have good compatibility with kerosene [17]. The Fe_3O_4 nanofluids prepared by phase-transfer method do not show the previously reported "time dependence of the thermal conductivity characteristic". The preparation of nanofluids with controllable microstructure is one of the key issues. It is well known that the properties of nanofluids strongly depend on the structure and shape of nanomaterials. The recent research shows that nanofluids synthesized by chemical solution method have both higher conductivity enhancement and better stability than those produced by the other methods [18]. This method is distinguished from the others by its controllability. The nanofluid microstructure can be varied and manipulated by adjusting synthesis parameters such as temperature, acidity, ultrasonic and microwave irradiation, types and concentrations of reactants and additives, and the order in which the additives are added to the solution.

3. The stability of nanofluids

The agglomeration of nanoparticles results in not only the settlement and clogging of microchannels but also the decreasing of thermal conductivity of nanofluids. So the investigation on stability is also a key issue that influences the properties of nanofluids for application, and it is necessary to study and analyze influencing factors to the dispersion stability of nanofluids. This section will contain: A) the stability evaluation methods for nanofluids; B) the ways to enhance the stability of nanofluids and C) the stability mechanisms of nanofluids.

3.1 The stability evaluation methods for nanofluids

3.1.1 Sedimentation and centrifugation methods

Many methods have been developed to evaluate the stability of nanofluids. The simplest method is sedimentation method [19, 20]. The sediment weight or the sediment volume of nanoparticles in a nanofluid under an external force field is an indication of the stability of the characterized nanofluid. The variation of concentration or particle size of supernatant particle with sediment time can be obtained by special apparatus [5]. The nanofluids are considered to be stable when the concentration or particle size of supernatant particles keeps constant. Sedimentation photograph of nanofluids in test tubes taken by a camera is also a usual method for observing the stability of nanofluids [5]. Zhu et al. used a sedimentation balance method to measure the stability of the graphite suspension [21]. The tray of sedimentation balance immersed in the fresh graphite suspension. The weight of sediment nanoparticles during a certain period was measured. The suspension fraction of graphite nanoparticles at a certain time could be calculated. For the sedimentation method, long period for observation is the defect. Therefore centrifugation method is developed to evaluate the stability of nanofluids. Singh et al. applied the centrifugation method to observe the stability of silver nanofluids prepared by the microwave synthesis in ethanol by reduction of $AgNO_3$ with PVP as stabilizing agent [12]. It has been found that the obtained nanofluids are stable for more than 1 month in the stationary state and more than 10 h under centrifugation at 3,000 rpm without sedimentation. Excellent stability of the obtained nanofluid is due to the protective role of PVP as it retards the growth and agglomeration of nanoparticles by steric effect. Li et al. prepared the aqueous polyaniline colloids, and used the centrifugation method to evaluate the stability of the colloids [22]. Electrostatic repulsive forces between nanofibers enabled the long-term stability of the colloids.

3.1.2 Zeta potential analysis

Zeta potential is electric potential in the interfacial double layer at the location of the slipping plane versus a point in the bulk fluid away from the interface, and it shows the potential difference between the dispersion medium and the stationary layer of fluid attached to the dispersed particle. The significance of zeta potential is that its value can be related to the stability of colloidal dispersions. So, colloids with high zeta potential (negative or positive) are electrically stabilized while colloids with low zeta potentials tend to coagulate or flocculate. In general, a value of 25 mV (positive or negative) can be taken as the arbitrary value that separates low-charged surfaces from highly-charged surfaces. The colloids with zeta potential from 40 to 60 mV are believed to be good stable, and those with more than 60 mV have excellent stability. Kim et al. prepared Au nanofluids with an outstanding stability even after 1 month although no dispersants were observed [23]. The stability is due to a large negative zeta potential of Au nanoparticles in water. The influence of pH and sodium dodecylbenzene sulfonate (SDBS) on the stability of two water-based nanofluids was studied [24], and zeta potential analysis was an important technique to evaluate the stability. Zhu et al. [25] measured the zeta potential of Al_2O_3-H_2O nanofluids under different pH values and different SDBS concentration. The Derjaguin-Laudau-Verwey-Overbeek (DLVO) theory was used to calculate attractive and repulsive potentials. Cationic gemini surfactant as stabilizer was used to prepare stable water based nanofluids containing MWNTs [26]. Zeta potential measurements were employed to study the absorption mechanisms of the surfactants on the MWNT surfaces with the help of Fourier transformation infrared spectra.

3.1.3 Spectral absorbency analysis

Spectral absorbency analysis is another efficient way to evaluate the stability of nanofluids. In general, there is a linear relationship between the absorbency intensity and the concentration of nanoparticles in fluid. Huang et al. evaluated the dispersion characteristics of alumina and copper suspensions using the conventional sedimentation method with the help of absorbency analysis by using a spectrophotometer after the suspensions deposited for 24 h [27]. The stability investigation of colloidal FePt nanoparticle systems was done via spectrophotometer analysis [28]. The sedimentation kinetics could also be determined by examining the absorbency of particle in solution [25].

If the nanomaterials dispersed in fluids have characteristic absorption bands in the wavelength 190-1100 nm, it is an easy and reliable method to evaluate the stability of nanofluids using UV-vis spectral analysis. The variation of supernatant particle concentration of nanofluids with sediment time can be obtained by the measurement of absorption of nanofluids because there is a linear relation between the supernatant nanoparticle concentration and the absorbance of suspended particles. The outstanding advantage comparing to other methods is that UV-vis spectral analysis can present the quantitative concentration of nanofluids. Hwang et al. [29] studied the stability of nanofluids with the UV-vis spectrophotometer. It was believed that the stability of nanofluids was strongly affected by the characteristics of the suspended particles and the base fluid such as particle morphology. Moreover, addition of a surfactant could improve the stability of the suspensions. The relative stability of MWNT nanofluids [26] could be estimated by measuring the UV-vis absorption of the MWNT nanofluids at different sediment times. From the above relation between MWNT concentration and its UV-vis absorbance value the concentration of the MWNT nanofluids at different sediment times could be obtained. The above three methods can be united to investigate the stability of nanofluids. For example, Li et al. evaluated the dispersion behavior of the aqueous copper nano-suspensions under different pH values, different dispersant type and concentration by the method of zeta potential, absorbency and sedimentation photographs [20].

3.2 The ways to enhance the stability of nanofluids

3.2.1 Surfactants used in nanofluids

Surfactants used in nanofluids are also called dispersants. Adding dispersants in the two-phase systems is an easy and economic method to enhance the stability of nanofluids. Dispersants can markedly affect the surface characteristics of a system in small quantity. Dispersants consists of a hydrophobic tail portion, usually a long-chain hydrocarbon, and a hydrophilic polar head group. Dispersants are employed to increase the contact of two materials, sometimes known as wettability. In a two-phase system, a dispersant tends to locate at the interface of the two phases, where it introduces a degree of continuity between the nanoparticles and fluids. According to the composition of the head, surfactants are divided into four classes: non-ionic surfactants without charge groups in its head (include polyethylene oxide, alcohols, and other polar groups); anionic surfactants with negatively charged head groups (anionic head groups include long-chain fatty acids, sulfosuccinates, alkyl sulfates, phosphates, and sulfonates); cationic surfactants with positively charged head groups (cationic surfactants may be protonated long-chain amines and long-chain quaternary ammonium compounds); and amphoteric surfactants with zwitterionic head groups (charge depends on pH. The class of amphoteric surfactants is represented by betaines and certain lecithins). How to select suitable dispersants is a key issue. In general, when the base fluid of nanofluids is polar solvent, we should select water soluble surfactants, otherwise we will select oil soluble. For nonionic surfactants, we can evaluate the solubility through the term hydrophilic/lipophilic balance (HLB) value. The lower the HLB number the more oil soluble the surfactants, and in turn the higher the HLB number the more water-soluble the surfactants is. The HLB value can be obtained easily by many handbooks.

3.2.2 Surface modification techniques-surfactant free method

Although surfactant addition is an effective way to enhance the dispersibility of nanoparticles, surfactants might cause several problems [30]. For example, the addition of surfactants may contaminate the heat transfer media. Surfactants may produce foams when heating, while heating and cooling are routinely processes in heat exchange systems. Furthermore surfactant molecules attaching on the surfaces of nanoparticles may enlarge the thermal resistance between the nanoparticles and the base fluid, which may limit the enhancement of the effective thermal conductivity. Use of functionalized nanoparticles is a promising approach to achieve long-term stability of nanofluid. It represents the surfactant free technique. Yang et al. presented a work on the synthesis of functionalized silica (SiO_2) nanoparticles by grafting silanes directly to the surface of silica nanoparticles in original nanoparticle solutions [31]. One of the unique characteristics of the nanofluids was that no deposition layer formed on the heated surface after a pool boiling process. Chen *et al.* introduced hydrophilic functional groups on the surface of the nanotubes by mechanochemical reaction [29]. The prepared nanofluids, with no contamination to medium, good fluidity, low viscosity, high stability, and high thermal conductivity, would have potential applications as coolants in advanced thermal systems. A wet-mechanochemical reaction was applied to prepare surfactant-free nanofluids containing double- and single-walled CNTs. Results from the infrared spectrum and zeta potential measurements showed that the hydroxyl groups had been introduced onto the treated CNT surfaces [32]. Plasma treatment was used to modify the surface characteristics of diamond nanoparticles [33]. Through plasma treatment using gas mixtures of methane and oxygen, various polar groups were imparted on the surface of the diamond nanoparticles, improving their dispersion property in water. A stable dispersion of titania nanoparticles in an organic solvent of diethylene glycol dimethylether (diglyme) was successfully prepared using a ball milling process [34]. In order to enhance dispersion stability of the solution, surface modification of dispersed titania particles was carried out during the centrifugal bead mill process. Surface modification was utilized with silane coupling agents, (3-acryl-oxypropyl) trimethoxysilane and trimethoxypropylsilane. Zinc oxide nanoparticles could be modified by polymethacrylic acid (PMAA) in aqueous system [35]. The hydroxyl groups of nano-ZnO particle surface could interact with carboxyl groups of PMAA and form poly (zinc methacrylate) complex on the surface of nano-ZnO. PMAA enhanced the dispersibility of nano-ZnO particles in water. The modification did not alter the crystalline structure of the ZnO nanoparticles.

3.3 Stability mechanisms of nanofluids

Particles in dispersion may adhere together and form aggregates of increasing size which may settle out due to gravity. Stability means that the particles do not aggregate at a significant rate. The rate of aggregation is in general determined by the frequency of collisions and the probability of cohesion during collision. Derjaguin, Verway, Landau and Overbeek (DVLO) developed a theory which dealt with colloidal stability [36, 37]. DLVO theory suggests that the stability of a particle in solution is determined by the sum of van der Waals attractive and electrical double layer repulsive forces that exist between particles as they approach each other due to the Brownian motion they are undergoing. If the attractive force is larger than the repulsive force, the two particles will collide, and the suspension is not stable. If the particles have a sufficient high repulsion, the suspensions will exist in stable state. For stable nanofluids or colloids, the repulsive forces between particles must be dominant. According to the types of repulsion, the fundamental mechanisms that affect colloidal stability are divided into two kinds, one is steric repulsion, and another is electrostatic (charge) repulsion. For steric stabilization, polymers are always involved into the suspension system, and they will adsorb onto the particles surface, producing an additional steric repulsive force. For example, Zinc oxide nanoparticles modified by PMAA have good compatibility with polar solvents [35]. Silver nanofluids are very stable due to the protective role of PVP as it retards the growth and agglomeration of nanoparticles by steric effect. PVP is an efficient agent to improve the stability of graphite suspension [21]. The steric effect of polymer dispersant is determined by the concentration of the dispersant. If the PVP concentration is low, the surface of the graphite particles is gradually coated by PVP molecules with the increase of PVP. Kamiya et al. studied the effect of polymer dispersant structure on electrosteric interaction and dense alumina suspension behavior [38]. An optimum hydrophilic to hydrophobic group ratio was obtained from the maximum repulsive force and minimum viscosity. For electrostatic stabilization, surface charge will be developed through one or more of the following mechanisms: 1) preferential adsorption of ions; 2) dissociation of surface charged species; 3) isomorphic substitution of ions; 4) accumulation or depletion of electrons at the surface and 5) physical adsorption of charged species onto the surface.

4. Application of nanofluids

4.1 Heat transfer Intensification

Since the origination of the nanofluid concept about a decade ago, the potentials of nanofluids in heat transfer applications have attracted more and more attention. Up to now, there are some review papers, which present overviews of various aspects of nanofluids [1, 3-6, 39-44], including preparation and characterization, techniques for the measurements of thermal conductivity, theory and model, thermophysical properties, convective heat transfer. In this part, we will summarize the applications of nanofluids in heat transfer enhancement.

4.1.1 Electronic applications

Due to higher density of chips, design of electronic components with more compact makes heat dissipation more difficult. Advanced electronic devices face thermal management challenges from the high level of heat generation and the reduction of available surface area for heat removal. So, the reliable thermal management system is vital for the smooth operation of the advanced electronic devices. In general, there are two approaches to improve the heat removal for electronic equipment. One is to find an optimum geometry of cooling devices; another is to increase the heat transfer capacity. Recent researches illustrated that nanofluids could increase the heat transfer coefficient by increasing the thermal conductivity of a coolant. Jang et al. designed a new cooler, combined microchannel heat sink with nanofluids [45]. Higher cooling performance was obtained when compared to the device using pure water as working medium. Nanofluids reduced both the thermal resistance and the temperature difference between the heated microchannel wall and the coolant. A combined microchannel heat sink with nanofluids had the potential as the next generation cooling devices for removing ultra-high heat flux. Nguyen et al. designed a closed liquid-circuit to investigate the heat transfer enhancement of a liquid cooling system, by replacing the base fluid (distilled water) with a nanofluid composed of distilled water and Al_2O_3 nanoparticles at various concentrations [46]. Measured data have clearly shown that the inclusion of nanoparticles within the distilled water has produced a considerable enhancement in convective heat transfer coefficient of the cooling block. With particle loading 4.5 vol%, the enhancement is up to 23% with respect to that of the base fluid. It has also been observed that an augmentation of particle concentration has produced a clear decrease of the junction temperature between the heated component and the cooling block. Silicon microchannel heat sink performance using nanofluids containing Cu nanoparticles was analyzed [47]. It was found nanofluids could enhance the performance as compared with that using pure water as the coolant. The enhancement was due to the increase in thermal conductivity of coolant and the nanoparticle thermal dispersion effect. The other advantage was that there was no extra pressure drop since the nanoparticle was small and particle volume fraction was low.

The thermal requirements on the personal computer become much stricter with the increase in thermal dissipation of CPU. One of the solutions is the use of heat pipes. Nanofluids, employed as working medium for conventional heat pipe, have shown higher thermal performances, having the potential as a substitute for conventional water in heat pipe. At a same charge volume, there is a significant reduction in thermal resistance of heat pipe with nanofluid containing gold nanoparticles as compared with water [48]. The measured results also show that the thermal resistance of a vertical meshed heat pipe varies with the size of gold nanoparticles. The suspended nanoparticles tend to bombard the vapor bubble during the bubble formation. Therefore, it is expected that the nucleation size of vapor bubble is much smaller for fluid with suspended nanoparticles than that without them. This may be the major reason for reducing the thermal resistance of heat pipe. Chen et al. studied the effect of a nanofluid on flat heat pipe (FHP) thermal performance [49], using silver nanofluid as the working fluid. The temperature difference and the thermal resistance of the FHP with the silver nanoparticle solution were lower than those with pure water. The plausible reasons for enhancement of the thermal performance of the FHP using the nanofluid can be explained by the critical heat flux enhancement by higher wettability and the reduction of the boiling limit. Nanofluid oscillating heat pipe with ultrahigh-performance was developed by Ma et al. [50]. They combined nanofluids with thermally excited oscillating motion in an oscillating heat pipe, and heat transport capability significantly increased. For example, at the input power of 80.0 W, diamond nanofluid could reduce the temperature difference between the evaporator and the condenser from 40.9 to 24.3°C. This study would accelerate the development of a highly efficient cooling device for ultrahigh-heat-flux electronic systems. The thermal performance investigation of heat pipe indicated that nanofluids containing silver or titanium nanoparticles could be used as an efficient cooling fluid for devices with high energy density. For a silver nanofluid, the temperature difference decreased 0.56-0.65°C compared to water at an input power of 30-50 W [51]. For the heat pipe with titanium nanoparticles at a volume concentration of 0.10%, the thermal efficiency is 10.60% higher than that with the based working fluid [52]. These positive results are promoting the continued research and development of nanofluids for such applications.

4.1.2 Transportation

Nanofluids have great potentials to improve automotive and heavy-duty engine cooling rates by increasing the efficiency, lowering the weight and reducing the complexity of thermal management systems. The improved cooling rates for automotive and truck engines can be used to remove more heat from higher horsepower engines with the same size of cooling system. Alternatively, it is beneficial to design more compact cooling system with smaller and lighter radiators. It is in turn benefit the high performance and high fuel economy of car and truck. Ethylene glycol based nanofluids have attracted much attention in the application as engine coolant [53-55], due to the low-pressure operation compared with a 50/50 mixture of ethylene glycol and water, which is the nearly universally used automotive coolant. The nanofluids has a high boiling point, and it can be used to increase the normal coolant operating temperature and then reject more heat through the existing coolant system [56]. Kole et al. prepared car engine coolant (Al_2O_3 nanofluid) using a standard car engine coolant (HP KOOLGARD) as the base fluid [57], and studied the thermal conductivity and viscosity of the coolant. The prepared nanofluid, containing only 3.5% volume fraction of Al_2O_3 nanoparticles, displayed a fairly higher thermal conductivity than the base fluid, and a maximum enhancement of 10.41% was observed at room temperature. Tzeng et al. [58] applied nanofluids to the cooling of automatic transmissions. The experimental platform was the transmission of a four-wheel drive vehicle. The used nanofluids were prepared by dispersing CuO and Al_2O_3 nanoparticles into engine transmission oil. The results showed that CuO nanofluids produced the lower transmission temperatures both at high and low rotating speeds. From the thermal performance viewpoint, the use of nanofluid in the transmission has a clear advantage.

The researchers of Argonne National Laboratory have assessed the applications of nanofluids for transportation [59]. The use of high-thermal conductive nanofluids in radiators can lead to a reduction in the frontal area of the radiator up to 10%. The fuel saving is up to 5% due to the reduction in aerodynamic drag. It opens the door for new aerodynamic automotive designs that reduce emissions by lowering drag. The application of nanofluids also contributed to a reduction of friction and wear, reducing parasitic losses, operation of components such as pumps and compressors, and subsequently leading to more than 6% fuel savings. In fact, nanofluids not only enhance the efficiency and economic performance of car engine, but also will greatly influence the structure design of automotives. For example, the engine radiator cooled by a nanofluid will be smaller and lighter. It can be placed elsewhere in the vehicle, allowing for the redesign of a far more aerodynamic chassis. By reducing the size and changing the location of the radiator, a reduction in weight and wind resistance could enable greater fuel efficiency and subsequently lower exhaust emissions. Computer simulations from the US department of energy's office of vehicle technology showed that nanofluid coolants could reduce the size of truck radiators by 5%. This would result in a 2.5% fuel saving at highway speeds.

The practical applications are on the road. In USA, car manufacturers GM and Ford are running their own research programs on nanofluid applications. A €8.3 million FP7 project, named NanoHex (Nanofluid Heat Exchange), began to run. It involved 12 organizations from Europe and Israel ranging from Universities to SME's and major companies. NanoHex is overcoming the technological challenges faced in development and application of reliable and safe nanofluids for more sophisticated, energy efficient, and environmentally friendly products and services [60].

4.1.3 Industrial cooling applications

The application of nanofluids in industrial cooling will result in great energy savings and emissions reductions. For US industry, the replacement of cooling and heating water with nanofluids has the potential to conserve 1 trillion Btu of energy [39, 61]. For the US electric power industry, using nanofluids in closed loop cooling cycles could save about 10-30 trillion Btu per year (equivalent to the annual energy consumption of about 50,000–150,000 households). The associated emissions reductions would be approximately 5.6 million metric tons of carbon dioxide, 8,600 metric tons of nitrogen oxides, and 21,000 metric tons of sulfur dioxide.

Experiments were performed using a flow-loop apparatus to explore the performance of polyalphaolefin nanofluids containing exfoliated graphite nanoparticle fibers in cooling [63]. It was observed that the specific heat of nanofluids was found to be 50% higher for nanofluids compared with polyalphaolefin and it increased with temperature. The thermal diffusivity was found to be 4 times higher for nanofluids. The convective heat transfer was enhanced by ~10% using nanofluids compared with using polyalphaolefin. Ma et al. proposed the concept of nano liquid-metal fluid, aiming to establish an engineering route to make the highest conductive coolant with about several dozen times larger thermal conductivity than that of water [64]. The liquid metal with low melting point is expected to be an idealistic base fluid for making super conductive solution which may lead to the ultimate coolant in a wide variety of heat transfer enhancement area. The thermal conductivity of the liquid-metal fluid can be enhanced through the addition of more conductive nanoparticles.

4.1.4 Heating buildings and reducing pollution

Nanofluids can be applied in the building heating systems. Kulkarni et al. evaluated how they perform heating buildings in cold regions [65]. In cold regions, it is a common practice to use ethylene or propylene glycol mixed with water in different proportions as a heat transfer fluid. So 60:40 ethylene glcol/water (by weight) was selected as the base fluid. The results showed that using nanofluids in heat exchangers could reduce volumetric and mass flow rates, resulting in an overall pumping power savings. Nanofluids necessitate smaller heating systems, which are capable of delivering the same amount of thermal energy as larger heating systems, but are less expensive. This lowers the initial equipment cost excluding nanofluid cost. This will also reduce environmental pollutants because smaller heating units use less power, and the heat transfer unit has less liquid and material waste to discard at the end of its life cycle.

4.1.5 Nuclear systems cooling

The Massachusetts Institute of Technology has established an interdisciplinary center for nanofluid technology for the nuclear energy industry. The researchers are exploring the nuclear applications of nanofluids, specifically the following three [66]: 1) main reactor coolant for pressurized water reactors (PWRs). It could enable significant power uprates in current and future PWRs, thus enhancing their economic performance. Specifically, the use of nanofluids with at least 32% higher critical heat flux (CHF) could enable a 20% power density uprate in current plants without changing the fuel assembly design and without reducing the margin to CHF; 2) coolant for the emergency core cooling systems (ECCSs) of both PWRs and boiling water reactors. The use of a nanofluid in the ECCS accumulators and safety injection can increase the peak-cladding-temperature margins (in the nominal-power core) or maintain them in uprated cores if the nanofluid has a higher post-CHF heat transfer rate; 3) coolant for in-vessel retention of the molten core during severe accidents in high-power-density light water reactors. It can increase the margin to vessel breach by 40% during severe accidents in high-power density systems such as Westinghouse APR1000 and the Korean APR1400. While there exist several significant gaps, including the nanofluid thermal-hydraulic performance at prototypical reactor conditions and the compatibility of the nanofluid chemistry with the reactor materials. Much work should be done to overcome these gaps before any applications can be implemented in a nuclear power plant.

4.1.6 Space and defense

Due to the restriction of space, energy and weight in space station and aircraft, there is a strong demand for high efficient cooling system with smaller size. You et al. [67] and Vassalo et al. [68] have reported order of magnitude increases in the critical heat flux in pool boiling with nanofluids compared to the base fluid alone. Further research of nanofluids will lead to the development of next generation of cooling devices that incorporate nanofluids for ultrahigh-heat-flux electronic systems, presenting the possibility of raising chip power in electronic components or simplifying cooling requirements for space applications. A number of military devices and systems require high-heat flux cooling to the level of tens of MW/m^2. At this level, the cooling of military devices and system is vital for the reliable operation. Nanofluids with high critical heat fluxes have the potential to provide the required cooling in such applications as well as in other military systems, including military vehicles, submarines, and high-power laser diodes. Therefore, nanofluids have wide application in space and defense fields where power density is very high and the components should be smaller and weight less.

4.2 Mass transfer enhancement

Several researches have studied the mass transfer enhancement of nanofluids. Kim et al. initially examined the effect of nanoparticles on the bubble type absorption for NH_3/H_2O absorption system [69]. The addition of nanoparticles enhances the absorption performance up to 3.21 times. Then they visualized the bubble behavior during the NH_3/H_2O absorption process and studied the effect of nanoparticles and surfactants on the absorption characteristics [70]. The results show that the addition of surfactants and nanoparticles improved the absorption performance up to 5.32 times. The addition of both surfactants and nanoparticles enhanced significantly the absorption performance during the ammonia bubble absorption process. The theoretical investigations of thermodiffusion and diffusionthermo on convective instabilities in binary nanofluids for absorption application were conducted. Mass diffusion is induced by thermal gradient. Diffusionthermo implies that heat transfer is induced by concentration gradient [71]. Ma et al. studied the mass transfer process of absorption using CNTs-ammonia nanofluids as the working medium [72, 73]. The absorption rates of the CNTs-ammonia binary nanofluids were higher than those of ammonia solution without CNTs. The effective absorption ratio of the CNTs-ammonia binary nanofluids increased with the initial concentration of ammonia and the mass fraction of CNTs. Komati et al. studied CO_2 absorption into amine solutions, and the addition of ferrofluids increased the mass transfer coefficient in gas/liquid mass transfer [74], and the enhancement extent depended on the amount of ferrofluid added. The enhancement in mass transfer coefficient was 92.8% for a volume fraction of the fluid of about 50% (solid magnetite volume fraction of about 0.39%). The research about the influence of Al_2O_3 nanofluid on the falling film absorption with ammonia-water showed that the sorts of nanoparticles and surfactants in the nanofluid and the concentration of ammonia in the basefluid were the key parameters influencing the absorption effect of ammonia [75].

4.3 Energy applications

4.3.1 Energy storage

The temporal difference of energy source and energy needs made necessary the development of storage system. The storage of thermal energy in the form of sensible and latent heat has become an important aspect of energy management with the emphasis on efficient use and conservation of the waste heat and solar energy in industry and buildings [76]. Latent heat storage is one of the most efficient ways of storing thermal energy. Wu et al. evaluated the potential of Al_2O_3-H_2O nanofluids as a new phase change material (PCM) for the thermal energy storage of cooling systems. The thermal response test showed the addition of Al_2O_3 nanoparticles remarkably decreased the supercooling degree of water, advanced the beginning freezing time and reduced the total freezing time. Only adding 0.2 wt% Al_2O_3 nanoparticles, the total freezing time of Al_2O_3-H_2O nanofluids could be reduced by 20.5%. Liu et al. prepared a new sort of nanofluid phase change materials (PCMs) by suspending small amount of TiO_2 nanoparticles in saturated $BaCl_2$ aqueous solution [77]. The nanofluids PCMs possessed remarkably high thermal conductivities compared to the base material. The cool storage/supply rate and the cool storage/supply capacity all increased greatly than those of $BaCl_2$ aqueous solution without added nanoparticles. The higher thermal performances of nanofluids PCMs indicate that they have a potential for substituting conventional PCMs in cool storage applications. Copper nanoparticles are efficient additives to improve the heating and cooling rates of PCMs [78]. For composites with 1 wt % copper nanoparticle, the heating and cooling times could be reduced by 30.3 and 28.2%, respectively. The latent heats and phase-change temperatures changed very little after 100 thermal cycles.

4.3.2 Solar absorption

Solar energy is one of the best sources of renewable energy with minimal environmental impact. The conventional direct absorption solar collector is a well established technology, and it has been proposed for a variety of applications such as water heating; however the efficiency of these collectors is limited by the absorption properties of the working fluid, which is very poor for typical fluids used in solar collectors. Recently this technology has been combined with the emerging technologies of nanofluids and liquid-nanoparticle suspensions to create a new class of nanofluid-based solar collectors. Otanicar et al. reported the experimental results on solar collectors based on nanofluids made from a variety of nanoparticles (CNTs, graphite, and silver) [79]. The efficiency improvement was up to 5% in solar thermal collectors by utilizing nanofluids as the absorption media. In addition they compared the experimental data with a numerical model of a solar collector with direct absorption nanofluids. The experimental and numerical results demonstrated an initial rapid increase in efficiency with volume fraction, followed by a leveling off in efficiency as volume fraction continues to increase. Theoretical investigation on the feasibility of using a nonconcentrating direct absorption solar collector showed that the presence of nanoparticles increased the absorption of incident radiation by more than nine times over that of pure water [80]. Under the similar operating conditions, the efficiency of an absorption solar collector using nanofluid as the working fluid was found to be up to 10% higher (on an absolute basis) than that of a flat-plate collector. Otanicar et al. evaluated the overall economic and environmental impacts of the technology in contrast with conventional solar collectors using the life cycle assessment methodology [81]. Results showed that for the current cost of nanoparticles the nanofluid based solar collector had a slightly longer payback period but at the end of its useful life has the same economic saving as a conventional solar collector. Sani et al. investigated the optical and thermal properties of nanofluids consisting in aqueous suspensions of single wall carbon nanohorns [82]. The observed nanoparticle-induced differences in optical properties appeared promising, leading to a considerably higher sunlight absorption. Both these effects, together with the possible chemical functionalization of carbon nanohorns, make this new kind of nanofluids very interesting for increasing the overall efficiency of the sunlight exploiting device.

4.4 Mechanical applications

4.4.1 Friction reduction

Advanced lubricants can improve productivity through energy saving and reliability of engineered systems. Tribological research heavily emphasizes reducing friction and wear. Nanoparticles have attracted much interest in recent years due to their excellent load-carrying capacity, good extreme pressure and friction reducing properties. Zhou et al. evaluated the tribological behavior of Cu nanoparticles in oil on a four-ball machine. The results showed that Cu nanoparticles as an oil additive had better friction-reduction and antiwear properties than zinc dithiophosphate, especially at high applied load. Meanwhile, the nanoparticles could also strikingly improve the load-carrying capacity of the base oil [83]. Dispersion of solid particles was found to play an important role, especially when a slurry layer was formed. Water-based Al2O3 and diamond nanofluids were applied in the minimum quantity lubrication (MQL) grinding process of cast iron. During the nanofluid MQL grinding, a dense and hard slurry layer was formed on the wheel surface and could benefit the grinding performance. Nanofluids showed the benefits of reducing grinding forces, improving surface roughness, and preventing workpiece burning. Compared to dry grinding, MQL grinding could significantly reduce the grinding temperature [84]. Wear and friction properties of surface modified Cu nanoparticles as 50CC oil additive were studied. The higher the oil temperature applied, the better the tribological properties of Cu nanoparticles were. It could be inferred that a thin copper protective film with lower elastic modulus and hardness was formed on the worn surface, which resulted in the good tribological performances of Cu nanoparticles, especially when the oil temperature was higher [85]. Wang et al. studied the tribological properties of ionic liquid-based nanofluids containing functionlized MWNTs under loads in the range of 200-800 N [86], indicating that the nanofluids exhibited preferable friction-reduction properties under 800 N and remarkable antiwear properties with use of reasonable concentrations. Magnetic nanoparticle $Mn_{0.78}Zn_{0.22}Fe_2O_4$ was also an efficient lubricant additive. When used as a lubricant additive in 46# turbine oil, it could improve the wear resistance, load-carrying capacity, and antifriction ability of base oil, and the decreasing percentage of wear scar diameter was 25.45% compared to the base oil. This was a typical self-repair phenomenon [87]. Chen et al. reported on dispersion stability enhancement and self-repair principle discussion of ultrafine-tungsten disulfide in green lubricating oil [88]. Ultrafine-tungsten disulfide particulates could fill and level up the furrows on abrasive surfaces, repairing abrasive surface well. What is more, ultrafine-tungsten disulfide particulates could form a WS2 film with low shear stress by adsorbing and depositing in the hollowness of abrasive surface, making the abrasive surface be more smooth, and the FeS film formed in tribology reaction could protect the abrasive surface further, all of which realize the self-repair to abrasive surface. The tribological properties of liquid paraffin with SiO_2 nanoparticles additive made by a sol-gel method was investigated by Peng et al. [89]. The optimal concentrations of SiO_2 nanoparticles in liquid paraffin was associated with better tribological properties than pure paraffin oil, and an anti-wear ability that depended on the particle size, and oleic acid surface-modified SiO_2 nanoparticles with an average diameter of 58 nm provided better tribological properties in load-carrying capacity, anti-wear and friction-reduction than pure liquid paraffin. Nanoparticles can easily penetrate into the rubbing surfaces because of their nanoscale. During the frictional process, the thin physical tribofilm of the nanoparticles forms between rubbing surfaces, which cannot only bear the load but also separates the rubbing faces. The spherical SiO_2 nanoparticles could roll between the rubbing faces in sliding friction, the originally pure sliding friction becomes mixed sliding and rolling friction. Therefore, the friction coefficient declines markedly and then remains constant.

4.4.2 Magnetic sealing

Magnetic fluids (Ferromagnetic fluid) are kinds of special nanofluids. They are stable colloidal suspensions of small magnetic particles such as magnetite (Fe_3O_4). The properties of the magnetic nanoparticles, the magnetic component of magnetic nanofluids, may be tailored by varying their size and adapting their surface coating in order to meet the requirements of colloidal stability of magnetic nanofluids with non-polar and polar carrier liquids [90]. Comparing with the mechanical sealing, magnetic sealing offers a cost-effective solution to environmental and hazardous-gas sealing in a wide variety of industrial rotation equipment with high speed capability, low friction power losses and long life and high reliability [91]. A ring magnet forms part of a magnetic circuit in which an intense magnetic field is established in the gaps between the teeth on a magnetically permeable shaft and the surface of an opposing pole block. Ferrofluid introduced into the gaps forms discrete liquid rings capable of supporting a pressure difference while maintaining zero leakage. The seals operate without wear as the shaft rotates because the mechanical moving parts do not touch. With these unique characteristics, sealing liquids with magnetic fluids can be applied in many application areas. It is reported that an iron particle dispersed magnetic fluids was utilized in the sealing of a high rotation pump. The sealing holds pressure of 618 kPa with a 1800 r/min [92]. Mitamura et al. studied the application of a magnetic fluid seal to rotary blood pumps. The developed magnetic fluid seal worked for over 286 days in a continuous flow condition, for 24 days (on-going) in a pulsatile flow condition and for 24 h (electively terminated) in blood flow [93]. Ferro-cobalt magnetic fluid was used for oil sealing, and the holding pressure is 25 times as high as that of a conventional magnetite sealing [94].

4.5 Biomedical application

4.5.1 Antibacterial activity

Organic antibacterial materials are often less stable particularly at high temperatures or pressures. As a consequence, inorganic materials such as metal and metal oxides have attracted lots of attention over the past decade due to their ability to withstand harsh process conditions. The antibacterial behaviour of ZnO nanofluids shows that the ZnO nanofluids have bacteriostatic activity against [95]. Electrochemical measurements suggest some direct interaction between ZnO nanoparticles and the bacteria membrane at high ZnO concentrations. Jalal et al. prepared ZnO nanoparticles via a green method. The antibacterial activity of suspensions of ZnO nanoparticles against Escherichia coli (E. coli) has been evaluated by estimating the reduction ratio of the bacteria treated with ZnO. Survival ratio of bacteria decreases with increasing the concentrations of ZnO nanofluids and time [96]. Further investigations have clearly demonstrated that ZnO nanoparticles have a wide range of antibacterial effects on a number of other microorganisms. The antibacterial activity of ZnO may be dependent on the size and the presence of normal visible light [97]. Recent research showed that ZnO nanoparticles exhibited impressive antibacterial properties against an important foodborne pathogen, E. coli O157:H7, and the inhibitory effects increased as the concentrations of ZnO nanoparticles increased. ZnO nanoparticles changed the cell membrane components including lipids and proteins. ZnO nanoparticles could distort bacterial cell membrane, leading to loss of intracellular components, and ultimately the death of cells, considered as an effective antibacterial agent for protecting agricultural and food safety [98].

The antibacterial activity research of CuO nanoparticles showed that they possessed antibacterial activity against four bacterial strains. The size of nanoparticles was less than that of the pore size in the bacteria and thus they had a unique property of crossing the cell membrane without any hindrance. It could be hypothesized that these nanoparticles formed stable complexes with vital enzymes inside cells which hampered cellular functioning resulting in their death [99]. Bulk equivalents of these products showed no inhibitory activity, indicating that particle size was determinant in activity [100]. Lee et al. reported the antibacterial efficacy of nanosized silver colloidal solution on the cellulosic and synthetic fabrics [101]. The antibacterial treatment of the textile fabrics was easily achieved by padding them with nanosized silver colloidal solution. The antibacterial efficacy of the fabrics was maintained after many times laundering. Silver colloid is an efficient antibacterial agent. The silver colloid prepared by a one-step synthesis showed high antimicrobial and bactericidal activity against Gram-positive and Gram-negative bacteria, including highly multiresistant strains such as methicillin-resistant staphylococcus aureus. The antibacterial activity of silver nanoparticles was found to be dependent on the size of silver particles. A very low concentration of silver gave antibacterial performance [102]. The aqueous suspensions of fullerenes and nano-TiO_2 can produce reactive oxygen species (ROS). Bacterial (E. coli) toxicity tests suggestted that, unlike nano-TiO_2 which was exclusively phototoxic, the antibacterial activity of fullerene suspensions was linked to ROS production. Nano-TiO_2 may be more efficient for water treatment involving UV or solar energy, to enhance contaminant oxidation and perhaps for disinfection. However, fullerol and PVP/C_{60} may be useful as water treatment agents targeting specific pollutants or microorganisms that are more sensitive to either superoxide or singlet oxygen [103]. Lyon et al. proposed that C_{60} suspensions exerted ROS-independent oxidative stress in bacteria, with evidence of protein oxidation, changes in cell membrane potential, and interruption of cellular respiration. This mechanism requires direct contact between the nanoparticle and the bacterial cell and differs from previously reported nanomaterial antibacterial mechanisms that involve ROS generation (metal oxides) or leaching of toxic elements (nanosilver) [104].

4.5.2 Nanodrug delivery

Over the last few decades, colloidal drug delivery systems have been developed in order to improve the efficiency and the specificity of drug action [105]. The small size, customized surface, improved solubility, and multi-functionality of nanoparticles open many doors and create new biomedical applications. The novel properties of nanoparticles offer the ability to interact with complex cellular functions in new ways [106]. Gold nanoparticles provide non-toxic carriers for drug and gene delivery applications. With these systems, the gold core imparts stability to the assembly, while the monolayer allows tuning of surface properties such as charge and hydrophobicity. Another attractive feature of gold nanoparticles is their interaction with thiols, providing an effective and selective means of controlled intracellular release [107]. Nakano et al. proposed the drug delivery system using nano-magnetic fluid [108], which targetted and concentrated drugs using a ferrofluid cluster composed of magnetic nanoparticles. The potential of magnetic nanoparticles stems from the intrinsic properties of their magnetic cores combined with their drug loading capability and the biochemical properties that can be bestowed on them by means of a suitable coating. CNT has emerged as a new alternative and efficient tool for transporting and translocating therapeutic molecules. CNT can be functionalised with bioactive peptides, proteins, nucleic acids and drugs, and used to deliver their cargos to cells and organs. Because functionalised CNT display low toxicity and are not immunogenic, such systems hold great potential in the field of nanobiotechnology and nanomedicine [109, 110]. Pastorin et al. have developed a novel strategy for the functionalisation of CNTs with two different molecules using the 1,3-dipolar cycloaddition of azomethine ylides [111]. The attachment of molecules that will target specific receptors on tumour cells will help improve the response to anticancer agents. Liu et al. have found that prefunctionalized CNTs can adsorb widely used aromatic molecules by simple mixing, forming "forest–scrub"-like assemblies on CNTs with PEG extending into water to impart solubility and aromatic molecules densely populating CNT sidewalls. The work establishes a novel, easy-to-make formulation of a SWNT-doxorubicin complex with extremely high drug loading efficiency [112].

In recent years, graphene based drug delivery systems have attracted more and more attention. In 2008, Sun et al. firstly reported the application of nano-graphene oxide (NGO) for cellular imaging and drug delivery [113]. They have developed functionalization chemistry in order to impart solubility and compatibility of NGO in biological environments. Simple physicosorption via π-stacking can be used for loading doxorubicin, a widely used cancer drug onto NGO functionalized with antibody for selective killing of cancer cells in vitro. Functional nanoscale graphene oxide is found to be a novel nanocarrier for the loading and targeted delivery of anticancer drugs [114]. Controlled loading of two anticancer drugs onto the folic acid-conjugated NGO via п–п stacking and hydrophobic interactions demonstrated that NGO loaded with the two anticancer drugs showed specific targeting to MCF-7 cells (human breast cancer cells with folic acid receptors), and remarkably high cytotoxicity compared to NGO loaded with either doxorubicin or camptothecin only. The PEGylated (PEG: polyethylene glycol) nanographene oxide could be used for the delivery of water-insoluble cancer drugs [115]. PEGylated NGO readily complexes with a water insoluble aromatic molecule SN38, a camptothecin analogue, via noncovalent van der Waals interaction. The NGO-PEG-SN38 complex exhibits excellent aqueous solubility and retains the high potency of free SN38 dissolved in organic solvents. Yang et al. found $GO-Fe_3O_4$ hybrid could be loaded with anti-cancer drug doxorubicin hydrochloride with a high loading capacity [116]. This $GO-Fe_3O_4$ hybrid showed superparamagnetic property and could congregate under acidic conditions and be redispersed reversibly under basic conditions. This pH-triggered controlled magnetic behavior makes this material a promising candidate for controlled targeted drug delivery.

4.6 Other applications

4.6.1 Intensify microreactors

The discovery of high enhancement of heat transfer in nanofluids can be applicable to the area of process intensification of chemical reactors through integration of the functionalities of reaction and heat transfer in compact multifunctional reactors. Fan et al. studied a nanofluid based on benign TiO_2 material dispersed in ethylene glycol in an integrated reactor-heat exchanger [117]. The overall heat transfer coefficient increase was up to 35% in the steady state continuous experiments. This resulted in a closer temperature control in the reaction of selective reduction of an aromatic aldehyde by molecular hydrogen and very rapid change in the temperature of reaction under dynamic reaction control.

4.6.2 Nanofluids as vehicular brake fluids

A vehicle's kinetic energy is dispersed through the heat produced during the process of braking and this is transmitted throughout the brake fluid in the hydraulic braking system [39], and now there is a higher demand for the properties of brake oils. Copper-oxide and aluminum-oxide based brake nanofluids were manufactured using the arc-submerged nanoparticle synthesis system and the plasma charging arc system, respectively [118, 119]. The two kinds of nanofluids both have enhanced properties such as a higher boiling point, higher viscosity and a higher conductivity than that of traditional brake fluid. By yielding a higher boiling point, conductivity and viscosity, the nanofluid brake oil will reduce the occurrence of vapor-lock and offer increased safety while driving.

4.6.3 Nanofluids based microbial fuel cell

Microbial fuel cells (MFC) that utilize the energy found in carbohydrates, proteins and other energy rich natural products to generate electrical power have a promising future. The excellent performance of MFC depends on electrodes and electron mediator. Sharma et al. constructed a novel microbial fuel cell (MFC) using novel electron mediators and CNT based electrodes [120]. The novel mediators are nanofluids which were prepared by dispersing nanocrystalline platinum anchored CNTs in water. They compared the performance of the new E. coli based MFC to the previously reported E. coli based microbial fuel cells with Neutral Red and Methylene Blue electron mediators. The performance of the MFC using CNT based nanofluids and CNT based electrodes has been compared against plain graphite electrode based MFC. CNT based electrodes showed as high as ~6 fold increase in the power density compared to graphite electrodes. The work demonstrates the potential of noble metal nanoparticles dispersed on CNT based MFC for the generation of high energies from even simple bacteria like E. coli.

4.6.4 Nanofluids with unique optical properties

Optical filters are used to select different wavelengths of light. The ferrofluid based optical filter has tunable properties. The desired central wavelength region can be tuned by an external magnetic field. Philip et al. developed a ferrofluid based emulsion for selecting different bands of wavelengths in the UV, visible and IR regions [121]. The desired range of wavelengths, bandwidth and percentage of reflectivity could be easily controlled by using suitably tailored ferrofluid emulsions. Mishra et al. developed nanofluids with selective visible colors in gold nanoparticles embedded in polymer molecules of polyvinyl pyrrolidone (PVP) in water [122]. They compared the developments in the apparent visible colors in forming the Au-PVP nanofluids of 0.05, 0.10, 0.50, and 1.00 wt% Au-contents. The surface plasmon bands, which occurs over 480-700 nm, varies sensitively in its position as well as the intensity when varying the Au-content 0-1 wt%.

5. Conclusions

Many interesting properties of nanofluids have been reported in the past decades. This paper presents an overview of the recent developments in the study of nanofluids, including the preparation methods, the evaluation methods for their stability, the ways to enhance their stability, the stability mechanisms, and their potential applications in heat transfer intensification, mass transfer enhancement, energy fields, mechanical fields and biomedical fields, etc.

Although nanofluids have displayed enormously exciting potential applications, some vital hinders also exist before commercialization of nanofluids. The following key issues should receive greater attention in the future. Firstly, further experimental and theoretical researches are required to find the major factors influencing the performance of nanofluids. Up to now, there is a lack of agreement between experimental results from different groups, so it is important to systematically identify these factors. The detailed and accurate structure characterizations of the suspensions may be the key to explain the discrepancy in the experimental data. Secondly, increase in viscosity by the use of nanofluids is an important drawback due to the associated increase in pumping power. The applications for nanofluids with low viscosity and high conductivity are promising. Enhancing the compatibility between nanomaterials and the base fluids through modifying the interface properties of two phases may be one of the solution routes. Thirdly, the shape of the additives in nanofluids is very important for the properties, therefore the new nanofluid synthesis approaches with controllable microscope structure will be an interesting research work. Fourthly, Stability of the suspension is a crucial issue for both scientific research and practical applications. The stability of nanofluids, especially the long term stability, the stability in the practical conditions and the stability after thousands of thermal cycles should be paid more attention. Fifthly, there is a lack of investigation of the thermal performance of nanofluids at high temperatures, which may widen the possible application areas of nanofluids, like in high temperature solar energy absorption and high temperature energy storage. At the same time, high temperature may accelerate the degradation of the surfactants used as dispersants in nanofluids, and may produce more foams. These factors should be taken into account. Finally, the properties of nanofluids strongly depend on the shape and property of the additive. Therefore nanofluid research can be richened and extended through exploring new nanomaterials. For example, the newly discovered 2-D monatomic sheet graphene is a promising candidate material to enhance the thermal conductivity of the base fluid [123, 124]. The concept of nanofluids is extended by the use of phase change materials, which goes well beyond simply increasing the thermal conductivity of a fluid [125]. It is found that the indium/polyalphaolefin phase change nanofluid exhibits simultaneously enhanced thermal conductivity and specific heat.

6. Acknowledgment

The work was supported by New Century Excellent Talents in University (NECT-10-883), the Program for Professor of Special Appointment (Eastern Scholar) at Shanghai Institutions of Higher Learning, and partly by National Natural Science Foundation of China (51106093).

7. References

[1] V. Trisaksri, S. Wongwises, Renew. Sust. Energ. Rev. 11, 512 (2007).
[2] S. Özerinç, S. Kakaç, A.G. Yazıcıoğlu, Microfluid Nanofluid 8, 145 (2009).
[3] X. Wang, A.S. Mujumdar, Int. J. Therm. Sci. 46, 1 (2007).
[4] X. Wang, A.S. Mujumdar, Brazilian J. Chem. Eng. 25, 613 (2008).
[5] Y. Li, J. Zhou, S. Tung, E. Schneider, S. Xi, Powder Technol. 196, 89 (2009).
[6] S. Kakaç, A. Pramuanjaroenkij, Int. J. Heat Mass Transf. 52, 3187 (2009).
[7] J.A. Eastman, S.U. Choi, S. Li, W. Yu, L.J. Thompson, Appl. Phys. Lett. 78, 718 (2001).
[8] C. Lo, T. Tsung, L. Chen, J. Cryst. Growth 277, 636 (2005).
[9] C. Lo, T. Tsung, L. Chen, C. Su, H. Lin, J. Nanopart. Res. 7, 313 (2005).
[10] H. Zhu, Y. Lin, Y. Yin, J. Colloid Interface Sci. 277, 100 (2004).
[11] H. Bönnemann, S.S. Botha, B. Bladergroen, V.M. Linkov, Appl. Organomet. Chem. 19, 768 (2005).
[12] A.K. Singh, V.S. Raykar, Colloid Polym. Sci. 286, 1667 (2008).
[13] A. Kumar, J. Colloid Interface Sci. 264, 396 (2003).
[14] H. Zhu, C. Zhang, Y. Tang, J. Wang, J. Phys. Chem. C 111, 1646 (2007).
[15] Y. Chen, X. Wang, Mater. Lett. 62, 2215 (2008).
[16] X. Feng, H. Ma, S. Huang, W. Pan, X. Zhang, F. Tian, C. Gao, Y. Cheng, J. Luo, J. Phys. Chem. 110, 12311 (2006).
[17] W. Yu, H. Xie, L. Chen, Y. Li, Colloids Surf. A: Physicochemical And Engineering Aspects 355, 109 (2010).
[18] L.Q. Wang, J. Fan, Nanoscale Res. Lett. 5, 1241 (2010).
[19] X. Wei, L. Wang, Particuology 8, 262 (2010).
[20] X. Li, D. Zhu, X. Wang, J. Colloid Interface Sci. 310, 456 (2007).
[21] H. Zhu, C. Zhang, Y. Tang, J. Wang, B. Ren, Y. Yin, Carbon 45, 226 (2007).

[22] D. Li, R.B. Kaner, Chem. Commun. 14, 3286 (2005).

[23] H.J. Kim, I.C. Bang, J. Onoe, Opt. Laser Eng. 47, 532 (2009).

[24] X. Wang, X. Li, S. Yang, Energy Fuels 23, 2684 (2009).

[25] D. Zhu, X. Li, N. Wang, X. Wang, J. Gao, H. Li, Curr. Appl. Phys. 9, 131 (2009).

[26] L. Chen, H. Xie, Thermochim. Acta 506, 62 (2010).

[27] J. Huang, X. Wang, Q. Long, X. Wen, Y. Zhou, L. Li, Symposium On Photonics And Optoelectronics 1-4 (2009).

[28] M. Farahmandjou, S.A. Sebt, S.S. Parhizgar, P. Aberomand, M. Akhavan, Chin. Phys. Lett. 26, 027501 (2009).

[29] Y. Hwang, J. Lee, C. Lee, Y. Jung, S. Cheong, B. Ku, S. Jang, Thermochim. Acta 455, 70 (2007).

[30] L. Chen, H. Xie, Y. Li, W. Yu, Thermochim. Acta 477, 21 (2008).

[31] X. Yang, Z. Liu, Nanoscale Res. Lett. 5, 1324 (2010).

[32] L. Chen, H. Xie, Thermochim. Acta 497, 67 (2010).

[33] Q. Yu, Y.J. Kim, H. Ma, Appl. Phys. Lett. 92, 103111 (2008).

[34] I.M. Joni, A. Purwanto, F. Iskandar, K. Okuyama, Ind. Eng. Chem. Res. 48, 6916 (2009).

[35] E. Tang, G. Cheng, X. Ma, X. Pang, Q. Zhao, Appl. Surf. Sci. 252, 5227 (2006).

[36] T. Missana, A. Adell, J. Colloid Interface Sci. 230, 150 (2000).

[37] I. Popa, G. Gillies, G. Papastavrou, M. Borkovec, J. Phys. Chem. B 114, 3170 (2010).

[38] H. Kamiya, Y. Fukuda, Y. Suzuki, M. Tsukada, T. Kakui, M. Naito, J. Am. Ceram. Soc. 82, 3407 (1999).

[39] K.V. Wong, O. De Leon, Adv. Mech. Eng. 2010, 1 (2010).

[40] G. Donzelli, R. Cerbino, A. Vailati, Phys. Rev. Lett. 102, 1 (2009).

[41] M. Arruebo, R. Fernández-pacheco, M.R. Ibarra, J. Santamaría, Rev. Literature Arts Am. 2, 22 (2007).

[42] W. Yu, D.M. France, D. Singh, E.V. Timofeeva, D.S. Smith, J.L. Routbort, J. Nanosci. Nanotechnol. 10, 4824 (2010).

[43] K. Ma, J. Liu, Phys. Lett. A 361, 252 (2007).

[44] G. Paul, M. Chopkar, I. Manna, P.K. Das, Renew.Sust. Energ Rev. 14, 1913 (2010).

[45] S.P. Jang, S.U.S. Choi, Appl. Therm. Eng. 26, 2457 (2006).

[46] C.T. Nguyen, G. Roy, N. Galanis, S. Suiro, Proceedings of the 4th WSEAS Int. Conf. on heat transfer, thermal engineering and environment, Elounda, Greece, August 21-23, 103-108 (2006).

[47] H. Shokouhmand, M. Ghazvini, J. Shabanian, Proceedings of the World Congress on Engineering 2008 Vol III, WCE 2008, July 2 - 4, 2008, London, U.K.

[48] C.Y. Tsaia, H.T. Chiena, P.P. Dingb, B. Chanc, T.Y. Luhd, P.H. Chena, Mater. Lett. 58, 1461 (2004).

[49] Y.T. Chen, W.C. Wei, S.W. Kang, C.S. Yu, 24th IEEE SEMI-THERM Symposium 16 (2008).

[50] H.B. Ma, C. Wilson, B. Borgmeyer, K. Park, Q. Yu, S.U.S. Choi, M. Tirumala, Appl. Phys. Lett. 88, 143116 (2006).

[51] S.W. Kang, W.C. Wei, S.H. Tsai, C.C. Huang, Appl. Therm. Eng. 29, 973 (2009).

[52] P. Naphon, P. Assadamongkol, T. Borirak, Int. Commun. Heat Mass Transf. 35, 1316 (2008).

[53] H.Q. Xie, L.F. Chen, Phys. Lett. A 373, 1861 (2009).

[54] H.Q. Xie, W. Yu, Y. Li, J. Phys. D: Appl. Phys. 42, 095413 (2009).

[55] W. Yu, H.Q. Xie, Y. Li, L.F. Chen, Thermochim. Acta, 491, 92 (2009).

[56] W. Yu, D.M. France, S.U.S. Choi and J.L. Routbort, "Review and Assessment of Nanofluid Technology for Transportation and Other Applications,"Argonne National Laboratory Technical Report, ANL/ESD/07-9, April 2007, 78 (2007).

[57] M. Kole, T.K. Dey, J. Phys. D: Appl. Phys. 43 315501(2010).

[58] S.Z. Tzeng, C.W. Lin, K.D. Huang, Acta Mechanica, 179, 11 (2005).

[59] D. Singh, J. Toutbort, G. Chen, et al., "Heavy vehicle systems optimization merit review and peer evaluation," Annual Report, Argonne National Laboratory, 2006.

[60] http://www.labnews.co.uk/feature_archive.php/5449/5/keeping-it-cool

[61] J. Routbort, et al., Argonne National Lab, Michellin North America, St. Gobain Corp., 2009,
http://www1.eere.energy.gov/industry/nanomanufacturing/pdfs/nanofluids industrial cooling.pdf.

[62] http://96.30.12.13/execsumm/VU0319--Nanofluid%20for%20Cooling%20Enhancement %20of%20Electrical%20Power%20Equipment.pdf

[63] I.C. Nelson, D. Banerjee, R. Ponnappan, J. Thermophys. Heat Transf. 23, 752 (2009).

[64] K.Q. Ma, J. Liu, Phys. Lett. A 361, 252 (2007).

[65] D.P. Kulkarni, D.K. Das, R.S. Vajjha, Appl. Energy, 86, 2566 (2009).

[66] J. Buongiorno, L.W. Hu, J.K. Sung, R. Hannink, B. Truong, E. Forrest, Nucl. Technol. 162, 80 (2008).

[67] S.M. You, J.H. Kim, and K.H. Kim, Appl. Phys. Lett. 83, 3374 (2003).

[68] P. Vassallo, R. Kumar, S.D. Amico. Int. J. Heat Mass Transf. 47, 407 (2004).

[69] J. Kim, J. Jung, Y. Kang, Int. J. Refrig. 29, 22 (2006).

[70] J. Kim, J. Jung, Y. Kang, Int. J. Refrig. 30, 50 (2007).

[71] J. Kim, Y. Kang, C. Choi, Int. J. Refrig. 30, 323 (2007).

[72] X. Ma, F. Su, J. Chen, Y. Zhang, J. Mech. Sci. Technol. 21, 1813 (2007).

[73] X. Ma, F. Su, J. Chen, T. Bai, Z. Han, Int. Commun. Heat Mass Transf. 36, 657 (2009).

[74] S. Komati, A.K. Suresh, Chem. Technol. 1100, 1094 (2008).

[75] L. Yang, K. Du, B. Cheng, Y. Jiang, 2010 Asia-Pacific Power And Energy Engineering Conference 1-4 (2010).

[76] M.F. Demirbas, Energy Source. Part B: Economics, Planning, and Policy 1, 85 (2006).

[77] S. Wu, D. Zhu, X. Zhang, J. Huang, Energy Fuels 24, 1894 (2010).

[78] Y. Liu, Y. Zhou, M. Tong, X. Zhou, Microfluid. Nanofluid. 7, 579 (2009).

[79] T.P. Otanicar, P.E. Phelan, R.S. Prasher, G. Rosengarten, R.a. Taylor, Renew.Sust. Energ Rev. 2, 033102 (2010).

[80] H. Tyagi, P. Phelan, R. Prasher, J. Solar Energy Eng. 131, 041004 (2009).

[81] T.P. Otanicar, J.S. Golden, Environ. Sci. Technol. 43, 6082 (2009).

[82] E. Sani, S. Barison, C. Pagura, L. Mercatelli, P. Sansoni, D. Fontani, D. Jafrancesco, F. Francini, Opt. Express 18, 4613 (2010).

[83] J. Zhou, Z. Wu, Z. Zhang, W. Liu, Q. Xue, Tribol. Lett. 8, 213 (2000).

[84] B. Shen, A. Shih, S. Tung, Tribology Transactions 51, 730 (2008).

[85] H. Yu, Y. Xu, P. Shi, B. Xu, X. Wang, Q. Liu, T. Nonferr. Metal. Soc. China 18, 636 (2008).

[86] B. Wang, X. Wang, W. Lou, J. Hao, J. Phys. Chem. C 114, 8749 (2010).

[87] W. Li-jun, G. Chu-wen, R. Yamane, J. Trib. 130, 031801 (2008).

[88] S. Chen, D.H. Mao, Adv. Tribol. 995 (2010).

[89] D. Peng, C. Chen, Y. Kang, Y. Chang, S. Chang, Ind. Lubr. Trib. 62, 111 (2010).

[90] L. Vekas, China Particuology 5, 43 (2007).

[91] R.E. Rosensweig, Ann. Rev. Fluid Mech. 19, 437 (1987).

[92] Y.S. Kim, K. Nakatsuka, T. Fujita, T. Atarashi, J. Magn. Magn. Mater. 201, 361 (1999).

[93] Y. Mitamura, S. Arioka, D. Sakota, K. Sekine, M. Azegami, J Phys.: Condens. Mater. 20, 204145 (2008).

[94] Y. Kim, J. Magn. Magn. Mater. 267, 105 (2003).

[95] L. Zhang, Y. Jiang, Y. Ding, M. Povey, D. York, J. Nanopart. Res. 9, 479 (2006).

[96] R. Jalal, E.K. Goharshadi, M. Abareshi, M. Moosavi, A. Yousefi, P. Nancarrow, Mater. Chem. Phys. 121, 198 (2010).

[97] N. Jones, B. Ray, K.T. Ranjit, A.C. Manna, FEMS Microbiol. Lett. 279, 71 (2008).

[98] Y. Liu, L. He, a. Mustapha, H. Li, Z.Q. Hu, M. Lin, J. Appl. Microbiol. 107, 1193 (2009).

[99] O. Mahapatra, M. Bhagat, C. Gopalakrishnan, K. Arunachalam, J. Exp. Nanosci. 3, 185 (2008).

[100] P. Gajjar, B. Pettee, D.W. Britt, W. Huang, W.P. Johnson, A.J. Anderson, J. Biol. Eng. 3, 9 (2009).

[101] H.J. Lee, S.Y. Yeo, S.H. Jeong, Polym. 8, 2199 (2003).

[102] A. Panacek, L. Kvítek, R. Prucek, M. Kolar, R. Vecerova, N. Pizúrova, V.K. Sharma, T. Nevecna, R. Zboril, J. Phys. Chem. B 110, 16248 (2006).

[103] L. Brunet, D.Y. Lyon, E.M. Hotze, P.J. Alvarez, M.R. Wiesner, Environ. Sci. Technol. 43, 4355 (2009).

[104] D.Y. Lyon, P.J. Alvarez, Environ. Sci. Technol. 42, 8127 (2008).

[105] A. Vonarbourg, C. Passirani, P. Saulnier, J. Benoit, Biomater. 27, 4356 (2006).

[106] R. Singh, J.W. Lillard, Exp. Mol. Pathol. 86, 215 (2009).

[107] P. Ghosh, G. Han, M. De, C.K. Kim, V.M. Rotello, Adv. Drug Deliver. Rev. 17 1307 (2008).

[108] M. Nakano, H. Matsuura, D. Ju, T. Kumazawa, S. Kimura, Y. Uozumi, N. Tonohata, K. Koide, N. Noda, P. Bian, M. Akutsu, K. Masuyama, K. Makino, Proceedings of the 3rd International Conference on Innovative Computing, Information and Control, ICICIC2008, Dalian, China, June 18-20 (2008).

[109] A. Bianco, K. Kostarelos, M. Prato, Curr. Opin. Chem. Biol. 9, 674 (2005).

[110] C. Tripisciano, K. Kraemer, a. Taylor, E. Borowiak-Palen, Chem. Phys. Lett. 478, 200 (2009).

[111] G. Pastorin, W. Wu, S. Wieckowski, J. Briand, K. Kostarelos, M. Prato, A. Bianco, Chem. Commun. 1, 1182 (2006).

[112] Z. Liu, X. Sun, N. Nakayama-Ratchford, H. Dai, ACS Nano 1, 50 (2007).

[113] X. Sun, Z. Liu, K. Welsher, J.T. Robinson, A. Goodwin, S. Zaric, H. Dai, Nano Res. 1, 203 (2008).

[114] L. Zhang, J. Xia, Q. Zhao, L. Liu, Z. Zhang, Small 6, 537 (2010).

[115] Z. Liu, J.T. Robinson, X. Sun, H. Dai, J. Am. Chem. Soc. 130, 10876 (2008).

[116] X. Yang, X. Zhang, Y. Ma, Y. Huang, Y. Wang, Y. Chen, J. Mater. Chem. 19, 2710 (2009).

[117] X. Fan, H. Chen, Y. Ding, P.K. Plucinski, A.a. Lapkin, Green Chem. 10, 670 (2008).

[118] M.J. Kao, C.H. Lo, T.T. Tsung, Y.Y. Wu, C.S. Jwo, H. M. Lin, J. Alloys Compd. 434, 672 (2007).

[119] M. J. Kao, H. Chang, Y.Y. Wu, T.T. Tsung, H. M. Lin, J. Chin. Soc. Mech. Eng. 28, 123 (2007).

[120] T. Sharma, A.L.M. Reddy, T.S. Chandra, S. Ramaprabhu, Int. J. Hydrogen Energy, 33, 6749 (2008).

[121] J. Philip, T. Jaykumar, P. Kalyanasundaram, B. Raj. Meas. Sci. Technol. 14, 1289 (2003).

[122] A. Mishra, P. Tripathy, S. Ram, H.J. Fecht. J. Nanosci. Nanotechnol. 9, 4342 (2009).

[123] W. Yu, H.Q. Xie, D. Bao, Nanotechnology, 21, 055705 (2010).

[124] W. Yu, H.Q. Xie, W. Chen, J. Appl. Phys. 107, 094317 (2010).

[125] Z.H. Han, F.Y. Cao, B. Yang, Appl. Phys. Lett. 92, 243104 (2008).

6

Nanoparticle Dynamics in Polymer Melts

Giovanni Filippone and Domenico Acierno

Dept. of Materials and Production Engineering, University of Naples Federico II

Italy

1. Introduction

Adding solid particles to polymeric materials is a common way to reduce the costs and to impart desired mechanical and transport properties. This makes polymers potential substitutes for more expensive non-polymeric materials. The advantages of filled polymers are normally offset by the increased complexity in the rheological behaviour of the resulting composite. Usually, a compromise has to be made between the benefits ensured by the filler, the increased difficulties in melt processing, the problems in achieving a uniform dispersion of the solid particulate, and the economics of the process due to the added step of compounding [Shenoy, 1999]. Filled polymers can be described as a suspension of particles and particle aggregates dispersed in the polymer matrix. Interactions between individual particles or aggregates and the matrix, as well as between particles, hinder the material deformability modifying both the solid- and melt-state behaviour of the host polymer. In polymer-based microcomposites, these effects only become significant at relatively high filler contents, i.e. when the filler particles are sufficiently close to each other to form a network that spans large sections of the polymer matrix. Over the last fifteen years, the same reinforcing and thixotropic effects have been observed with the use of very small amounts of inorganic nanoparticles, which has resulted in extensive research in the field of polymer-based nanocomposites (PNCs) [Usuki et al., 1993; Kojima et al., 1993]. In order to fully understand the exceptional properties of PNCs, the morphological and structural implications stemming from the nanometric sizes of the filler have to be taken into account. With respect to traditional microcomposites, nanocomposites show very high specific interface area, typically of order of ~10^2 m^2 g^{-1}. The matrix properties are significantly affected in the vicinity of the reinforcement, varying continuously from the interface towards the bulk polymer. As a consequence, the large amount of reinforcement surface area means that a relatively small amount of nanoscale reinforcement can have remarkable effects on the macroscale properties of the composite material.

A noticeable consequence of the nanometric dimensions of the filler is the extremely high numerical density of particles, or alternatively the very small inter-particles distances. If N monodisperse spherical particles with radius r are randomly distributed in a volume V, the distance between the centres of the particles can be approximated to $h=(V/N)^{1/3}$. Introducing the particle volume fraction $\Phi=Nv/V$, where $v=4\pi r^3/3$ is the volume of the single particle, the wall-to-wall distance between contiguous particles is:

$$h = \left[\sqrt[3]{\frac{4\pi}{3\Phi}} - 2 \right] r \tag{1}$$

Once fixed the filler content, h linearly scales with r. In addition, we observe that, for diluted systems ($\Phi<0.1$) such those we are interested in, Eq. 1 gives $h\sim2r$. This means that, if the filler particles are well dispersed within the host polymer, nanometric inter-particles distances are expected for nanocomposites. In such systems a large fraction of polymer is in contact with the filler. At the most, if the particle radius is of the same order as the mean radius of gyration of host polymer chains, R_g, each single chain potentially interacts with more than one nanoparticle, and there is no bulk polymer [Jancar & Recman, 2010]. Two scenarios are possible when inter-particles distances are so small: if a good affinity exists between the polymer and the filler, then a polymer-mediated transient network between the particles set up [Ozmusul et al., 2005; Saint-Michel et al., 2003; Zhang & Archer, 2002]; on the other hand, if the polymer-filler interactions are weak, the nanoparticles aggregate forming flocs, which eventually assemble into a space-spanning filler network [Filippone et al. 2009; Inoubli et al., 2006; Ren et al., 2000]. In both cases, the presence of a three-dimensional mesostructure has a profound effect on the composite rheology.

The formation of the network, either polymer-mediated or formed by bare nanoparticles, originates from local rearrangements of the filler occurring in the melt both during flow and at rest. Nanoparticles, in fact, are subjected to relevant Brownian motion even in highly viscous mediums such as polymer melts. To get an idea about the relevance of such a phenomenon, we estimate the self-diffusion time of a spherical particle of radius r, τ_s, that is the time required for the particle moves of a length equal to its radius [Russel et al., 1989]:

$$\tau_s = \frac{6\pi\eta_s r^3}{k_B T} \tag{2}$$

Here η_s is the viscosity of the suspending medium, k_B is the Boltzmann's constant and T is the temperature. For a simple low viscosity ($\eta_s\sim10^{-2}$ Pa*s) Newtonian liquid at room temperature, Eq. 2 gives the well-known result that particles of a few microns in size experience appreciable Brownian motions. Setting $T=400°K$ and $\eta_s\sim10^3$ Pa*s as typical values for melted polymers, we obtain the noteworthy result that particles of a few tens of nanometers display Brownian motions on timescales of order of a $10^1\div10^2$ seconds. Such durations are typically accessed in many transforming processes of the polymer industry, as well as during rheological analyses. The result is that, unlike polymer microcomposites, PNCs can be depicted as "living systems", in which the particles are free to move and rearrange in the melt, both in flow and even at rest, towards more favourable thermodynamic states. In this sense, PNCs are reminiscent of colloidal suspensions. Therefore, these simpler systems can be considered as the natural starting point to understand the much more complex rheological behaviour of PNCs.

The simplest case of colloidal dispersion is represented by a Newtonian suspension of hard spheres. The inter-particles interaction is zero at all separations and infinitely repulsive at contact. Coupled with thermal fluctuations, this kind of interaction results in a wide variety of possible structures. The suspension may behave like a gas, a liquid, a crystal or a glass depending on the particle volume fraction Φ [Pusey & van Megen, 1986]. In the presence of interactions, this phase behaviour is modified due to interplay between Φ and the energy of interaction, U. We are mainly interested in weakly attractive colloidal dispersions, which are reminiscent of a large number of PNCs in which Van der Waals forces between nanoparticles and aggregates are of major importance. In such systems, aggregation results in disordered clusters of particles. These mesostructures may or may not span the whole space depending on Φ and U [Prasad, 2003]. The rheological behaviour of weakly interacting colloidal dispersions can be rationalized with a simple two-phase model that combines the elasticity of the disordered particle network and the Newtonian viscosity of the suspending liquid [Cipelletti et al., 2000; Trappe & Weitz, 2000; Trappe et al., 2001]. Despite the complexities stemming from the intrinsic non-Newtonian feature of polymer matrices, in this chapter we show that a similar approach can be successfully applied to a series of model PNCs with weak polymer-filler interaction. We emphasize that many PNCs of technological interest fall in this family. The two-phase model is validated through the building of a master curve of the elastic modulus of samples at different composition. A refinement of the model is also presented, which accounts for hydrodynamic effects. The dynamics of de-structuring and reforming of the filler network are studied by analysing the effects of large amplitude deformations. Besides simplifying the viscoelastic analysis of complex systems such as PNCs, the proposed approach can be extended to a wide variety of complex fluids where the elasticity of the components can be superimposed. In particular, the elastic modulus has been recently suggested to follow a universal behaviour with volume fraction also in case of interacting systems in which polymer bridging mechanisms exist [Surve et al., 2006]. This suggests a possible general feature for the proposed approach.

2. Viscoelasticity and structure of PNCs

We start our analysis dealing with the implications of Brownian motion in simple model systems constituted by polymer melts filled with small amounts of different kinds of spherical particles. Specifically, we discuss the effect of particle size and matrix viscosity on the ability of the filler to aggregate and eventually assemble in a three-dimensional network. Then, a two-phase model firstly proposed for weakly attractive particles suspended in a Newtonian medium is presented [Trappe & Weitz, 2000]. The physical picture of an elastic particle network interspersed in a background fluid qualitatively accounts for the viscoelastic behaviour of the suspension. Afterwards, the relations between structure and viscoelasticity of two model PNCs is described in the framework of the two-phase model. A refinement of the model is therefore presented, which accounts for hydrodynamic effects successfully capturing the frequency dependent viscoelastic behaviour of simple PNCs. Finally, the dynamics of de-structuring and reforming of the filler network are studied by analysing the effects of large amplitude deformations.

2.1 Brownian motion in polymer melts filled with nanoparticles – Gelation and ageing

2.1.1 Preliminary considerations

Untreated inorganic particles are difficult to disperse in polymer matrices due to the typically poor polymer-filler affinity. Such incompatibility clearly emerges in the case of PNCs, where the specific surface of the particles is very high. The hydrodynamic forces developed during intense melt mixing processes breakup the large aggregates down to clusters of few tens of particles [Baird & Collias, 1998]. Above the melting or glass transition temperature of the polymer matrix, however, these aggregates may reassemble into bigger structures because of the inter-particles attraction forces. Since the refractive indexes of polymers and inorganic fillers are typically very different, Van der Waals forces becomes of major importance leading to formation of aggregates and particle gels. The two most simple experimental techniques to follow the rearrangements of the filler in the melt are: (i) direct visualization of the particles through electron microscopy performed on solid samples; (ii) monitoring of rheological parameters sensitive to the material internal structure. Both these techniques have been applied to several model PNC systems constituted by polymer matrices filled with different kinds of inorganic nanoparticles with spherical symmetry. The rational for selecting such model systems originates from the intrinsic high complexity of other technologically relevant PNCs. The properties of such systems, often based on layered or tubular nanoparticles, are too sensitive to the state of dispersion of the filler and the wide variety of the possible nanostructures achievable during processing. The materials, the compounding procedures and the experimental details about the characterization of the composites are described in detail in the following paragraphs. Many of the results have been taken from papers previously published by our group, wherein further experimental details can be found [Acierno et al., 2007a, 2007b; Romeo et al., 2008; Filippone et al., 2010; Romeo et al., 2009].

2.1.2 Materials and sample preparation

Nano- and microcomposites were prepared using two different polymeric matrices. The first one is polypropylene (PP Moplen HP563N by Basell; weight average molecular weight M_w=245 KDa; zero shear viscosity η_0=1.9*10³ Pa*s at 190°C; terminal relaxation time τ_t≈0.4 s) with glass transition temperature T_g=6°C and melting temperature T_m=169°C. The second polymer matrix is atactic polystyrene (PS, kindly supplied by Polimeri Europa). In particular, we used two PS matrices at different molecular weight, coded as PS-low (M_w=125 KDa; η_0=1.7*10³ Pa*s at 200°C; τ_t≈0.1 s) and PS-high (M_w=268 KDa; η_0=2.1*10⁴ Pa*s at 200°C; τ_t≈100 s), both having glass transition temperatures T_g=100°C.

Three kinds of nanoparticles were used as fillers: titanium dioxide (TiO₂ by Sigma Aldrich; density ρ=3.9 g/mL; surface area ~190÷290 m²/g; average primary particles diameter d=15 nm) and alumina nanospheres (Al₂O₃ by Sigma Aldrich; ρ=4.00 g/mL; surface area: 35–43 m²/g; d≈40 nm) were used to prepare PP/TiO₂ and PP/Al₂O₃ nanocomposites with filler volume fractions up to Φ=0.064; fumed silica (SiO₂ by Degussa; ρ=2.2 g/mL; surface area ~135-165 m²/g; d=14 nm) was mixed with the two PS matrices up to Φ=0.041. PP/TiO₂ microcomposites were also prepared by using titanium dioxide microparticles (TiO₂ by Sigma Aldrich; ρ=3.9 g/mL; surface area ~0.14-0.04 m²/g; d≈4 μm).

Nano- and microcomposites were prepared by melt compounding the constituents using a co-rotating extruder (Minilab Microcompounder, ThermoHaake) equipped with a capillary die (diameter 2 mm). The extrusions were carried out at 190°C. The screw speed was set to ~100 rpm, corresponding to an average shear rate of order of 50 s⁻¹ inside the extrusion chamber. A feedback chamber allowed an accurate control of the residence time, which was set to 250 s for all the samples. The polymer and the filler were previously dried under vacuum for sixteen hours at 70°C. The neat polymers used as reference materials were extruded in the same conditions to allow for an accurate comparison with the filled samples.

2.1.3 Characterization

The morphology of the composites was examined by transmission electron microscopy (TEM mod. EM 208, Philips). The observations were performed on slices with thickness ~150 nm, which were randomly cut from the extruded pellets using a diamond knife at room temperature.

Rheological tests were carried out by means of either a strain-controlled rotational rheometer (ARES L.S, Rheometric Scientific) or a stress-controlled rotational rheometer (ARG2, TA Instruments). The tests were carried out using parallel plates with diameter of 25 mm for the nanocomposites, while plates of 50 mm were used for the neat polymers because of their low viscosity. All measurements were performed in an atmosphere of dry nitrogen. The testing temperature was T=190°C for the PP/TiO$_2$ samples and T=200°C for the PS/SiO$_2$ samples. The viscoelastic moduli display a range of strain-independence, i.e. a range of linear viscoelasticity, which depends on the filler content. In order to determine the limits of the linear viscoelastic regime, oscillatory strain scans were performed on each sample at a fixed frequency of 0.063 rad s^{-1}. Low-frequency (ω=0.063 rad s^{-1}) time-sweep experiments were performed to study the evolution of the linear viscoelastic properties during time. The frequency-dependent elastic (G') and viscous (G'') moduli of the samples were measured by oscillatory shear scans in the linear regime. To account for the marked sensitivity of the rheological response on filler content, we evaluated the effective amount of filler of each sample used for the rheological experiments through thermogravimetrical analyses (TGA). The filler volume fraction Φ was estimated as:

$$\Phi = \frac{c\rho_p}{\rho_f + c(\rho_p - \rho_f)} \tag{3}$$

where c is filler weight fraction as deduced from TGA and ρ_p and ρ_f are the densities of polymer and filler, respectively.

2.1.4 Effect of the filler mobility on the linear viscoelasticity

The internal structure of the as extruded sample PP/TiO$_2$ at Φ=0.038 is shown in the TEM micrograph of Figure 1.a. Well distributed nanoparticle aggregates of a few hundreds of nanometers are suspended in the polymer matrix. The magnification of one of these aggregates is reported in Figure 1.b. A few hundreds of individual TiO$_2$ nanoparticles are closely packed into dense clusters with irregular shape.

The sample was subjected to a thermal annealing at 190°C in quasi-quiescent conditions, i.e. by submitting it to shear oscillations in the rheometer at small strain amplitude (γ=2%) and frequency (ω=0.063 rad s^{-1}). This allows to monitor the evolutions during time of slow dynamic populations, relaxing in timescales of order of τ=1/ω≈16 sec, without altering the internal structure of the sample.

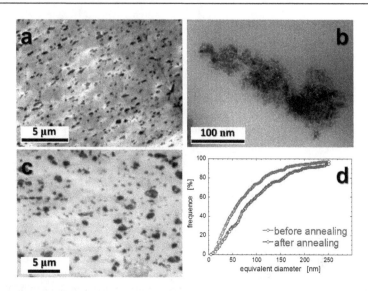

Fig. 1. (a) TEM micrograph of the as extruded sample PP/TiO$_2$ at Φ=0.045. (b) Magnification of an aggregate of TiO$_2$ nanoparticles. (c) TEM micrograph of the same sample as in (a) after a three-hours thermal annealing at T=190°C. (d) CSD of the samples shown in (a) and (c) (images taken from Acierno et al., 2007a).

The microstructure of the annealed sample is shown in Figure 1.c. A visual comparison with the morphology of the as extruded sample reveals the presence of bigger aggregates and the disappearance of the smaller ones. An analysis of the TEM micrographs was carried out to quantify the effect of the thermal annealing. An equivalent diameter for the aggregates was defined as $D_i=(\pi A_i)^{0.5}$, where A_i is the measured area of the i-th cluster. Once the sizes of the aggregates are available, the cumulative size distribution (CSD) and the number average size of the TiO$_2$ aggregates, $D_n=\Sigma n_i D_i / \Sigma n_i$ (n_i clusters with size D_i), was determined for each sample. The comparison between the CSDs is shown in Figure 1.d. The lowering of the cumulative CSD curves indicates an increase of the cluster sizes occurred during the thermal conditioning. In particular, the average size of the TiO$_2$ aggregates increases from $D_n \approx 125$ nm to $D_n \approx 170$ nm.

The coarsening of the microstructure is a consequence of the diffusion of the clusters under the push of Van der Waals attraction. Rheological parameters such as the linear viscoelastic moduli are extremely sensitive to the internal microstructure. Thus, we use them to follow such internal rearrangements. The time evolutions of G' and G'' at ω=0.063 rad s^{-1} are shown in Figure 2.

The elastic modulus, which at the beginning is lower than the viscous one, increases during the first stage and then it reaches a plateau after a certain time; on the other hand, the loss modulus remains essentially constant. Preliminary investigations revealed that the neat matrices display a constant value of the moduli in the analysed time window. Thus, the increase in the sample elasticity is related to the structuring of the inorganic phase. The characteristic timescale for such a phenomenon can be roughly estimated as the Smoluchowski time for two clusters of radius R to come in contact, τ_a [Russel et al., 1989]. This characteristic time depends on on the self-diffusion time of each aggregate, given by Equation 2, and on the average inter-aggregates distance, inversely proportional to the filler amount:

$$\tau_a = \frac{\pi \eta_s R^3}{\overline{\Phi} k_B T} \qquad (4)$$

$\overline{\Phi}$ is the actual filler volume fraction, i.e. the volume of the particles in a cluster plus the free volume enclosed between them. These regions are actually inaccessible to the polymer, and depends on how primary particles are assembled together inside the aggregates. As shown in Figure 1.b, the TiO_2 clusters appear rather compact. As a consequence, we can reasonably assume that the primary particles are packed at a volume fraction of ~60%, which is close to random close packing. The actual filler volume fraction of the PP/TiO_2 nanocomposites can be consequently estimated as $\overline{\Phi} \approx \Phi / 0.6$. Assuming $R = D_n / 2 \approx 65$ nm, Equation 4 gives $\tau_a \sim 4*10^3$ s, in good agreement with the data shown in Figure 2. This result suggests that the increasing of the sample elasticity during time is related to cluster-cluster aggregation. In order to support the previous conclusion, we increase τ_a by increasing either the size of primary particles or aggregates or the viscosity of the suspending medium. According to Equation 2, in these conditions we expect that the elasticity of the samples cannot increase significantly because of the reduced particle mobility.

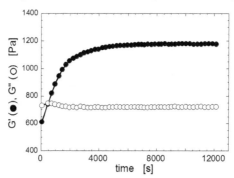

Fig. 2. Time evolution of G' (full) and G'' (empty) at $\omega = 0.063$ rad s⁻¹ and $T = 190°C$ for the nanocomposite PP/TiO_2 at $\Phi = 0.038$ (image taken from Romeo et al., 2009).

As first test, we investigate the time evolutions of the linear viscoelastic moduli at $\omega = 0.063$ rad s⁻¹ for a PP/TiO_2 microcomposite (particles radius $R \approx 1$ µm) at $\Phi \approx 0.035$. Based on Equation 4, we expect that two micron-sized particles should come at contact after timescales of order of ~10^7 s. As a matter of fact, the results shown in Figure 3.a indicate that both moduli remain stable during the aging test until ~10^4 s.

As second test, we monitor the moduli of a nanocomposite based on a high viscosity matrix such as PS-high ($\eta_0 = 2.1*10^4$ Pa*s at 200°C) filled with SiO_2 particles at $\Phi \approx 0.035$. Silica aggregates exhibits the typical open and branched structure of fractal objects. In such systems the mass M scales with length L as $M \sim L^{d_f}$, d_f being the fractal dimension [Weitz & Oliveira, 1984]. The actual filler volume fraction thus becomes [Wolthers et al., 1997]:

$$\overline{\Phi} = \Phi \times (L / d)^{3-d_f} \qquad\qquad (5)$$

Setting $L=D_n\approx125$ nm as emerged from the analysis of many TEM micrographs, and taking d_f=2.2 as a typical fractal dimension of fumed silica aggregates [Kammler et al., 2004], Equation 4 gives $\tau_a\sim10^5$ s. This is in agreement with the results of the time sweep experiment shown in Figure 3.b, which indicate that cluster assembling phenomena, if any, are negligible in the timescale of the test. Obviously, the not structured sample keeps a predominantly viscous connotation, i.e. $G''>>G'$.

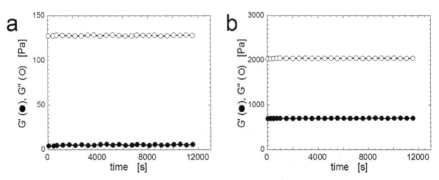

Fig. 3. Time evolution of G' (full) and G'' (empty) at ω=0.063 rad s^{-1} for the microcomposite PP/TiO$_2$ at Φ=0.035 and T=190°C (a), and the nanocomposite PS-high/SiO$_2$ at Φ=0.035 and T=200°C (b) (images taken from Romeo et al., 2009).

Particle rearrangements eventually give rise to mesoscopic structures, such as branched aggregates or space-spanning filler network, which strongly alter the frequency response of the sample. The ω-dependent G' and G'' of two PP/TiO$_2$ samples filled with micro- and nanoparticles both at $\Phi\approx0.035$ are compared in Figure 4.a. In both cases the matrix governs the high-frequency response. This suggests that the relaxation modes of the polymer chains and sub-chains are only slightly affected by the presence of the filler at these high frequencies. The presence of microparticles negligibly affects the whole response of the composite. On the contrary, the nanoparticles significantly alter the low-frequency moduli of the material, and in particular the elastic one.

The flattening of G' over long timescales is a general feature characterizing different kinds of PNCs [Krishnamoorti & Yurekli, 2001; Du et al., 2004]. Such a behaviour, however, is not a direct consequence of the nanometric size of the particles, but rather it originates from particle mobility. To emphasize this point, in Figure 4.b the frequency response of the three-hours aged samples PS-low/SiO$_2$ and PS-high/SiO$_2$ at $\Phi\approx0.035$ are compared. The nanocomposite with high viscosity matrix displays a liquid-like behaviour at low frequency reminiscent of that of the neat polymer (not shown). Differently, the PS-low/SiO$_2$ sample exhibits a predominant elastic connotation, the low-frequency plateau of G' being indicative of the presence of a space-spanning filler network formed during the ageing process.

To summarize, the viscoelastic response of a filled polymer is greatly affected by particle mobility. When the characteristic diffusion time of the particles and/or aggregates resulting from the extrusion process is too high, either because the matrix is too viscous or the particle size is too big, the filler is unable to rearrange and only produce a small perturbation of the composite viscoelastic response. Conversely, when mobility of the inorganic phase is high enough, random motion and attractive Van der Waals forces lead to the structuring of the primary aggregates. This eventually results in the formation of a whole space-spanning filler network. Since this network exhibits the connotation of an elastic solid, a drastic slowing down of relaxation dynamics occurs at low frequencies.

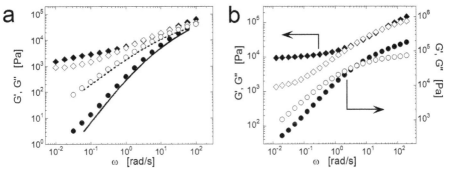

Fig. 4. (a) G' (full) and G'' (empty) for the samples PP/TiO$_2$ at $\Phi \approx 0.035$ filled with micrometric (circles) and nanometric (diamonds) particles. Solid and dashed lines represent the elastic and viscous modulus of the neat PP, respectively. (b) G' (full) and G'' (empty) for the nanocomposite samples PS-low/SiO$_2$ (diamonds, left axis) and PS-high/SiO$_2$ (circles, right axis) at $\Phi \approx 0.035$ (image taken from Romeo et al., 2009).

2.2 Linear viscoelasticity of PNCs

2.2.1 Weakly attractive particles suspended in Newtonian fluids – A two-phase model

Colloids are typically nanometer to micron sized particles forming a dispersed phase in a suspending medium. Colloidal dispersions exhibit a wide spectrum of rheological properties, ranging from simply viscous fluids to highly elastic pastes depending on the amount of particles and the sign and magnitude of inter-particles interactions. Here we are interested in weakly attractive systems, where the particles are inclined to assemble together into more or less branched flocs. In these systems, the ω-dependent storage and loss modulus typically exhibit a strong dependence on both Φ and U. Such a high variability makes extremely difficult a general description of the viscoelastic behaviour of colloidal dispersions. A drastic simplification has been introduced by Trappe and Weitz, which showed that modelling the suspension above the particle percolation threshold as an elastic filler network interspersed in a background fluid (two-phase model) qualitatively accounts for the viscoelasticity of their samples [Trappe & Weitz, 2000]. The authors studied a dispersion of carbon black in base stock oil as a function of particle volume fraction and interaction potential. The U was tuned by adding a dispersant that acts as a surfactant. Without dispersant carbon black particles are rather strongly attractive, and flocs of ~100 μm in size form even at very low amounts of particles. The linear viscoelastic moduli of samples at different Φ were measured as a function of frequency with a strain-controlled rheometer with Couette geometry. The authors found that a rheological transition occurs at $\Phi_c = 0.053$: at $\Phi > \Phi_c$ the suspension is clearly elastic and G' is nearly independent on ω; at

$\Phi<\Phi_c$ the viscous feature definitely prevails over the elastic one, and the suspension rheology looks like that of the suspending fluid. Microscopic analyses reveal that the rheological transition reflects the state of dispersion of the filler: isolated carbon black flocs are suspended in the background fluid below Φ_c, whereas above this threshold the aggregates assemble in a three-dimensional space spanning network.

Despite their marked differences, the moduli of the samples at $\Phi>\Phi_c$ can be scaled onto a single pair of master curves. The authors qualitatively accounted for the observed scaling by assuming that the carbon black forms a solid but tenuous network with a purely elastic, ω-independent modulus. The elasticity of this network, G'_0, increases with Φ as the network becomes more and more robust. Interspersed throughout this structure is the purely viscous suspending fluid, which G'' linearly increases with ω and is substantially independent of Φ. Consequently, the elasticity of the network prevails at low ω, while the viscosity of the fluid dominates at high ω. Within this simplified picture, scaling the elasticity of each sample along the viscosity of the matrix results in the collapse of data of samples at different composition onto a single pair of master curves.

Although the proposed approach can account for the basic scaling behaviour, many issue remain unresolved. For example, the behaviour of the weaker of the two moduli in each regime is not addressed. At low frequencies, $G''(\omega)$ must be determined by the loss modulus of the network, which is larger than that of the suspending fluid. Similarly, at the highest frequencies $G'(\omega)$ must reflect the storage component of the suspending fluid with the solid network in it. In addition, the model does not take into account hydrodynamic effects. Despite these limitations, the good quality of the scaling supports the reliability of the approach, indicating that there is a strong similarity in the structures of the networks that form at different Φ. This also implies some predictive feature of the model: the tiny elasticity of samples at low Φ (as long as greater than Φ_c), which networks are too tenuous to be appreciated through direct dynamic-mechanical analyses, can be predicted with good approximation by simply referring to the master curve of G'.

2.2.2 Weakly attractive nanoparticles suspended in non-Newtonian mediums – Recovering the two-phase model

The relatively high mobility of nanoparticles even in highly viscous fluids such as polymer melts makes PNCs similar to colloidal dispersions. The main difference with these simpler systems is the non-Newtonian feature of the suspending medium. According to Trappe and Weitz, the viscoelasticity of a colloidal suspension above Φ_c originates from the combination of the responses of an *elastic* particle gel that of the *purely viscous* background fluid. In the case of a PNC, instead, the suspending medium is viscoelastic by itself, and its response combines with that of the space-spanning network giving rise to a more complex ω- and Φ-dependent viscoelastic behaviour. It follows that a separation of the effects of the solid and fluid phases is no more possible in the case of PNCs. However, we argue that a recovery of the two-phase model is possible if the elasticity of the polymer is neglected with respect to that of the filler network. Under this assumption, the viscoelasticity of the PNC can be split into the independent responses of an *elastic* particle network and that of the *predominantly viscous* polymer. The former contribution depends on the filler content and governs the long timescale response of the composite, whereas the latter is responsible for the high-frequency behaviour (Figure 5).

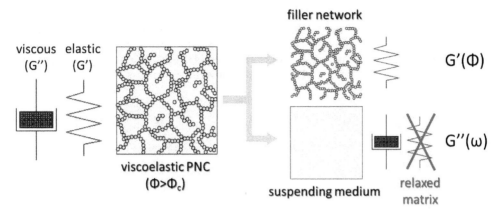

Fig. 5. Scheme of the viscoelasticity of a PNC at $\Phi > \Phi_c$. For fully relaxed polymer matrix, the filler network is the only responsible for the elastic connotation of the system.

To test the validity of the previous considerations, we focus on the ω-dependence of the moduli of PP/TiO$_2$ and PS-low/SiO$_2$ nanocomposite samples at $\Phi > \Phi_c$, i.e. in which the filler rearranges in experimentally accessible timescales forming a space-spanning network. All these samples share a similar pseudo solid-like behaviour at low frequency, with weak ω-dependences of both moduli and G' greater than G''. Since the filler mainly affects the elastic modulus of the samples, G' increases with Φ more rapidly than G''. As a consequence, a further crossover between G' and G'' occurs at intermediate frequencies in addition to that at high ω related to the relaxation of the neat polymer. The coordinates of such additional crossing point, $(\omega_c; G_c)$, shift towards higher and higher frequencies and moduli with increasing the filler content. This is shown in Figure 6 for three samples PS-low/SiO$_2$ at different composition.

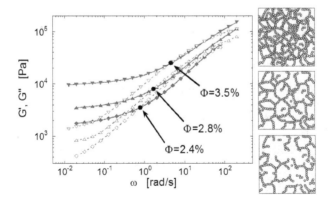

Fig. 6. G' (full, red) and G'' (empty, blu) for the nanocomposite samples PS-low/SiO$_2$ at increasing filler content. The additional crossover is indicated by the arrows.

The additional low-frequency crossover can be interpreted as the point at which the network elasticity equals the viscous contribution of the polymer. As a consequence, normalizing the moduli of samples at different Φ by their elasticity, and doing so along the background fluid viscosity, the curves should collapse onto a single pair of master curves. Accordingly, the scaling has to be done by shifting the curves both horizontally and vertically using as shift factors $a=1/\omega_c$ and $b=1/G_c$, respectively. The resulting master curves are shown in Figure 7 for both the PP/TiO$_2$ and PS-low/SiO$_2$ nanocomposites.

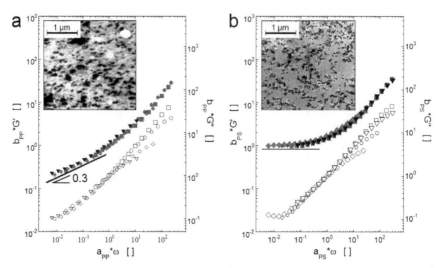

Fig. 7. Master curves of G' (full, left axis) and G" (empty, right axis) for the systems PP/TiO$_2$ (a) and PS-LOW/SiO$_2$ (b). Each colour corresponds to a composition. Note that only curves at $\Phi>\Phi_c$ have been used to build the master curves. The TEM micrographs shown in the insets represent the microstructures of samples at $\Phi\approx0.035$ (image taken from Romeo et al., 2009).

The scaled moduli lie on top of each other in about five decades of frequencies, supporting the validity of the adopted approach. Deviations are observed for the viscous moduli at high scaled frequencies. This is not unexpected, since the relaxation modes of the polymers are independent on the filler content and cannot be scaled using a and b as scaling factors.

Once the master curves are built, the differences in elasticity and dynamic of the particle networks become evident. The SiO$_2$ network is characterized by an ω-independent elastic modulus at low frequency, which emphasizes its truly solid-like feature. Differently, the TiO$_2$ network displays a slow relaxation dynamic with $G'\sim\omega^{0.3}$. These differences are related to the differences in network structures formed in the two composites. The TEM images reported in the insets of Figure 7 show that the SiO$_2$ nanoparticles form a tenuous, fractal network of sub-micron sized, branched flocs interspersed within the host PS. Differently, the TiO$_2$ nanoparticles are assembled into dense clusters, which mobility is presumably slowed down by the surrounding aggregates. The transient character of the latter network emerges as a glassy-like decrease of G', which reflects the internal rearrangements of the TiO$_2$ clusters. Such slow relaxation dynamics are characteristic of colloidal glasses [Shikata & Pearson, 1994; Mason & Weitz, 1995] and has been observed in many other soft materials [Sollich et al., 1997].

2.2.3 Refining the two-phase model – Role of the hydrodynamic effects

Despite the good quality of the scaling shown in Figure 7, unresolved issues exist regarding the physical meaning of the shift factors. The underlying physics of the model lies on the independent rheological responses of the *neat polymer* and the particle network. Actually, the coordinates of the crossover point of the moduli of the nanocomposite, identified by Trappe and Weitz as the shift factors for their system, do not rigorously reflect the properties of the two pristine phases of the model. In addition, the presence of the particles implies hydrodynamic effects, which cannot be eluded for a correct scaling of the data. To account for these issues, the procedure to get the shift factors for the building of the master curve has to be revisited. For this aim, hereinafter we only refer to the system PS-low/SiO$_2$, which particle network exhibits a truly solid-like behaviour at low frequency.

Hydrodynamic effects reflect the perturbation of the flow lines in proximity of the filler. In a liquid filled with a solid particulate, the suspending fluid flows in the narrow gap between contiguous particles or aggregates, locally experiencing a greater flow rate than what externally imposed or measured. Gleissle and Hochstein quantitatively accounted for hydrodynamic effects in oscillatory shear experiments by introducing an empiric amplification factor, representing the ratio between the complex moduli of the filled sample over that of the neat matrix: $B(\Phi) = G^*(\Phi)/G^*_{PS}$ [Gleissle & Hochstein, 2003]. In the case of microparticles, $B(\Phi)$ well describe the increase of G^* of the suspension in the whole range of accessible frequencies. Differently, non-continuum effects emerge over long timescales in the case of PNCs. Consequently, the hydrodynamic effects only are appreciable at high frequencies, i.e. where the rheological response is governed by the polymer matrix. This is shown in Figure 8, where the complex moduli of various PS-low/SiO$_2$ nanocomposites at different composition are reported together with the resulting $B(\Phi)$.

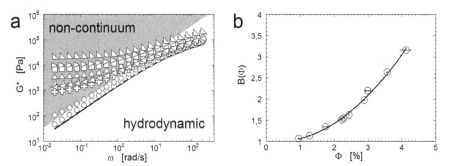

Fig. 8. (a) Complex modulus of PS-low/SiO$_2$ nanocomposite at various filler content. The regions in which non-continuum and hydrodynamic effects are dominant are emphasized. (b) Amplification factor for the data shown in (a) (images taken from Filippone et al., 2010).

After the hydrodynamic contribution has been quantified for each sample, then new and more rigorous shift factors can be identified. Specifically, we now refer to the point at which the elasticity of the filler network, given by the plateau modulus of the nanocomposite, $G'_0(\Phi)$, equals the viscous modulus of the neat matrix amplified by $B(\Phi)$ to account for hydrodynamic effects, $B(\Phi) \cdot G^*$.

The comparison between the old $(a; b)$ and new $(a'; b')$ shift factors is shown in Figure 9.a for the sample at Φ=2.9%; in Figure 9.b the new master curve is reported.

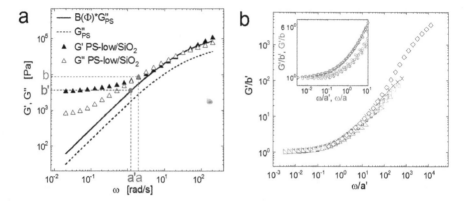

Fig. 9. (a) Comparison of the shift factors for the samples-low/SiO₂ at Φ=2.8%. (b) Master curve of G' built using a' and b' as shift factors; the inset shows a magnifications at low scaled frequencies of the master curves obtained using as shift factors (a'; b') (red) and (a; b) (blu) (images taken from Filippone et al., 2010).

The elastic moduli scaled using a' and b' as shift factors lies on top of each other over about seven decades of scaled frequencies, confirming the validity of the adopted approach. Again, the slight deviations at ω/a' greater than ~10^1 do not invalidate the consistency of the scaling, being a consequence of the intrinsic viscoelastic feature of the suspending fluid.

Besides exactly capturing the underlying physics of the two-phase model, the refined model guarantees a better scaling of the elasticity of samples at different composition. This is shown in the inset of Figure 9.b, where the master curves built using the two pairs of shift factors are compared. The lower scattering of the data scaled using a' and b' confirms the importance of properly accounting for hydrodynamic contributions when dealing with PNCs.

2.3 Strength and reversibility of the filler network in PP/Al₂O₃ PNCs

Aim of this paragraph is the study of the relationships between the rheology and structure of PP/Al₂O₃ nanocomposites. The structuring (during a quiescent annealing process) and de-structuring (promoted by large amplitude shear flows) of the filler network are investigated by means of both rheological and TEM analyses. The internal morphology of the sample PP/Al₂O₃ at Φ=4.2% at the end of the extrusion process is shown in Figure 10.

Although a homogeneous distribution can be observed on microscale, the presence of aggregates of a few hundred nanometers is noticed at higher magnifications. The aggregates appear as open structures formed of tens of nanospheres of different sizes. Such non-equilibrium structures rearrange towards a more favourable thermodynamic state during a subsequent aging above the PP melting temperature. The morphology of a sample at Φ=4.2% after a 3-hours annealing at T=190°C is shown in Figure 11.

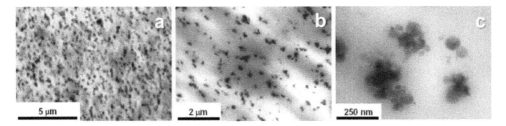

Fig. 10. TEM micrographs of the as extruded PP/Al$_2$O$_3$ sample at Φ=4.2% at various magnifications (image taken from Acierno et al., 2007b).

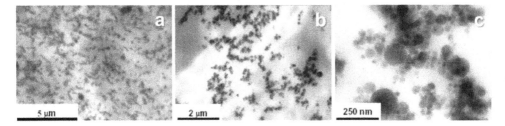

Fig. 11. TEM micrographs of the 3-hours thermal annealed PP/Al$_2$O$_3$ sample at Φ=4.2% at various magnifications (image taken from Acierno et al., 2007b).

Pristine individual aggregates are now assembled into a disordered network that spans large sections of the sample. The particles and aggregates are essentially kept together by Van der Waals attractions and/or other kinds of weak bonds between the functional sites located at the particle surfaces. The application of large strains provides an excess energy to overcome such attractive interactions, thus destroying the network. After that, the particles may or may not aggregate again depending on the strength of inter-particle interactions.

The relaxation dynamics of a viscoelastic fluid can be indifferently monitored by frequency scans or stress relaxation tests. In the latter kind of experiment, a constant strain, γ_0, is imposed to the sample in the linear regime, and the transient stress, $\sigma(t)$, is measured as a function of time. The stress relaxation modulus, $G(t)=\sigma(t)/\gamma_0$, is shown in Figure 12 after the application of large amplitude oscillatory shear (LAOS) at a constant frequency ω=0.0628 rad s^{-1} and different γ_0 on the 3-hours aged sample at Φ=4.2%.

Large deformations have a drastic effect on the relaxation spectrum: the bigger the strain amplitude, the faster the relaxation dynamics. Time sweep tests in linear regime were performed after each LAOS to test the viscoelastic behaviour of the sheared sample, and the results are shown in the inset of Figure 12. The elastic feature progressively vanishes with increasing the deformation amplitude. Interestingly, the steadiness of the elastic modulus during time suggests an irreversibility of the network break-up process, at least within the experimental time window. Moreover, a polymer-like behaviour is recovered after the LAOS at the largest amplitude (γ_0=500%). In such case, the inorganic phase does not affect the rheological response of the nanocomposite at all.

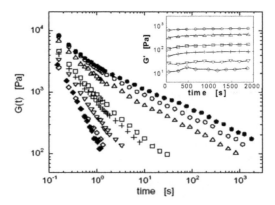

Fig. 12. $G(t)$ of the 3-hours aged PP/Al$_2$O$_3$ sample at Φ=4,5% after LAOS at different strain amplitudes γ_0: 0.8% (solid circles), 10% (open circles), 25% (triangles), 50% (squares), 100% (crosses), 250% (reverse triangles), 500% (diamonds). Solid diamonds represents the $G(t)$ of the neat polymer. The time evolutions of G' after each LAOS are shown in the inset. Symbols are the same of stress relaxation moduli (image taken from Acierno et al., 2007b).

The morphology of the 3-hours aged sample after the LAOS at γ_0=500% is reported in Figure 13. The network formed during aging is no more visible, and the presence of many small clusters characterizes the sheared system. The flocs show a more open structure than that of a not sheared sample, either aged or not, suggesting a weaker tendency to the clustering for the Al$_2$O$_3$ nanoparticles after the large deformations. The cumulative cluster size distributions were determined through the analysis of TEM micrographs. The results are shown in Figure 14 for the as extruded, 3-hours annealed and sheared after aging samples at Φ=4.2%. The number average equivalent diameters of the clusters are reported in the same figure.

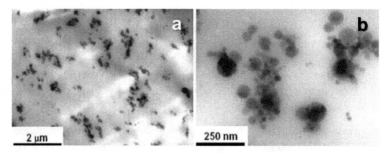

Fig. 13. TEM morphology of the 3-hours aged sample at Φ=4.2% after the LAOS at γ=500% (image taken from Acierno et al., 2007b).

The CSD of the as extruded sample is rather sharp, indicating a good dispersion efficiency of the extrusion process. The thermal annealing results in a significant widening of the CSD, with the appearance of very large clusters (D_n greater than 800 nm). This confirms the metastable feature of the samples, which quickly evolve toward states of less free energy under the push of the inter-particle attractive interactions. The resulting filler network breaks up when the sample is subjected to LAOS, and a remarkable sharpening of the CSD is observed.

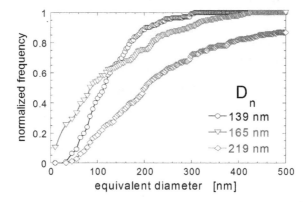

Fig. 14. CSDs for the PP/Al$_2$O$_3$ sample at Φ=4.2% as extruded (black circles), 3-hours aged (blue diamonds) and 3-hours aged after LAOS at γ_0=500% (red triangles) (image taken from Acierno et al., 2007b).

Interestingly, the strength of the filler network depends on whether the LAOS is applied before or after the thermal annealing. This is shown in Figure 15, where the loss factor $tan\delta$=G''/G' (15.a) and the complex moduli (15.b) of the samples at Φ=4.2% submitted to LAOS (γ_0=500%) are reported before and after the ageing; the curves of the 3-hours aged but not sheared sample are also reported for comparison.

If the LAOS is applied to the sample before the formation of the particle network, the system quickly evolves to a more elastic structure and the $tan\delta$ asymptotically reaches values close to those of the not sheared sample. However, the comparison between the G* shown in

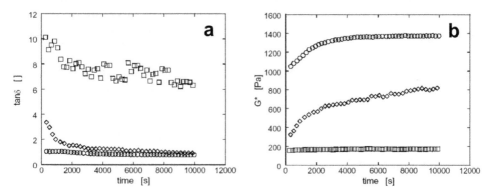

Fig. 15. Loss factor (a) and complex moduli (b) of the samples at Φ=4.2% submitted to LAOS (γ_0=500%) before (diamonds) and after (squares) the thermal annealing. The curves of the 3-hours aged and not sheared sample (circles) are also reported for comparison (image taken from Acierno et al., 2007b).

Figure 15.b reveals that the strength of the network formed after the LAOS is much lower than that of the not sheared sample.

Such a result can be explained by assuming some rearrangement of the reactive sites of the particle surfaces after the network break-up, which may weaken the surface activity of the particles. This reduces the intensity of inter-particle interactions and, as a consequence, the strength of the filler network [Bicerano et al., 1999]. On the other hand, if a "strong" network forms and then it is destroyed by LAOS, the restoration of new bonds required for the reformation of the network can results inhibited. This could explain the irreversibility of the structuring process noticed after the LAOS performed on the aged sample.

3. Conclusions

The effect of small amounts of nanoparticles on the melt-state linear viscoelastic behaviour has been investigated for different polymer-nanoparticles model systems characterized by poor polymer-particles interactions and low particle contents. The drastic increase of the rheological properties with respect to the matrices has been related to the formation of a filler network above a critical particles volume fraction. This is a consequence of particles and clusters rearrangements taking place during a thermal annealing. The filler mobility depends on both particle size and viscosity of the suspending medium. Once formed, the filler network exhibits an elastic feature that mixes with the intrinsic viscoelastic response of the polymer matrix, resulting in a complex Φ- and ω-dependent viscoelastic response of the nanocomposite. However, starting from a two-phase model proposed for colloidal suspensions in Newtonian fluids, we have shown that the contributions of filler network and suspending medium can be decoupled due to the weak polymer-particle interactions and the differences in temporal relaxation scales. The adopted approach has been validated through the building of a master curve of the moduli, which reflects the scaling of the elasticity of composites along the viscosity of the suspending medium. The two-phase model well works irrespective of the structure of the filler network, making evident the strict interrelationships between the structure, both on nano- and micro-scale, and the melt-state behaviour of the studied PNCs. The physical meaning of the two-phase model clearly emerges once hydrodynamic effects have been properly taken into account. Besides clarifying the various timescales of PNCs, the proposed model allows for predicting the modulus of particle networks which are too tenuous to be appreciated through simple frequency scans. The application of a large amplitude oscillatory shear flows provides an excess energy for the system to escape from the metastable configuration in which it is trapped. This destroys the network formed during the thermal annealing, leading to a more tenuous structure which is unable to significantly contribute to the system elasticity. After the network has been destroyed the sample cannot recover its previous solid-like feature during a subsequent thermal annealing. This is probably due to some rearrangement of the reactive sites of the particle surfaces occurring after the rupture of the inter-particles bonds formed during annealing.

Besides well describing the behaviour of PNCs in the framework of simpler systems such as Newtonian colloidal suspensions, the analysis proposed in this chapter is expected to be useful to understand a wide variety of complex fluids in which a superposition of the elasticity of the components is possible. The generalization of our approach to such systems and to other technologically relevant PNCs, such as nanocomposites based on layered silicates or carbon nanotubes, still remains to be proved.

4. References

Acierno, D.; Filippone, G.; Romeo, G.; Russo, P. (2007). Rheological aspects of PP-TiO$_2$ nanocomposites: a preliminary investigation. *Macromol. Symp.*, Vol. 247, pp. 59-66

Acierno, D.; Filippone, G.; Romeo, G.; Russo, P. (2007). Dynamics of stress bearing particle networks in poly(propilene)/alumina nanohybrids, *Macromol. Mater. Eng.*, Vol. 292, pp. 347-353

Baird, D. G.; Collias D. I. (1995). In: *Polymer Processing Principles and Design*, John Wiley and Sons, Inc., ISBN: 978-0-471-25453-9, New York (USA)

Bicerano, J.; Douglas, J. F.; Brune, D. A. (1999). Model for the viscosity of particle dispersions. *J. Macromol. Sci.: Rev. Macromol. Chem. Phys.* Vol. 39, No. 4, pp. 5611-642

Cipelletti, L.; Manley, S.; Ball, R. C.; Weitz, D. A. (2000). Universal aging features in restructuring of fractal colloidal gels. *Phys. Rev. Lett.*, Vol. 84, No. 10, pp. 2275-2278

Du, F.; Scogna, R. C.; Zhou, W.; Brand, S.; Fischer, J. E.; Winey, K. I. (2004). Nanotube networks in polymer nanocomposites: rheology and electrical conductivity, *Macromolecules*, Vol. 37, pp. 9048-9055

Filippone, G.; Romeo, G.; Acierno, D. (2010). Viscoelasticity and structure of polystyrene/fumed silica nanocomposites: Filler Network and Hydrodynamic Contributions. *Langmuir*, Vol. 26, No. 4, pp. 2714-2720

Gleissle, W.; Hochstein, B. (2003). Validity of the Cox–Merz rule for concentrated suspensions. *J. Rheol.*, Vol. 47, pp. 897-910

Inoubli, R.; Dagréou, S.; Lapp, A.; Billon, L.; Peyrelasse, J. (2006). Nanostructure and mechanical properties of polybutylacrylate filled with grafted silica particles. *Langmuir*, Vol. 22, No. 15, pp. 6683-6689

Israelachvili, J. (1991). In: *Intermolecular and surface forces*, Academic Press, ISBN 0-12-375181-0, London, UK

Jancar, J.; Recman, L. (2010). Particle size dependence of the elastic modulus of particulate filled PMMA near its T$_g$, *Polymer*, Vol. 51, No. 17, (August 2010), pp. 3826-3828

Kammler, H. K.; Beucage, G.; Mueller, R.; Pratsinis S. E. (2004). Structure of flame-made silica nanoparticles by ultra-small-angle X-ray scattering. *Langmuir*, Vol. 20, pp. 1915-1921

Kojima Y, Usuki A, Kawasumi M, Okada A, Fukushima Y, Kurauchi T, Kamigaito O. (1993) Mechanical properties of nylon 6-clay hybrid, *J Mat Res* Vol. 8, pp. 1185-1189

Krishnamoorti, R.; Yurekli, K. (2001). Rheology of polymer layered silicate nanocomposites, *Curr. Opin. Colloid Interface Sci.* Vol. 6, No. 5-6, pp. 464-470

Mason, T. G.; Weitz, D. A. (1995) Linear viscoelasticity of colloidal hard sphere suspensions near the glass transition. *Phys. Rev. Lett.*, Vol. 75, pp. 2770-2773

Ozmusul, M. S.; Picu, R. C.; Sternstein, S. S.; Kumar, S. (2005). Lattice Monte Carlo simulations of chain conformations in polymer nanocomposites , *Macromolecules*, Vol. 38, No. 10, pp. 4495-4500

Prasad, V.; Trappe, V.; Dinsmore, A. D.; Segrè, P. N.; Cipelletti, L.; Weitz, D. A. (2003) Universal features of the fluid to solid transition for attractive colloidal particles. *Faraday Discuss.*, Vol. 123, pp. 1-12

Pusey, P. N.; van Megen, W. (1986). Phase behaviour of concentrated suspensions of nearly hard colloidal spheres. *Nature*, Vol. 320, pp. 340-342

Ren, J.; Silva, A. S.; Krishnamoorti, R. (2000). Linear viscoelasticity of disordered polystyrene–polyisoprene block copolymer based layered-silicate nanocomposites. *Macromolecules*, Vol. 33, No. 10, pp. 3739-3746

Romeo, G.; Filippone, G.; Fernández-Nieves, A.; Russo, P.; Acierno, D. (2008). Elasticiy and dynamics of particle gels in non-Newtonian melts. *Rheol. Acta*, Vol. 47, pp. 989-997

Romeo, G.; Filippone, G.; Russo, P.; Acierno, D. (2010). Effects of particle dimension and matrix viscosity on the colloidal aggregation in weakly interacting polymer-nanoparticle composites: a linear viscoelastic analysis. *Polym. Bull.*, Vol. 63, No.6, pp. 883-895

Russel, W. B.; Saville, D. A.; Schowalter, W. R. (1989). In: *Colloidal dispersions*, Cambridge University Press, ISBN 0-521-34188-4, Cambridge, UK

Saint-Michel, F.; Pignon, F.; Magnin, A. (2003). Fractal behavior and scaling law of hydrophobic silica in polyol. *J. Colloid space face Sci.*, Vol. 267, No. 2, pp. 314-319

Shenoy, A. V. (1999). In: *Rheology of filled polymer systems*, Kluwer Academic Publishers, ISBN 0-4112-83100-7, Dordrecht, The Netherlands

Shikata, T.; Pearson, D.S. (1994). Viscoelastic behavior of concentrated spherical suspensions. *J. Rheol.*, Vol. 38, pp. 601-613

Sollich, P.; Lequeux, F.; Hébraud, P.; Cates, M. E. (1997). Rheology of soft glassy materials. *Phys. Rev. Lett.* 78, 2020-2023

Surve, M.; Pryamitsyn, V.; Ganesan, V. (2006). Universality in structure and elasticity of polymer-nanoparticle gels. *Phys. Rev. Lett.*, Vol. 96, No. 17, pp. 1778051-1778054

Trappe, V.; Weitz, D. A. (2000). Scaling of the viscoelasticity of weakly attractive particles. *Phys. Rev. Lett.*, Vol. 85, No. 2, pp. 449-452

Trappe, V.; Prasad, V.; Cipelletti, L.; Segrè, P.N.; Weitz, D. A. (2001). Jamming phase diagram for attractive particles. *Nature*, Vol. 411, pp. 772-775

Usuki A.; Kojima Y.; Kawasumi M.; Okada A.; Fukushima Y.; Kurauchi T.; Kamigaito O. (1993). Synthesis of nylon 6-clay hybrid. *J Mat Res*, Vol. 8, pp. 1179-1184

Weitz, D. A.; Oliveira, M. (1984). Fractal structures formed by kinetic aggregation of aqueous gold colloids. *Phys. Rev. Lett.*, Vol. 52, pp. 1433-1436

Wolthers, W.; van den Ende, D.; Bredveld, V.; Duits, M. H. G.; Potanin, A. A.; Wientjens, R. H. W.; Mellema, J. (1997). Linear viscoelastic behavior of aggregated colloidal dispersions. *Phys. Rev. E*, Vol. 56, pp. 5726-5733

Zhang, Q.; Archer, L. A. (2002). Poly(ethylene oxide)/silica nanocomposites: structure and rheology. *Langmuir*, Vol. 18, No. 26, pp. 10435-10442

7

Dielectric and Transport Properties of Thin Films Deposited from Sols with Silicon Nanoparticles

Nickolay N. Kononov[1] and Sergey G. Dorofeev[2]
[1]Prokhorov General Physics Institute, Russian Academy of Sciences, Moscow,
[2]Faculty of Chemistry, Moscow State University, Moscow,
Russia

1. Introduction

Currently, there is steady scientific interest in structures formed by nanocrystalline silicon particles (nc-Si). This interest is to a large extent caused by the fact that efficient methods for fabricating silicon nanoparticles capable of bright and stable photoluminescence in the visible region of the spectrum with high quantum yield were developed over the last decade (Jurbergs at al., 2006). The main carriers of such nanoparticles are colloidal solutions (sols) based on methanol, chloroform, hexane, etc. Such sols are very promising objects for developing technologies for applying highly uniform thin nc-Si films onto various substrates. The use of such films seems very promising for developing light emitting elements based on nc-Si electroluminescence (Anopchenko at al., 2009). Furthermore, nc-Si films are very promising as elements of solar panels (De la Torre at al., 2006), thin film transistors (Min at al., 2002), and single electronic devices (Tsu, 2000). In the case in which films consist of nanoparticles with a diameter smaller than 10 nm, their total characteristics are controlled not only by their material, but also by properties of atoms on the surface of these particles. In other words, in general, such films should be considered as a multicomponent medium the properties of which are controlled by both crystalline cores of nanoparticles and surface atoms and molecules and air voids being a film component.

In the modern scientific literature, most papers are devoted to the study of properties of amorphous silicon (a-Si) films with introduced silicon nanocrystals (Conte at al., 2006; Wang at al., 2003). Such films can be deposited, e.g., in the high frequency discharge in a mixture of gases SiH_4, Ar, or H_2 (PECVD method), followed by high temperature annealing (Saadane at al., 2003).

Recently, we showed that homogeneous thin films (with a thickness up to 30 nm) can be grown by size selective precipitation sols containing nanocrystalline silicon particles (Dorofeev at al., 2009). Such films (nc-Si) are formed by closely adjacent crystalline Si nanoparticles; therefore, their physical characteristics to a certain extent should be similar to characteristics of films based on porous silicon (por-Si). The optical absorption and photoluminescence ability of por-Si films have been very comprehensively studied to date (see, e.g., Kovalev at al., 1996; Brus at al., 1995); however, the number of studies of transport and dielectric properties of such films in an ac electric field is extremely small. Here we indicate papers (Axelrod at al., 2002; Ben-Chorin at al., 1995; Urbach at al., 2007) devoted to such studies of por-Si.

A similar situation exists as applied to nc-Si films; however, we are not aware of results of studies on the conductivity in an ac electric field (ac conductivity and dielectric relaxation in such films).

In this chapter we analyze the dielectric and transport properties of nc-Si films deposited on a glass and quartz substrates from the sol containing nanoparticles of silicon. Silicon nanoparticles were synthesized in the process of laser pyrolysis of silane and placed in ethanol or methanol, repeatedly centrifuged resulting in a colloidal solution (sol) in which the silicon nanoparticles could be a long time (over two years). We analyze three kinds of films. The films deposited on a substrate by centrifugation of sols of nanoparticles in a week after their synthesis. Films deposited on a substrate of sols in which the nanoparticles were 2 years after their synthesis and films deposited from two-year-old sol in which has been added the conductive tetra-aniline. More circumstantial experimental details we will present in the following sections. In the future of the films deposited on a substrate of silicon nanoparticles in a week after their synthesis we call films I, films obtained from similar nanoparticles, but two years after their synthesis (aged nanoparticles) - films II and films deposited of sols with aged nanoparticles and with the tetra aniline addition - films III.

For films I we present measurements of the nc-Si film permittivity in the optical range ($5 \times 10^{14} \leq v \leq 10^{15}$ Hz) and in the frequency range of $10 \leq v \leq 10^6$Hz. In the latter range, the ac conductivity (σ_{ac}) of nc-Si films is also determined.

In the optical region, the real ε' and imaginary ε'' components of the complex permittivity were determined from an ellipsometric analysis of light beams incident and reflected from the free boundary of the nc-Si film. In the frequency range of $10 \leq v \leq 10^6$ Hz, the ε' and ε'' spectra, were determined from an analysis of the frequency dependence of the nc-Si film impedance.

In an optical spectral region, ε' and ε'' varied within 2.1–1.1 and 0.25–0.75, respectively, as the frequency increased. We attribute such low values of ε' and ε'' to the nc-Si film structure. The nc-Si particles forming such films consist of crystalline cores surrounded by a SiO_x shell ($0 \leq x \leq 2$). The SiO_x shell results from the interaction of the Si nanoparticle surface with ambient air. On the basis of the analysis of the Raman spectra, it is suggested that the amorphous component is involved in the nc-Si powders and films due to oxygen atoms arranged at the nanoparticle surface.

Using the Bruggeman effective medium approximation (EMA) (Bruggeman, 1935), the structural composition of nc-Si film was simulated. It was shown that good agreement between the frequency dependences of ε' and ε'' obtained from the EMA and the ε' and ε'' spectra determined from ellipsometric data is achieved when nc-Si films are considered as a two component medium consisting of SiO and air voids existing in it. In the frequency range of 10–10^6 Hz, the ε' and ε'' dispersion was determined from an analysis of the frequency dependences of the capacitance of nc-Si films and their impedance spectra. It was found that ε' and ε'' vary within 6.2–3.4 and 1.8–0.08, respectively, as the frequency increases.

It is found that the function $\varepsilon'(\omega)$ in this frequency range is well approximated by the semiempirical Cole–Cole dependence (Cole–Cole dielectric relaxation) (Cole, K. S. & Cole, R. H., 1941). At the same time, the $\varepsilon''(\omega)$ spectra of nc-Si films are well approximated by the Cole–Cole dependence only at frequencies higher than 2×10^2 Hz. In the low frequency spectral region, good approximation is achieved by combining the Cole–Cole dependence and the term associated with the presence of free electric charges. From analysis of the approximating dependences, the average room temperature relaxation times of dipole moments in nc-Si films were determined as 6×10^{-2} s.

The conductivity σ_{ac} of the studied films I in an ac electric field depends only on its frequency according to the power law; the exponent is 0.74 in the entire frequency range under study. Such behavior of σ_{ac} suggests that the electrical transport mechanism in films is hopping. Comparison of the measured frequency dependence $\sigma_{ac}(v)$ with similar dependences following from various models of hopping conductivity shows that the $\sigma_{ac}(v)$ behavior is most accurately described in the diffusion cluster approximation (DCA) (Dyre & Schrøder, 2000; Schrøder & Dyre, 2002; Schrøder & Dyre, 2008).

Analysis of the dependences of the dark conductivity of films on humidity of ambient air and the temperature dependence of absorption bands caused by associated Si–OH groups on the film surface allowed the conclusion to be drawn that conductivity at frequencies lower than 2×10^2 Hz is associated with proton transport through the hydrogen bound hydroxyl groups on the silicon nanoparticle surface.

For films II and III we present measurements of the nc-Si film permittivity and ac conductivity (σ_{ac}) in the frequency range of $1 \leq v \leq 10^6$Hz. The dielectric properties of the films II and III were studied by impedance spectroscopy only in the frequency range $1 \leq v \leq 10^6$Hz.

We found that in films II and III, a double dielectric relaxation exists and to adequately describe the spectra of ε' and ε'' of these films should use not only the Cole-Cole relationship, but and the law of Debye's dielectric relaxation.

By a total approximation of the experimental spectra of the films II and III the values of static dielectric constant ε_0 have obtained. These values are equal 11.5 and 67 respectively. Value $\varepsilon_0 \approx 11.5$ characteristic of film II is close to the static permittivity of crystalline silicon, but the magnitude of $\varepsilon_0 \approx 67$ of films III significantly higher than this value. Next, we analyze this fact.

In contrast to the conductivity of the films I σ_{AS} of films II and III are not subject to a power law over the entire range of measured frequencies. Next, we show that such a deviation from the law $\sigma_{AS} \sim \omega^s$ associated with the appearance in the spectra $\varepsilon''(\omega)$ of the films II and III Debye's components.

2. The films from silicon nanoparticles

2.1 Films deposited from freshly prepared sols of silicon nanoparticles (films I)

2.1.1 Samples and measurement procedures

The nc-Si films were deposited from silicon nanoparticles produced by CO_2 laser pyrolysis of silane. The system for synthesis of the nc-Si powders and conditions of the process are described in detail elsewhere (Kononov at al., 2005; Kuz'min at al., 2000). In what follows, we briefly outline the procedure of synthesis of the Si nanoparticles. In a reactor chamber filled with a buffer gas (helium or argon) to the pressure $P = 200$ Torr, a fine SiH_4 jet is formed and heated by focused cw CO_2 laser radiation beam crossing the jet. During pyrolysis of silane, the SiH_4 molecules are decomposed, and free Si atoms are produced. When colliding with each other and with the atoms of the buffer gas, the Si atoms form particles, whose average dimensions can be in the range from 10 to 100 nm, depending on the pressure of the buffer gas. The nc-Si powders produced in such a manner were dispersed by ultrasonic treatment in ethanol and centrifuged for 30 min with an acceleration of $2000g$ (g is the gravitational acceleration). As a result, almost all agglomerates of nc-Si particles are precipitated. After preliminary centrifugation, a stable colloidal solution (sol) of nc-Si in ethanol remains. No visible changes in the solution, including precipitations, were observed for two years. For the subsequent deposition of nanoparticles, a water solution of aluminum dihydrophosphate was added to the sol.

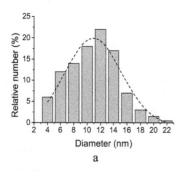 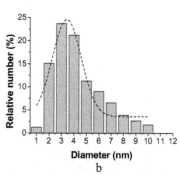

a   b

Fig. 1. (a) Histogram of the size distribution of particles, as obtained by processing of the TEM images of the nc-Si powder (b) Histogram of the size distribution of particles, as obtained by processing of the TEM images of the nc-Si powder etched in the (HF + HNO_3) acid mixture. The dushed lines represent the normal distribution functions for two different kinds of nc-Si particles

The size distribution of nc-Si particles was determined from images obtained using an LEO 912 AV OMEGA transmission electron microscope. The typical spectrum of silicon nanoparticles used for precipitation is shown in the Fig. 1. The nc-Si film thickness was determined using a Taly Step (Taylor-Hobbson) atomic force step profilometer. Ellipsometric spectra were measured using an Ellips 1891 ellipsometer (Institute of Semiconductor Physics, Siberian Branch, Russian Academy of Sciences). The transmission spectra were measured using a Lambda 900 (Perkin-Elmer) spectrophotometer. The Raman spectra of the films were recorded with a microlens equipped T 64000 (Jobin Ivon) Raman triple spectrograph in the backscattering layout of measurements at the power of the excitation argon laser 2 mW.

The impedance spectra were measured using an E7-20 immittance meter (Minsk Research Instrument Making Institute) and a Z-3000X (Elins) impedance meter. Samples for measuring impedance spectra were prepared as follows. First, aluminum electrodes separated by a rectilinear gap 1 mm wide were deposited on a glass substrate. Then *nc*-Si particles were precipitated from the sol on the substrate prepared in such a way, which formed a film. The third aluminum electrode was deposited on the obtained *nc*-Si film. As a result, a sandwich like structure similar to that shown in Fig. 2 was obtained. To achieve the ohmic lead contacts, the structure was annealed at a temperature of 400°C and a pressure of 10^{-5} Torr. Impedance spectra were measured at an amplitude voltage of 100 mV; however, the films under study can withstand a voltage to 15 V without electrical breakdown.

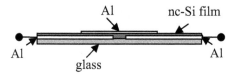

Fig. 2. Diagram of the sandwich like sample structures for measuring impedance spectra.

2.1.2 Experimental results

2.1.2.1 Raman scattering

To record the Raman spectra, we first deposited an aluminum film with the thickness ~300 μm onto the quartz substrate, and then, on top of the film, we deposited the *nc*-Si film from the sol. We proceeded in such manner in order to avoid the background scattering component produced by the quartz substrate. In this section, we analyze the Raman

spectra recorded for the initial *nc*-Si powder, for the films deposited at the second stage of centrifuging the sols of the initial *nc*-Si powder and film deposited from the sol with powder etched in the (5wt%HF+14wt%HNO$_3$) water mixture. The corresponding samples are identified as samples S_1, S_2, and S_3 respectively.

The typical Raman spectra recorded for these samples are shown in Fig. 3. All of the experimentally recorded spectra are very similar to the Raman spectra obtained for *p*-Si in (Tsang at al., 1992; Tsu at al., 1992) and for the *nc*-Si clusters in (Ehbrecht at al., 1995).

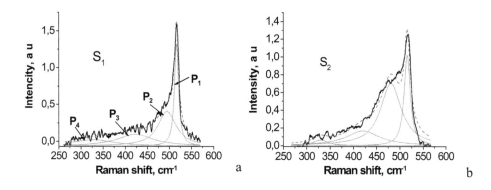

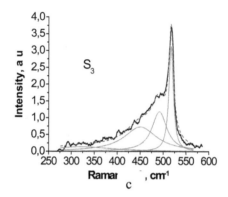

Fig. 3. Raman spectra of: a) nc-Si powder S_1, b) film deposited from the sol of the initial *nc*-Si powder-S_2 c) film deposited from the sol *nc*-Si powder etched in the (HF + HNO$_3$) acid mixture –S_3. The dotted line refers to the approximation of the spectrum with Lorentzian contours (the *P1*, *P2*, *P3*, and *P4* peaks).

The Raman spectra of all of the samples studied here can be fitted with four Lorentzian bands with a rather good accuracy (Fig. 3). In what follows, these bands are referred to as the P_1, P_2, P_3, and P_4 peaks. The Raman shift of the most intense P_1 peak with respect to the emission frequency of the probing laser is in the range of wave numbers from 515 to 517 cm^{-1} for all of the samples. The Raman shift of the similar peak for *c*-Si corresponds to the wave number 520.5 cm^{-1}. Thus, for all of the films studied here, the P_1 peak is shifted to smaller wave numbers with respect to the peak for *c*-Si (the red shift). The P_1 peak in the Raman spectra of the *nc*-Si particles is due to light scattering assisted by longitudinal optical (LO) and transverse optical (TO) phonons at the central point of the Brillouin zone for the *c*-Si crystal lattice. The red shift of the P_1 peak and its half width as functions of the nanoparticle dimensions are adequately described in the context of the phonon's confinement model (Campbell & Faushet 1986; Richter at al., 1981).

The result of application of this model to spherical nanoparticles is shown in Fig. 4. From Fig. 4, it can be seen that the average dimension of the *nc*-Si particles in the samples is in the range 4–6 nm, irrespective of whether the particles of the initial *nc*-Si powder were subjected to some treatment or not. For the sols of the *nc*-Si powders etched in the (HF + HNO$_3$) mixture, the average particle's dimensions determined in the phonon's confinement model are in good agreement with the particle dimensions corresponding to the peak of size distribution obtained for the particles by processing of the TEM images.

However, for the initial *nc*-Si powders, the average particle dimensions determined by the above mentioned two methods differ by a factor of about 2. There are two possible causes of the difference between the average particle's dimensions determined in the phonon's confinement model and by processing of the TEM images. One of the causes is associated with the fact that, in the phonon's confinement model, the nanoparticles are assumed to be single crystals. Therefore, the magnitude of the phonon wave's vector q in the nanoparticle can vary in the range $(0, 2\pi/L)$, where L is the particle diameter. However, if the nanoparticle core is polycrystalline and the average dimension of the elementary crystal lattice in the core is l, the confining condition $q \leq 2\pi/L$ should be replaced by the condition $q \leq 2\pi/l$. Thus, it is possible that the dimensions l = 4–6 nm calculated in the phonon's confinement model are related to the average dimensions of elementary lattices in the polycrystalline nanoparticle cores rather than to the average nanoparticles' dimensions in the initial nc-Si powder. From this assumption and the fact that, for nanoparticles subjected to etching, the average dimensions determined by the above two methods are the same, it follows that, on such etching of the nanoparticles, the remaining c-Si cores are single crystals. The other cause can follow from the well known low contrast of the finest nanoparticles (with the diameter 3 nm in the case under study) in the TEM images. Because of the low contrast, the processing of the TEM images always reduces the relative portion of the fine grained fraction of nanoparticles in the ensemble of particles under consideration.

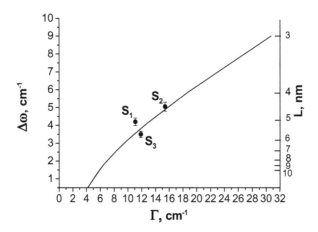

Fig. 4. The half-width and the red shift of the P_1 Raman peak versus the diameter of the spherical silicon nanoparticles, as obtained (solid line) in the context of the phonon's confinement model [19, 20] and (solid circles) from the approximation of the P_1 peak in samples S_1, S_2 and S_3, with the Lorentzian contours.

The Raman shift of the P_2 peak in the samples is in the range from 480 to 495 cm^{-1}. This peak corresponds to the TO-phonon assisted scattering in a-Si:H. Similarly to the P_2 peak, the P_3 and P_4 peaks are related to the amorphous component of the structure of the Si particles and result from scattering assisted by LO and longitudinal acoustic (LA) phonons.

From the comparison of the integrated intensities of the P_1 and P_2 peaks, Ic and Ia, we can determine the volume fraction of the crystalline phase, Xc, in the Si particles. To do this, we used the expression (Voutsas at al., 1995)

$$X_c = \frac{I_c}{I_c + \eta I_a} \tag{1}$$

where $\eta = \dfrac{\sigma_c}{\sigma_a}$ is the ratio between the integrated backscattering's cross sections in the crystalline and amorphous fractions (corresponding to the P_1 and P_2 peaks). According to (Kakinuma at al., 1991), the quantity η for silicon is η = 0.8–0.9. In the calculations, we set η = 0.8. For samples S_1, S_2 and S_3, the values of the parameter Xc are 0.45, 0.35 and 0.50 respectively.

From these values of X_c, it follows that almost a half of the volume of the particles is characterized by a high degree of disorder of the crystal lattice.

From comparison of the above values, it is evident that, in film S_2 deposited at the second stage of centrifuging from the sol with the initial nc-Si powder, the parameter X_c is smaller than X_c for the initial powder. The average particle's dimension in film S_2 is smaller than that in the initial powder. Correspondingly, the surface area to volume ratio for the particles in the S_2 film is larger than the corresponding ratio in the initial powder. Therefore, the effect of the nanoparticle surface on the general properties of the nanoparticles in film S_2 is bound to be more pronounced that the corresponding effect in the initial powder. Consequently, the smaller value of X_c (the higher degree of amorphization of the particles) in film S_2 in comparison with X_c in sample S_1 suggests that the disordered regionis at the nanoparticle surface rather than in the nanoparticle core. However, for film S_3 the value of X_c is larger than X_c for film S_2, although the average particle's dimensions in these films are comparable. Such difference suggests that the degree of disorder of particle surfaces in film S_3 is lower than that in film S_2.

Since film S_3 are deposited from the sols of the nc-Si powders subjected to etching, such lower degree of disorder in these films is due to the effect of the HF and HNO_3 acids on the particle surface. Here, it is reasonable to mention the studies (Luppi & Ossicini, 2005; Puzder at al., 2002), in which the effect of oxygen atoms on the structure of silicon clusters and on the degree of ordering of the Si crystal lattice in nanoparticles is analyzed, and the studies (Ma at al., 2000 Tsang at al., 1992;), in which the changes induced in the Raman peak similar to the P_2 peak (Fig. 3) by the effect of oxygen on the surface of p-Si passivated with hydrogen, are reported. The general idea of the above mentioned studies is that the crystal lattice of nanoparticles, whose surface is completely passivated with hydrogen, is practically the same as the lattice of the silicon crystal. However, if oxygen atoms appear at the nanoparticle surface, they can form the Si–O–Si and (Si=O) bonds and, thus, distort the lattice at the distances up to 0.5 nm. In this space region, the distortions of angles between the Si–Si bonds in the crystal lattice can be as large as 10° (Tsang at al., 1992). Therefore, if the surface of a nanoparticle of a diameter smaller than 3 nm is coated with the SiO_2 oxide, the crystal lattice is distorted within a noticeable volume fraction of such particle. As a consequence, if the p-Si surface is etched in the solution of HF, the Raman spectrum involves only one peak similar to the P_1 peak. If p-Si is exposed to oxygen in oxygen containing atmosphere, the Raman spectrum exhibits also the P_2 peak along with the P_1 peak. From the above mentioned studies and from the analysis of the Raman spectra discussed here, we can make the statement presented below. At the surface of nc-Si nanoparticles in all samples, there is a noticeable number of oxygen atoms, which distort the crystal lattice in these particles and bring about the appearance of the P_2 peak in the Raman spectra. Since the average nanoparticle's dimensions in film S_2 are smaller than those in powder S_1, the effect of these oxygen atoms on the crystal lattice structure in film S_2 is more pronounced than the effect in powder S_1. As a result, the volume fraction of the crystal phase in film S_2 is reduced compared to that in S_1.

Etching of the nc-Si particles in the solution of the (HF + HNO₃) acids results in a decrease in the particle dimensions. However, in this case, the total number of oxygen atoms at the nanoparticle's surface decreases, since a portion of oxygen atoms is replaced with hydrogen atoms. Therefore, in film S_3 two opposite processes are bound to occur. One process related to the decrease in the nanoparticle's dimensions yields a decrease in Xc, whereas the other process related to the decrease in the number of oxygen atoms at the nanoparticle surface brings about an increase in Xc. In film S_3, we experimentally observe the parameter X_c larger than X_c in film S_2; therefore, we can conclude that, on etching of the nc-Si particles, the latter process dominates over the former one.

2.1.2.2 Ellipsometric spectra

In the experiment, the ellipsometric angles ψ and Δ were measured as functions of the wavelength of a light beam incident at the angle Φ_0 on the free flat surface of the nc-Si film. The films under study were applied on glass and quartz substrates and on quartz substrates with preliminarily deposited aluminum films. The nc-Si film thicknesses (1–2 μm) were measured independently. When processing the ellipsometric data, the nc-Si films under study were considered as a 3D medium in air medium. The complex refractive index $N = n - ik$, where n is the film refractive index and k is the extinction coefficient, was determined by the expression (Azzam & Bashara, 1977)

$$N = N_0 \sin \Phi_0 \sqrt{1 + \left(\frac{1-\rho}{1+\rho} \right)^2 tg^2 \Phi_0} \qquad (2)$$

Here, ρ = e^{iΔ}· tgψ and N_0 are the complex refractive index of an ambient medium (air), which was equal to unity in the case at hand. It is known that formula (2) yields accurate values only when light is reflected from a semi-infinite medium with a boundary with an atomically clean surface. If impurities or an oxide filmare on the boundary, they introduce errors to the calculated values. In (Tompkins & Irene, 2005), the values n and k were compared for crystalline silicon (c-Si) in the absence and presence of the oxide film on its surface. It follows from this comparison that, in the presence of a SiO₂ film to 2 nm thick on the silicon surface, the value of n is almost identical to that of c-Si in the incident photon energy range of 1–3.4 eV; in the range of 3.4–5 eV, the refractive index differs from n of c-Si no more than by 20%, as well as k. However, in the range of 1–3.4 eV, the value of k in the presence of the SiO₂ film almost twice exceeds the c-Si extinction.

Since the real ε' and imaginary ε'' components of the medium permittivity are related to n and k by the known expressions $\varepsilon' = n_2 - k_2$ and $\varepsilon'' = 2nk$, it can be expected that the values of ε' calculated by Eq. (2) for nc-Si films will be slightly systematically underestimated, while the values of ε'' will be overestimated.

Nevertheless, representation of the pseudo dielectric functions by relation (2) is very convenient and is quite often used to study the dielectric properties of materials. For example, dielectric parameters of por-Si were studied using this equation in (Pickering, 1984). As applied to the present study, an analysis of the spectra obtained using formula (2) was limited by the energy range of incident photons, in which films strongly absorbed incident probe radiation, which could not reach the substrate surface in this case. If probe radiation reached the substrate surface with precipitated film, an interference structure arose in the spectra, which consisted of alternating minima and maxima. Such a structure at energies lower than 2 eV is easily seen in Fig. 5 (curves 3 and 3 ').

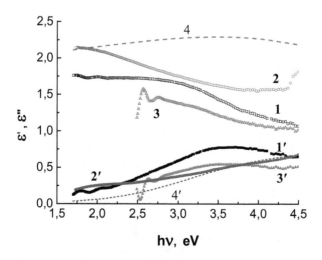

Fig. 5. Spectra of (1–3) real and imaginary (1'-3') permittivity components of *nc*-Si films precipitated on various substrates: (1, 1') film of initial (unetched) nanoparticles on the glass substrate; (2, 2') film of nanoparticles preliminarily etched in a HF/HNO_3 acid mixture on the quartz substrate; (3, 3') *nc*-Si film of initial nanoparticles on the glass substrate with a preliminary deposited aluminum film; and (4, 4') Bruggeman approximation for ε' and ε'', respectively.

We can see the spectra of pseudo dielectric functions ε' and ε'' of *nc*-Si films fabricated by precipitating initial silicon nanoparticles on the glass substrate and nanoparticles preliminary etched in a HF/HNO_3 acid mixture in a water for 30 min on the quartz substrate. Figure 5 also shows the ε' and ε'' spectra of *nc*-Si films precipitated on the glass substrate with a preliminarily deposited aluminum film.

It follows from this figure that the obtained values of ε' and ε'' are significantly lower than the similar values of *c*-Si.

Figure 6 shows the absorption spectra $\alpha(E)$ of the same films, obtained by the relation:

$$\alpha(E) = \frac{4\pi v}{c}k = \frac{4\pi E}{ch}k \tag{3}$$

where $E = hv$ is the energy of the incident photon and k is the experimentally measured extinction coefficient.

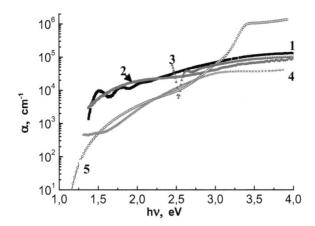

Fig. 6. *(1–3)* - absorption spectra of *nc*-Si films, obtained from ellipsometric data; *(4)* - absorption spectrum of film 1, obtained from its transmission spectrum; and *(5)* - absorption spectrum of crystalline silicon.

This figure also shows the absorption spectrum of the *nc*-Si film formed by unetched nanoparticles, which was calculated from its transmission spectrum. As a reference, the absorption spectrum of crystalline silicon (Aspens & Studna, 1983) is also shown. The size distribution of unetched and etched *nc*-Si particles used to precipitate films 1 and 3 are shown in Fig. 1.

A comparison of the absorption spectra of the film *nc*-Si grown from unetched nanoparticles, which were obtained from ellipsometric measurements and by processing the corresponding transmission spectrum, shows that the values of α obtained by ellipsometry are higher than the similar values calculated from transmission spectra, and this difference increases with decreasing the incident photon energies. As noted above, this difference is associated with the error of the extinction coefficient calculation by formula (2). At the same time, both spectra exhibit strong absorption of the *nc*-Si film in comparison with *c*-Si at energies lower than 1.5 eV. Such absorption enhancement in the low energy photon region is also inherent to the film grown by etched nanoparticles. At energies higher than 3 eV, all spectra exhibit absorption weaker than that of *c*-Si. In the Fig. 1, we can see that the diameter of an appreciable fraction of particles used to form films is smaller than 10 nm; therefore, the most probable cause of a decrease in the film absorption in the high-energy photon region is widening of the band gap in crystalline cores of silicon nanoparticles due to quantum confinement.

2.1.2.3 Dielectric dispersion

The permittivity spectra of *nc*-Si films were calculated from the measured frequency dependences of the capacitance of corresponding samples and their impedances, $Z(v) = Z' - iZ''$, $Z(v) = U(v) / I(v)$, where $U(v)$ is the potential difference at sample electrodes and $I(v)$ is the current flowing through the sample.

In what follows, we will analyze the dielectric properties of the Al–nc-Si–Al sandwich system in which the n-Si layer was precipitated from the sol with unetched nanoparticles. The thickness of this film was 2 μm; the geometrical capacitance of this system was $C_0 = 1.15 \times 10^{-10}$ F. The dielectric dispersion of this film is typical of other films obtained in a similar way from similar nc-Si particles.

Figure 7 shows the frequency dependence of the capacitance of this system. The capacitance was measured in parallel connection. The figure also shows the spectrum of the real component $\varepsilon'(v)$ of the film permittivity, calculated from the relation $\varepsilon' = C(v)/C_0$.

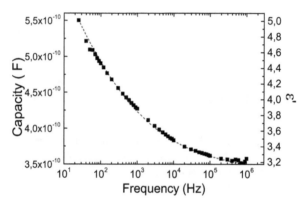

Fig. 7. Frequency dependence of the nc-Si film capacitance. The dashed curve is the approximation by function (3) (see text).

The $\varepsilon'(v)$ and $\varepsilon''(v)$ spectra of nc-Si films were also determined from the frequency dependence the film impedance by the expression

$$\varepsilon = \varepsilon' - i\varepsilon'' = \frac{1}{i2\pi v C_0 Z(v)} \tag{4}$$

Figure 8.a shows the dependences $\varepsilon'(v)$ and $\varepsilon''(v)$ calculated by the above method for the film under study. A comparison of the values of $\varepsilon'(v)$ obtained from $C(v)$ and $Z(v)$ measurements shows good quantitative and qualitative agreement of the values calculated in two different ways; in both cases, in the frequency range of $10 \leq v \leq 10^6$ Hz, the value of $\varepsilon'(v)$ is within 6–3.4 and decreases with frequency.

2.1.2.4 AC conductivity of films I

The ac conductivity of nc-Si films was determined by the known relation

$$\sigma_{AC}(v) - \sigma(0) = \varepsilon_0 \cdot \varepsilon''(v) \cdot 2\pi v$$

where $\sigma(0)$ is the dark conductivity of films in a dc electric field and $\varepsilon_0 = 8.85 \times 10^{-12}$ F/m is the permittivity of free space. The value of $\sigma(0)$ of the film under study at room temperature

(T = 297 K) was 9×10^{-10} Ω^{-1} m^{-1} and was used to calculate $\sigma_{AC}(\nu)$. The dependence $\sigma_{AC}(\nu)$ is shown in Fig. 9 on a log scale.

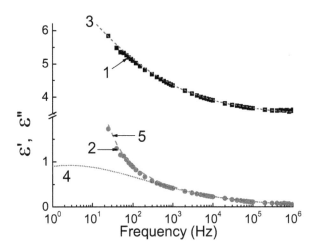

Fig. 8a. (*1, 2*) Frequency dependences of ε' and ε'', obtained from impedance spectra. Cole–Cole approximation of ε' - (3) and ε'' spectra (4) without and (5) with consideration of the contribution of free carriers.

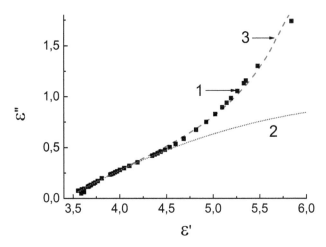

Fig. 8b. (*1*) Dependence $\varepsilon''(\varepsilon')$ for the *nc*-Si film. Cole–Cole approximation (2) without and (3) with consideration of the contribution of free carriers.

This figure suggests that $\sigma_{AC}(\nu)$ can be well approximated by the power law dependence with an exponent of 0.74.

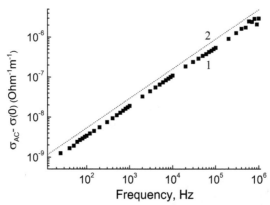

Fig. 9. (1) Frequency dependence of the ac conductivity of the *nc*-Si film; (2) dependence $\sigma_{ac}(v)$ defined by the DCA model (see text) using experimentally measured ε_s, ε_∞, and $\sigma(0)$.

2.2 Films deposited from aged nc-Si sols (Films II) and from aged nc-Si sols with tetraaniline (Films III)

2.2.1 Samples

The previous sections have presented experimental results of a study of thin films obtained from nanoparticles synthesized by one week prior to their deposition on the substrate. As already mentioned silicon nanoparticles could be no precipitation of sols for a long time. By the time of this writing, the silicon nanoparticles used for the deposition of films analyzed in the previous sections, were in sols over two years and in the next section we will report on the results of studies of the properties of the films deposited on substrates of these sols. It should be noted here that the film deposited on a substrate not as a result of centrifugation of sols, and with the spin coating method. Also in the following sections we will analyze the dielectric properties of films deposited from a 2-year nc-Si sols, in which the conductive tetramer – tetraaniline was added (Wang & MacDiarmid, 2002).

Because pure tetraaniline has low conductivity, for its increase the tetra aniline doped with p-toluensulfonic acid ($CH_3(C_6H_4)SO_3H$). Briefly the process of doping was as follows. A solution of tetraaniline and dimethyl sulfoxide (DMSO) as a solvent mixed with a DMSO solution of para-toluenesulfonic acid, so that the resulting solution in the molar ratio of tetra aniline and acid was 1.5. At the end of doping the color of resulting solution became green. The conductivity of the film which was deposited on a substrate of resulting tetraaniline solution was at room temperature 10^{-4} Ohm^{-1}m^{-1}. The resulting solution of tetraaniline in DMSO was added to the sol of silicon nanoparticles in ethanol in a mass ratio 1:10 before deposition of film on substrates.

As we have already reported, the films deposited on a substrate of silicon nanoparticles in a week after their synthesis we call the films I; films, obtained from the same nanoparticles, but two years after their synthesis (aged nanoparticles) - films II and films with aged nanoparticles and addition of tetraaniline - films III. Before the measurement the sandwich like structures of the films II and III with Mg or Al electrodes were heated to a temperature of 140^0 C and held at that temperature for 30 minutes.

2.2.2 Experimental results

2.2.2.1 Dielectric dispersion of the films II

The values of ε' and ε'' of films I and II in the frequency range below 10^4 Hz are quite close to each other. The main difference between the permittivity's spectra of these films observed in the frequency range higher of 10^5 Hz

In this frequencies region permittivity spectra of the films II reveal sharp decrease in ε' while the dielectric losses of ε'' has a form of enough narrow peak (see Figure 10). Such behavior of the spectrum is typical for the Debye dipole relaxation process, and will be discussed later.

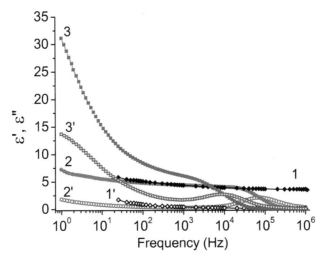

Fig. 10. The frequency dependence of ε' and ε'' for: Film I: - (1 and 1'), Film II - (2, 2'), Film III - (3, 3').

2.2.2.2 Dielectric dispersion of the films III

The $\varepsilon(\omega)$ spectra of the films III, similar to those of the films II, but for the films III decreasing of ε' is observed in the frequency of large 10^4 Hz. The frequency at which a maximum of dielectric loss ε'' observed in films III, is $v_{max} \approx 9{,}7 \cdot 10^3$ Hz, while for films II, this frequency is $7{,}5 \cdot 10^4$ Hz (see Figure 10). At frequencies $v \leq 10^4$ Hz the magnitude of ε' of films III reveal a sharp increase with decreasing frequency of the external electric field and greatly exceeds the corresponding values of the films I and II.

2.2.2.3 AC conductivity of the films II and III

In contrast to the film I conductivity of the films II and III may be approximated by a power law $\sigma(\omega) \sim \omega^s$ on the frequency of the alternating electric field only in a very limited range of frequencies. So for films II, this area is $1 \leq v \leq 4\cdot10^2$ Hz in which the exponent is 0.66, and for films III conductivity satisfactorily approximated by a power law with exponent s = 0,63 in the frequency range $5 \leq v \leq 5\cdot10^2$ Hz.

For both types of films, a significant increase of the growth rate of the conductivity is observed at frequencies exceeding $2 \cdot 10^3$ Hz, however, the conductivity of the films II and III begins very weakly dependent on frequency of external electric field (see Figure 11) at frequencies larger of 10^5 and $3 \cdot 10^4$ Hz respectively.

The conductivity of films III containing tetraaniline exceeds the conductivity of the films II in the frequency range $1 \leq v \leq 3\cdot10^4$ Hz ,while at higher frequencies observed the opposite picture in which the conductivity of the films II is higher then that of films III (see the same figure).

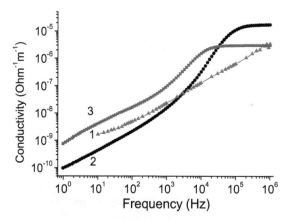

Fig. 11. Frequency dependence of AC conductivity of: films I - (1), film II - (2) and film III - (3)

2.3 Discussion

2.3.1 Ellipsometry of films I

Analysis of ellipsometric spectra shows that the value of ε' of the nc-Si films under study varies in the range of 2.1–1.1 in the energy range of 2–4.4 eV or in the frequency range of $5\times10^{14} – 1\times10^{15}$ Hz of the electromagnetic field, respectively, which is significantly below the values typical of c-Si in this range. In our opinion, there are two causes resulting in such low ε' and ε''. One is that nc-Si particles contacted with atmospheric oxygen for some time during film preparation; therefore, their surface was coated with a $SiO_x + SiO_2$ layer ($0 \leq x \leq 2$). Silicon nanoparticles oxidation was studied by Schuppler at al. (Schuppler at al., 1995), in that study the SiO_x layer thickness on their surface was determined as a function of the nanoparticle diameter. It was shown that the $SiO_x + SiO_2$ layer thickness in the nanoparticle diameter range of 10–3 nm is ~1 nm. However, this means that the ratio of the volume of the crystalline silicon core to the volume of the SiOx amorphous shell is from 100 to 40%. In other words, the oxidized crystalline silicon nanoparticle with a size smaller than 10 nm should exhibit amorphous properties to an appreciable extent. We have confirmed this statement previously based on an analysis of Raman spectra of nc-Si thin films (see section 2.A.2.1 and also Dorofeev at al., 2009). The second cause of a decrease in the permittivity is air gaps between nanoparticles, which appear during film formation.

To estimate the relation between crystalline and amorphous film components and their porosity, we use the Bruggeman EMA model. In the EMA approximation, the effective permittivity of the inhomogeneous medium consisting of spherical microobjects with permittivities $\varepsilon_1, \varepsilon_2, ...,\varepsilon_{N-1}$, immersed into a medium with ε_N ($\varepsilon_N \equiv \varepsilon_e$) is determined from the equation

$$\sum_{i=1}^{N} f_i \frac{(\varepsilon_i - \varepsilon_e)}{(\varepsilon_i + 2\varepsilon_e)} = 0 \quad \sum_{i=1}^{N} f_i = 1 \tag{5}$$

where: $f_i = \dfrac{V_i}{\sum\limits_{i=1}^{N} V_i}$

is the degree of medium volume filling with an element with permittivity ε_i and V_i is the volume occupied by this element.

Initially, to determine ε_e of the films under study, we assumed that the medium is two-phase and consists of purely crystalline silicon nanoparticles and the air gaps. In this case, Eq. (5) was reduced to a sum of two terms; knowing the dispersion relation of crystalline silicon, it was required to determine f_1 and f_2 so that approximating dispersion profiles would be identical to experimental $\varepsilon'(v)$ and $\varepsilon''(v)$. However, it was impossible to achieve satisfactory approximation at no values of f_1 and f_2.

Since the oxidation state of nanoparticles is unknown, we assumed that each particle in the two phase Bruggeman model behaves on average as a SiO_x medium (rather than as crystalline silicon), where $0 \leq x \leq 2$ was a fitting parameter, as well as f_1 and f_2. The $\varepsilon'(v)$ and $\varepsilon''(v)$ spectra for SiO_x in the entire range $0 \leq x \leq 2$ were taken from (Zuter, 1980), in which it was supposed that SiO_x is a mixture of Si–Si_yO_{4-y} tetrahedra; the random parameter takes values from 0 to 4 (random binding model (Hubner, 1980)). Using these spectra, it became possible to achieve good approximation of the experimental dependences $\varepsilon'(v)$ and $\varepsilon''(v)$ at $x = 1$ and $f_1 = f_2 = 0.5$. The approximating EMA spectra for these parameters are shown in Fig. 5 by dashed curves. Thus, it was shown that the nc-Si films under study on average behave as media consisting of SiO with a porosity of 0.5. Here we note already mentioned study (Pickering, 1984) in which the $\varepsilon'(v)$ and $\varepsilon''(v)$ spectra were measured and which are qualitatively and quantitatively rather similar to the spectra analyzed in the present study.

The absorption spectra of nc-Si films calculated from ellipsometric data are quite typical. As seen in Fig. 6, the film absorption at incident photon energies below 3 eV is stronger than that of c-Si; at higher energies, it is significantly lower. Such an absorption behavior shows that the SiOx shell with high density of states of defects near to the phase interface with the crystalline core mainly contributes to absorption for low energy photons; photons with energies above 3 eV are mostly absorbed by crystalline cores of nanoparticles with a wider band gap than that of c-Si due to quantum confinement.

2.3.2 Frequency dependence of the capacitance of films I

There are several models of the interpretation of the results of measurements of the ac conductivity of materials. For semiconductors, the model of (Goswami,A. & Goswami,A.P. 1973) is a good approximation, according to which a conductive material is a composition of a capacitor with capacitance C_1 and a resistor with conductance G_1 ($G_1 = 1/R$) connected in parallel. Furthermore, to take into account the effect of supplying contacts, a resistor with conductance G_2 ($G_2 = 1/r$) is connected in series with this group. According to this model, C_1 and G_1 are independent of the frequency of the applied ac electric field; however, G_1 depends on the conductive material temperature.

If the sample capacitance is measured in the mode of in parallel connected Cp, the measured value is related to C_1, G_1, and G_2 as

$$C_p = \frac{C_1 G_2^2}{\left(G_1 + G_2\right)^2 + \left(2\pi v C_1\right)^2}$$

We can see from this equality that the measured nc-Si film capacitance should satisfy the condition $Cp \sim v^{-2}$ while satisfying the conditions of the model of(27Goswami,A. & Goswami,A.P. 1973). However, it was impossible to approximate the experimental curve for $C_{nc\text{-}Si}$ shown in Fig. 7 by such power law dependence. Such a fact suggests that C_1 and G_1 should depend on frequency. Indeed, under experimental conditions, $G_2 \gg G_1$ and $G_2 \gg v C_1$, hence, $Cp \approx C_1$.

Therefore, for approximation, we used the following semi empirical function:

$$C_{nc-Si}(v) = C_\infty + \frac{C}{1 + \left(Av\right)^\beta} \qquad (6)$$

It follows from formula (6) that $C_{nc\text{-}Si} \rightarrow C_\infty$, at $v \rightarrow \infty$ and $C_{nc\text{-}Si} = C_\infty + C \equiv C(0)$ at $v = 0$.

Thus, the quantity C_∞ entering expression (6) is the film capacitance at an "infinitely high frequency" and $C(0) = C_\infty + C$ is the film static capacitance. The dimension of the fitting parameter A in the formula is time; the fitting parameter β defines the power law dependence of $C_{nc\text{-}Si}$ on the applied ac field frequency. Function (6) appeared to be a very good approximation of the experimental dependence $C_{nc\text{-}Si}(v)$ at the following coefficients: $C_\infty = 3.9 \times 10^{-10}$ F, $C(0) = 11.8 \times 10^{-10}$ F, $A = 0.5$, and $\beta = 0.32$.

The film capacitance is related to the real component of its permittivity by the relation $C_{nc\text{-}Si}(v) = \varepsilon'_{nc\text{-}Si}(v) \cdot C_0$. As noted above, $C_0 = 1.15 \times 10^{-10}$ F for the film under study; the static and optical permittivities $\varepsilon_s = \varepsilon(0) = 10.3$ and $\varepsilon_\infty = 3.4$ correspond to the determined capacitances $C(0)$ and C_∞.

The static permittivity of the film under study, which is 10.3, is significantly lower than the permittivity of crystalline silicon, which, as is known, is ~ 12. This result will be discussed below.

2.3.3 Dielectric relaxation in films I

The $\varepsilon'(\nu)$ and $\varepsilon''(\nu)$ frequency spectra obtained by measuring the nc-Si film impedance are shown in Fig. 8. The semi empirical Cole–Cole relation (Cole, K. S. & Cole, R. H., 1941; Moliton, 2007) appeared to be a good approximation for these spectra,

$$\varepsilon = \varepsilon_\infty + \frac{\varepsilon_s - \varepsilon_\infty}{1 + (i\omega\tau)^{1-h}} \quad 0 \leq h \leq 1 \tag{7}$$

where ε_s and ε_∞ are the static and optical permittivities determined above, $\omega = 2\pi\nu$ is the cyclic frequency, and τ is the dipole relaxation time.

As is known, the Cole–Cole relation is valid when a material simultaneously contains several types of dipoles each with a specific relaxation time. Therefore, the quantity τ entering Eq. (7) is the relaxation time averaged over the ensemble of dipole groups contained by the nc-Si film under study.

The approximating Cole–Cole curves are shown in Fig. 8.a by dashed curves. We can see that $\varepsilon'(\nu)$ is very well approximated in the entire measured frequency range; for $\varepsilon''(\nu)$, the Cole–Cole dependence exhibits good agreement only in the frequency range of $2 \times 10^2 \leq \nu \leq 10^6$ Hz. The values $\varepsilon_s = 10.8$, $\varepsilon_\infty = 3.43$, $\tau = 6 \times 10^{-2}$ s, and $h = 0.7$ correspond to the found approximation. It should be noted here that the value of ε_∞ is close to the values of ε' determined in the optical region by the ellipsometry method.

A comparison of the values of ε_s and ε_∞ corresponding to the Cole–Cole approximation with similar values determined from capacitance measurements shows the closeness of their numerical values. The value of $1 - h$ is also very close to the exponent β in formula (6). Furthermore, if we consider that A in formula (6) is the relaxation time multiplied by 2π, then $\tau = A/2\pi = 6.4 \times 10^{-2}$ s, which is also close to the average dipole relaxation time corresponding to the Cole–Cole approximation.

The static permittivity $\varepsilon_s = 10.8$ determined from the Cole–Cole relation is slightly larger than the similar value found from Eq. (6); however, it is also smaller than $\varepsilon_s = 12$ characteristic of crystalline silicon.

In our opinion, there are two causes resulting in a decrease in ε_s for the nc-Si film in comparison with ε_s of c-Si. The first cause is associated with air voids in the film body; the second cause is that the size distribution of nanoparticles composing the film includes a large fraction of particles with sizes smaller than 10 nm (see the Fig. 1). In (Tsu at al., 1997), the permittivity of silicon nanoparticles was calculated as a function of their size. According to these results, the static permittivity decreases as the particle diameter becomes smaller than 10 nm; for particles 10 nm in diameter, the permittivity is from 11.2 to 10.1, depending on the used calculation model.

In Fig. 8.a, in the frequency region $\nu \leq 2 \times 10^2$ Hz, we can see a notable disagreement between the Cole–Cole approximating function and the experimental dependence $\varepsilon''(\nu)$. This disagreement is caused by the fact that the Cole–Cole relation that describes dipole moment relaxation in dielectrics does not take into account the presence of free electric charges. However, free charges exist in the nc-Si film under study, which is indicated by the nonzero dc conductivity, which, as noted above, is $\sigma(0) = 9 \times 10^{-10}$ Ω^{-1} m^{-1} at temperature $T = 297$ K.

According to studies by Barton, Nakajima, and Namikawa (Barton 1966; Nakajima, 1972; Namikawa, 1975), the frequency v_m corresponding to the dispersion maximum for $\varepsilon''(v)$ is related to $\sigma(0)$ as $\sigma(0) = p(\varepsilon_s -\varepsilon_\infty).\varepsilon_0 2\pi v_m$, where the numerical coefficient p is approximately equal to unity. We can see in Fig. 8.a that the Cole–Cole approximating function reaches a maximum at the frequency v_m = 2.5 Hz, and this value is in good agreement with the experimental value of $\sigma(0)$ when using the Barton–Nakajima–Namikawa formula.

To take into account the conductivity associates with free electric charges, relation (4) should be written as

$$\varepsilon = \varepsilon_\infty + \frac{\varepsilon_s - \varepsilon_\infty}{1+(i\omega\tau)^{1-h}} + \frac{\sigma(0)}{\varepsilon_0\omega} \tag{8}$$

The approximation of the $\varepsilon''(v)$ spectrum of the film under study is shown by the dashed curve in Fig. 8.a (curve 5), from which it is obvious that function (8) is a good approach of the experimental dependence $\varepsilon''(v)$.

The effect of free electric charges on dielectric properties of the nc-Si film rather clearly appears in the Nyquist plot in which ε'' for each frequency is shown as a function of ε' (see Fig. 8.b).

It follows from the Cole–Cole approximation (see curve 2 in Fig. 8.b) that the $\varepsilon''(\varepsilon')$ should be shaped as a part of a semicircle whose center is below the horizontal axis ε''. The intersection of this circle with the ε' axis at ω = 0 and $\omega \to \infty$ yields the values of ε_s and ε_∞.

Figure 8.b shows only the semicircle part corresponding to the measured frequency range; therefore, the value ε_s = 10.8 is out of sight of the figure; the intersection of the semicircle with the ε' axis at $\omega \to \infty$ is clearly seen and corresponds to ε_∞ = 3.4. The same figure shows the approximation corresponding to function (8) (curve 3), similar to the approximation shown in Fig. 8.a.

2.3.4 AC conductivity of films I

To determine the nature of electric charge transport in nc-Si films, the frequency dependence of the conductivity $\sigma_{AC}(v)$, $\sigma_{AC}(v) - \sigma(0) = \varepsilon_0 \cdot 2\pi v \cdot \varepsilon''(v)$, was studied.

The $\sigma_{ac}(v)-\sigma(0)$ plot on a log scale for the film analyzed in this paper is shown in Fig. 10. We can see that $\sigma_{AC}(v)$ in the entire measured frequency range is well approximated by the power law function: $\sigma_{ac}(v) = \sigma(0) + Av^s$ with s = 0.74. Such $\sigma_{AC}(v)$ behavior means that the electric transport in the film has the hopping mechanism, which in turn is a manifestation of the structure disorder in that film region over which charge transport occurs.

Currently, there are several theoretical models describing hopping conductivity in unordered solids. All these models yield the power law dependence of the ac conductivity on the ac electric field frequency: $\sigma(v) \sim v^s$. However, the numerical values of the exponent s differ. For example, in the models (Austin & Mott, 1969; Hunt, 2001) according to which the conduction results from electric charge tunneling through energy barriers separating close localized states, the parameter s is given by

$$s = 1 + q \times \ln^{-1}\left(\frac{v}{v_{ph}}\right) \tag{9}$$

where q = 4 or 5, depending on the theoretical model, and $v_{ph} \approx 10^{12}$ Hz is the phonon frequency.

It follows from relation (9) that s should decrease with frequency. However, such behavior of s contradicts our experimental data and a large number of other experimental data (Dyre & Schrøder, 2000).

Currently, it has been sufficiently reliably determined that a large role in conduction processes in unordered solids is played by percolation processes with the result that electric transport occurs along trajectories with the lowest resistance (percolation trajectories) (Hunt, 2001; Isichenko, 1992). Conductive properties of percolation trajectories are controlled by the structure of (percolation) clusters composing the shell of solids.

In highly unordered solids, percolation trajectories at small scales exhibit a fractal structure with the result that their fractal dimension d_f appears larger than the topological one D (e.g., the fractal and topological dimensions of the Brownian particle trajectory is d_f = 2 and D = 1) (Isichenko, 1992).

In this regard, we note theoretical studies (Dyre & Schrøder, 2000; Schrøder & Dyre, 2002; Schrøder & Dyre, 2008) in which the diffusion cluster approximation (DCA) model is formulated. As these papers, it is argued that the so-called diffusion clusters with fractal dimensions of 1.1–1.7 make the largest contribution to the ac conductivity in the percolation mode. This statement means that the fractal structure of such clusters is simpler than the structure of multiply connected percolation clusters formed above the percolation threshold in conductive materials (backbone clusters), the fractal dimension of which is 1.7 (Isichenko, 1992). Simultaneously, the structure of diffusion clusters is more branched than the network of singly connected clusters and breaking of each results in disappearance of the current flowing through it (*redbonds*). The fractal dimension of *redbonds* clusters is 1.1 (Isichenko, 1992).

In these papers, the universal dependence of the dimensionless complex conductivity

$$\tilde{\sigma} = \frac{\sigma_{AC}(v) + i\sigma''(v)}{\sigma(0)}$$

on the dimensionless frequency

$$\tilde{\omega} = \frac{\varepsilon_0(\varepsilon_s - \varepsilon_\infty)}{\sigma(0)} 2\pi v$$

was derived. This dependence is given by

$$\ln\tilde{\sigma} = \left(\frac{i\tilde{\omega}}{\tilde{\sigma}}\right)^{\frac{d_f}{2}} \tag{10}$$

The fractal dimension d_f in formula (10) is a fitting parameter. Processing of a large number of experimental dependences in (Schrøder & Dyre, 2002; Schrøder & Dyre, 2008) showed that the best agreement in the frequency region $v > 1$ Hz is achieved at $d_f = 1.35$.

We compared the experimental dependence σ_{AC} (v) obtained in the present study with the values defined by formula (10). Here it should be noted that complex valued equation (10) has no analytical solution and should be solved numerically.

However, in the low frequency region $\omega \to 0$, Eq. (10) can be written as $\tilde{\sigma} - 1 = (i\tilde{\omega})^{\frac{d_f}{2}}$; accordingly (Kononov at al., 2011):

$$\sigma_{AC}(v) - \sigma(0) = \sigma(0)^{\left(1 - \frac{d_f}{2}\right)} \cos\left(\frac{\pi d_f}{4}\right) (2\pi\varepsilon_0 \Delta\varepsilon)^{\frac{d_f}{2}} v^{\frac{d_f}{2}} \tag{11}$$

where $\Delta\varepsilon = \varepsilon_s - \varepsilon_\infty$.

Substitution of experimentally determined values of ε_s, ε_∞, $\sigma(0)$, and $s \equiv d_f/2 = 0.74$ into formula (11) gives the approximating dependence for $\sigma_{ac}(v)$ (see Fig. 9) corresponding to the DCA model. We can see that the calculated dependence rather well approximates the experimental curve $\sigma_{ac}(v)$ in the entire measured frequency range. At the same time, the calculated dependence yields values of σ_{AC} larger than the experimental ones by a factor of ~1.5. We attribute such disagreement to possible errors when determining the numerical values of ε_s, ε_∞, and $\sigma(0)$.

2.3.5 Proton conductivity of films I

One of the possible causes that can result in $\sigma(0)$ measurement errors for nc-Si films is the dependence of $\sigma(0)$ on the ambient air humidity. We qualitatively determines the following systematic feature: the higher the laboratory air humidity, the higher (at a constant temperature) the conductivity $\sigma(0)$ of films similar to the film analyzed in this paper. On the contrary, if the film is preliminarily heated at a temperature of ~200°C for a time longer than 15 min and then it is cooled to its initial temperature, the film conductivity will decrease almost by two orders of magnitude. Thus the presence of water in an atmosphere surrounding the film changes its conductive properties significantly. In (Nogami & Abe, 1997; Nogami at al., 1998) a similar phenomenon was observed in the study of the ionic conductivity in fused silica glasses. It was shown that, in the presence of Si–O–H bonds on the glass surfaces, H_2O molecules form complexes with them, confined hydrogen bonds. These complexes can dissociate forming free H_3O^+ ions and bound Si–O– groups according to the scheme:

$$\text{Si-O-H} \cdots \text{OH}_2 \to \text{Si-O}^- + \text{H}^+ : \text{H}_2\text{O}$$

Here, dots denote the hydrogen bond between H and O atoms. In this case, the dissociated proton H+ can be trapped by a neighboring H_2O molecule,

$$\text{H}_2\text{O}: \text{H}_2\text{O}_{(1)}:\text{H}^+ + \text{H}_2\text{O}_{(2)} \to \text{H}_2\text{O}_{(1)} + \text{H}^+ + \text{H}_2\text{O}_{(2)} \to \text{H}_2\text{O}_{(1)} + \text{H}^+ : \text{H}_2\text{O}_{(2)} \text{ etc.}$$

Such a scheme allows implementation of proton transport near the glass surface. Returning to nc-Si films, we note that particles used to apply films represent hydrogenized nanocrystalline silicon. However, when exposing these particles to atmospheric air, a SiO_x shell ($0 \leq x \leq 2$) is formed on their surface. In (Du at al., 2003; Cao at al., 2007), the kinetics of the interaction of H_2O molecules with SiO_2 chain structures was calculated. It was shown that H_2O molecules very efficiently break Si–O–Si bonds during the interaction with SiO_2 surface groups with the formation of Si–O–H groups. The subsequent interaction of H_2O molecules and Si–O–H groups yields H_3O^+ ions, which, having high mobility, can appreciably contribute to the proton transport along the SiO_2 chain.

In addition to the above process, the collective proton conductivity caused by associated Si–O–H groups, i.e., groups linked by hydrogen bond, as shown in Fig. 12.a (Glasser, 1975). Arrows in the diagram indicate the direction of positive charge transport.

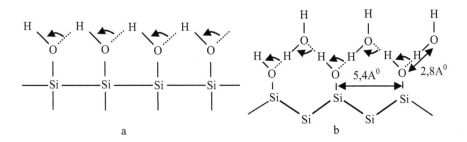

a   b

Fig. 12. Diagrams illustrating the mechanism of the collective proton conductivity, caused by (a) associated Si–O–H groups and (b) the interaction of water molecules with hydroxyl groups. Arrows indicate the direction of positive charge transport.

The collective proton conductivity is also possible during the interaction of water molecules with hydroxyl groups, which results in the surface structure shown in Fig. 12.b. Since the O···H—O group length is within 2.5-2.9 Å (Leite at al., 1998) and the angle between H-O-H bonds is ~104°, there is good spatial alignment between the element of this surface structure and the crystalline silicon lattice constant which, as is known, is 5.4 Å.

As applied to the nc-Si films analyzed in this paper, there is direct proof of the existence of such structures. Previously, in the investigations of IR transmission spectra of thin wafers (with thickness ≈50 µm) made by pressing ($P \sim 10^9$ Pa) from nc-Si powders similar to those used in this study, it was shown that the spectra contain a broad intense band with a maximum at ~3420 cm^{-1} (see Fig. 13 and Kononov at al., 2005). In papers (Wovchko at al., 1995; Stuart, 2004) this band is attributed to O-H vibrations in hydrogen bound hydroxyl groups. It was also shown that heating of nc-Si particles to 400°C causes an appreciably decrease in the intensity of the band near 3420 cm^{-1} and an increase in the intensity of the narrow band with a maximum near 3750 cm^{-1}, which is identified with

O–H vibrations in the isolated Si–O–H group (Kononov at al., 2005). Similar spectra are shown in Figure 13. Such behavior of the intensities of bands at 3420 and 3750 cm-1 means that associated Si–O–H groups become isolated upon heating of nc-Si particles. Accordingly, heating should decrease the proton conductivity associated with these groups.

Thus the dependence of the conductivity σ(0) of the *nc*-Si films under study on the ambient air humidity and the thermal behavior of the absorption bands associated with Si–O–H groups allows the conclusion to be drawn that the proton conductivity makes the main contribution to the dark dc conductivity of *nc*-Si films.

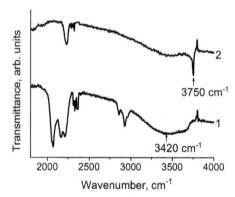

Fig. 13. Infrared transmittance spectra of: (1) - thin wafer from *nc*-Si particles prepared at a pressure of 5 × 10⁸ Pa at 20⁰C, (2) - the same *nc*-Si wafer but annealed at 400°C for 30 min.

2.3.6 Double dielectric relaxation in the films II and III

Earlier, we noted that the spectra of the $\varepsilon'(v)$ and $\varepsilon''(v)$ of the films II and III near a frequency $\approx 10^4$ Hz reveal the structure arising in the Debye dipole relaxation. Following this observation for the numerical approximation of the experimental spectra, we used not only semi-empirical law of Cole-Cole, but the law of Debye dipole relaxation. Thus, all experimental spectra were approximated by the following relation:

$$\varepsilon = \frac{\varepsilon_s - \varepsilon_\infty}{1+(i\omega\tau_1)^{1-h}} + \frac{\varepsilon_\infty}{1+(i\omega\tau_2)^2} + \frac{\sigma(0)}{\varepsilon_0\omega} \tag{12}$$

Here τ_1 and τ_2 is the relaxation times of dipole moments in the various structural components of the films. With the help of equation (12) was able to accurately approximate the dielectric spectra of the films studied; example of such an approximation for film II is shown in Figure 14.

Furthermore the approximation (12) allowed us to determine the static (ε_0), high-frequency (ε_∞) dielectric constants (v ~ 10⁵ Hz), the conductivity of the films at constant current σ (0), and relaxation times τ_1 and τ_2.

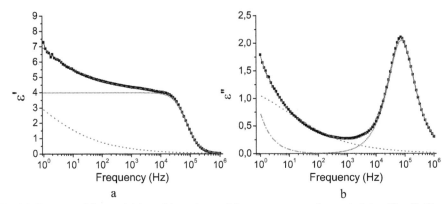

Fig. 14. Spectra of the real (a) and imaginary (b) components of permittivity film II. Short-dotted line shows the approximation of the Debye. Dotted line shows the approximation of the Cole-Cole. The dash-dotted line shows the approximation of free charges. The dashed line shows the complete approximation of the spectra.

№	ε_0	ε_∞	h	$\sigma(0)$, (Ohm·m)$^{-1}$	σ_{DC}, (Ohm·m)$^{-1}$	$\sigma_B(0)$, (Ohm·m)$^{-1}$	τ_1, s	τ_2, s
I	10,8	3,4	0,74	$1,4\cdot10^{-9}$	$9\cdot10^{-10}$	10^{-9}	0,06	-
II	11,5	4	0,66	$4\cdot10^{-11}$	$3,5\cdot10^{-11}$	$9\cdot10^{-11}$	0,72	$2,12\cdot10^{-6}$
III	66,9	4,9	0,5	$9\cdot10^{-11}$	$3,5\cdot10^{-9}$	$2\cdot10^{-9}$	0,27	$1,75\cdot10^{-5}$

Table 1. Fit parameters for the two dielectric relaxation lows of the films investigated in this study. σ_{DC} and $\sigma_B(0)$ - the conductivites at constant current received from direct measurements of the films resistance and from the Barton–Nakajima–Namikawa formula.

These values for films I, II and III are shown in the table 1. The table 1 also gives values of $\sigma(0)$ received from direct measurements of the films resistance at constant current at T = 297K, and those which obtained from the Barton–Nakajima–Namikawa formula. From table 1 it can be seen that the values of the static dielectric constant of films III are about 67, significantly higher than similar values of the films I and II, which are close to the values characteristic of crystalline silicon. However, the value of $\varepsilon_0 \approx 67$ is much lower quantities $\varepsilon_0 \sim 10^3$ typical for composites consisting of nanoparticles of tin dioxide and polyaniline which have been reported in (Kousik at al., 2007)

The authors of this work attributed so high ε_0 to an anomalously strong polarization of nanoparticle of SnO_2 which caused by inhomogeneity of the conductivity of its surface and core. However, the value of $\varepsilon_0 \approx 67$ which have been measured by us, is quite close to the values of the static dielectric constant of tetraaniline with different degrees of doping it with hydrochloric acid (Bianchi at al., 1999) and which, depending on the degree of doping lies in the range 35 - 80.

The presence in equation (12) two different laws of approximation indicates that there are two different dipole relaxation process associated with the various structural components of the studied films II and III. Very clear in understanding this phenomenon is a plot of ε'' vs ε' (Nyquist Plot), shown in Figure 15.

In the inset of Fig. 15 we can see that the dependence of ε'' vs of ε' for film III consists of two semicircles, which can be termed as high- and low- frequency components. The film II has a similar structure while the graph $\varepsilon''(\varepsilon')$ of the film I consists of only one semicircle (which is a low-frequency component) and low-frequency tail defined by the presence of free charges. Nanoparticles of silicon used for deposition of films II and III were in ethanol for two years after their synthesis, i.e., they were subjected to natural oxidation significantly longer than the nanoparticles of which consist film I. Therefore we can assume that oxidation of their surface is significantly higher than that of nanoparticles films I.

The previous sections have shown that the optical and electrical properties of films I greatly influenced by the surface of the nanoparticles from which these films are composed. It was found that the average properties of the surface similar to those of SiO and the component ε (ω) is determined by the Cole-Cole law related to the dipole relaxation in SiO_x shell of silicon nanoparticles.

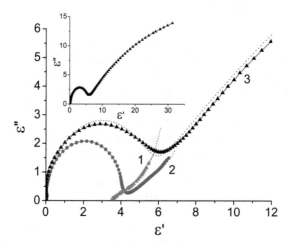

Fig. 15. The graph of dependence ε'' vs ε' for: film I - (1), film II - (2) and film III - (3) The inset shows an expanded plot $\varepsilon''(\varepsilon')$ for film III.

Since during the aging process of silicon nanoparticles the SiO_2 shell must increase, the appearance of high-frequency components of the Debye spectra ε (ω) of the films II and III gives reason to assume that the source of this component is the structure of SiO_2 with a narrow distribution of the dipole, which was formed on the surface of nanoparticles in two years of their presence in ethanol.

The fact that the Debye component of the spectrum ε (ω) as well as component Cole-Cole connected with the surface of the nanoparticles is confirmed by the fact noted earlier that the maxima of the Debye peak in the spectra of $\varepsilon''(\omega)$ of the films II and III correspond to different frequencies v_m.

The grains of silicon nanoparticles constituting the films II and III are similar to each other, so this difference frequency v_m can be attributed only to differences in the strength of interaction between the dipoles on the surface of the nanoparticles in these films. In other words, the presence of tetraaniline complexes on the surface of silicon nanoparticles leads to a weakening of the interaction between the dipoles are formed on the surface at the polarization of the particle.

2.3.7 AC conductivity of the films II and III

Dependence of the conductivity of the films I, II and III of the frequency of the applied electric field is shown in Figure 11. This figure shows that the conductivity of films I with good accuracy obey the law:

$$\sigma(\omega) = \sigma(0) + A\omega^s. \qquad (13)$$

In the entire range of measured frequencies, the exponent s = 0,74 equal to the value h which obtained from the approximation of the Cole-Cole and given in table 1.

Conductivity of the films II describe such an equation is possible only in very limited region, namely in the frequency range $v \leq 10^3$ Hz (let's call it a low-frequency component).

For low-frequency component of the conductivity of the films II as well as for the films I, the value of s coincides with that of h, shown in Table 1. For films of III this statement is incorrect. Indeed, as noted in Section B.2 conductivity of the films III is well approximated by a power law exponent with s = 0,63 only in small range of frequencies $5 \leq v \leq 5 \cdot 10^2$ Hz. As can be seen from Table 1, this value differs significantly from the values of h = 0,5 obtained from the approximation of the Cole-Cole.

The coincidence of the values of s and h for a film I is explained as follows circumstance. Spectrum $\varepsilon''(\omega)$ of the film over the entire range of measured frequencies is approximated by the Cole-Cole distribution which has the form:

$$\varepsilon'' = \frac{A(\omega\tau_1)^{1-h}}{1 + B(\omega\tau_1)^{1-h} + (\omega\tau_1)^{2(1-h)}} = \left(\frac{A}{B + \dfrac{1}{(\omega\tau_1)^{1-h}} + (\omega\tau_1)^{1-h}} \right)$$

Where A and B is constants and B≤2

If $\omega\tau_1 \gg 1$, this equation takes the form: $\varepsilon'' \approx \dfrac{A}{(\omega\tau_1)^{1-h}}$,

and hence the following relation is valid for the conductivity $\sigma(\omega) - \sigma(0) = \varepsilon_0 \cdot \omega \cdot \varepsilon'' \sim \omega^h$

As can be seen from Table 1 for the film I $\tau_1 = 0,06$ s, hence equation (13) is valid for it, at frequencies $v \geq 10$ Hz. A similar analysis is applicable also to the low frequency component of the film II. For film II $\tau_1 = 0,72$ s, therefore, the dependence (13) will be observed if $v \geq 1$ Hz. This fact is shown in Figure 16, where the conductivity of the films II and III is approximated by the sum of $\sigma(0)$ and two distributions of Cole-Cole and Debye.

From this figure it is clear that if the $\varepsilon''(\omega)$ spectrum of the films II describes only the distribution of the Cole-Cole, they would obey the conductivity relation (13) throughout the frequency range $1 \leq v \leq 10^6$ Hz as well as the conductivity of the films I.

For the film III observed more complicated situation, its spectrum is distorted with respect to relation (13), not only at high frequencies $v \geq 10^3$Hz, but also at frequencies $v \leq 10$ Hz (see Figure 16, b). According to the vast majority of experimental data, the frequency dependence of the conductivity of disordered media has kind of plateau (low-frequency plateaus) at low frequencies and is a power in excess of a certain critical frequency.

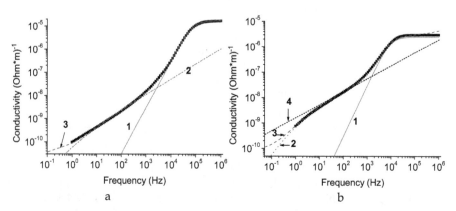

Fig. 16. The frequency dependence of AC conductivity of the films II (a) and III (b), as well as its approximation by: the Debye law - (1), the relation of Cole - Cole - (2) and the total approximation, which takes into account the dc conductivity - (3). (4) - power dependence with an exponent equal to the value of h at the Cole-Cole relation.

For films of III observes the opposite situation, instead, the appearance of a plateau at low frequencies, the conductivity $\sigma (\omega)$ begins to decrease more quickly with decreasing frequency of the external electric field. The reason for the absence of such low frequency plateau may be the existence of significant resistance at the interface of the film-electrode.

Comparison of $\sigma (0)$, σ_{DC} and $\sigma_B (0)$ from Table 1 shows their good agreement for film I. For films II are in good agreement the values $\sigma (0)$ and σ_{DC} but somewhat too high the value of $\sigma_B (0)$ with respect to them. For films III good agreement is observed for the values σ_{DC} and $\sigma_B (0)$ but $\sigma (0)$ is less than these quantities is about 20 times. The fact that σ_{DCI} more than 25 times higher then σ_{DCII} (see Table1) confirms our earlier assumption that the degree of surface oxidation of silicon nanoparticles of films II is significantly higher than that in films I.

At frequencies $v_{s1} \geq 1 \cdot 10^5$Hz for films II and $v_{s2} \geq 3 \cdot 10^4$Hz for films III conductivity begins to depend very weakly on the frequency of an external electric field. This behavior is usually associated with the manifestation of the nature of hopping conduction (Barsoukov & Macdonald, 2002), and the frequency v_c determined by the height of the barriers between potential wells, which are involved in the hopping transport of charge carriers. Because $v_{s1} > v_{s2}$, we can conclude that the presence of tetraaniline on the surface of silicon nanoparticles lowers the barriers separating localized states.

3. Conclusion

Dielectric and transport properties of thin films obtained by deposition of silicon nanoparticles from ethanol sols on a glass, quartz, and aluminum substrates were measured by optical ellipsometry and impedance spectroscopy methods. The real and imaginary permittivities of nc-Si films were measured in frequency ranges of 5×10^{14}–10^{15} and 10–10^6 Hz. It was found that the permittivity spectra depend on the time which has elapsed since the synthesis of nanoparticles until their deposition on the substrate.

Only one type of dipole relaxation, which can be described by semi-empirical Cole-Cole equation, exists in films prepared from sols with silicon nanoparticles, synthesized a week before their deposition on a substrate (film I). In films prepared from sols containing aged nanoparticles (film II) there is a double-dipole relaxation, which is revealed in the fact that for the approximation of the experimental spectra of these films not only Cole-Cole relation but the law of Debye dipole relaxation should be used. A similar confirmation is valid also for the films deposited from the sols with aged nanoparticles in which tetra aniline was added (film III).

In the measured frequency ranges, ε' and ε'' vary within 2.1–1.1, 3.4–6.2 and 0.25–0.75, 0.08–1.8, respectively. From the EMA analysis of the spectra, it was concluded that the nc-Si film in light reflection processes can on average be considered as a two component medium consisting of SiO and air gaps with a porosity of 50%.

It was shown that the complex dielectric dispersion of films in the frequency range of 10 – 2×10^6 Hz is well approximated by the semiempirical Cole–Cole relation, taking into account the effect of free charges controlling the dark dc conductivity of films.

An analysis of the frequency dependences of the ac conductivity of the studied films allowed the conclusion to be drawn that the ac conduction process is well described by the cluster diffusion approximation model.

The dependence of the dark conductivity of films on the ambient air humidity and the temperature dependence of absorption bands related to associated Si-O-H groups allows the conclusion to be drawn that the conductivity at frequencies lower than 2×10^2 Hz is controlled by proton transport through hydrogen bound hydroxyl groups on the surface of silicon nanoparticles.

Using Cole-Cole and Debye relations for approximation of experimental spectra $\varepsilon(\omega)$ the values of static permittivity ε_0 of films I, II and III have been found. For films I and II quantities ε_0 close to the values characteristic of crystalline silicon. For films of III $\varepsilon_0 \approx 67$, i.e. greatly exceeds ε_0 for c-Si. Such a high value ε_0 we attribute to increasing polarization of the silicon nanoparticles when the tetraaniline complexes are attached to their surface.

AC conductivity of the films II and III in the whole frequency range of 1-10^6 Hz can not be approximated by a power law, which is characteristic of the conductivity of the films I. We show that such deviation from the dependence $\sigma_{AS} \sim \omega^s$ is associated with a double-dielecrtic relaxation typical for films II and III and with the presence in the spectra $\varepsilon''(\omega)$ of these films Debye components.

4. Acknowledgment

We sincerely thank Dr. Helen Yagudayev, the senior researcher of the Shemyakin - Ovchinnicov Institute of Bioorganic Chemistry of RAS, for providing us the conductive tetraaniline solutions.

We also thank our colleagues prof. Plotnichenko V.G., prof. Kuz'min G.P., prof. Ischenko A.A., dr. Koltashev V.V., researcher Tikhonevich O.V. for the fruitful cooperation in investigation of the properties of nano-sized silicon.

5. References

[1] Anopchenko,A.; Marconi, A.; Moser, E.; Prezioso, S.; Wang; M., Pavesi; L., Pucker, G. & Bellutti. P. J. (2009) Low-voltage onset of electroluminescence in nanocrystalline-Si/SiO$_2$ multilayers *Journal of Applied Physics*, Vol.106, (2009), p.p. 033104, ISSN 0021-8979

[2] Aspens, D. E. & Studna, A. A. (1983) Dielectric functions and optical parameters of Si, Ge, Gap, GaAs, GaSb, InP, InAs, and InSb from 1.5 to 6.0 eV, *Physical Review* B, Vol. 27,№2, (January 1983), p.p.985 - 1009

[3] Austin, I. G. & Mott, N. F. (1069). Polarons in crystalline and non-crystalline materials. *Advances in physics*, Vol.18, №71, (January 1969), pp. 41-102

[4] Axelrod, E.; Givant, A.; Shappir, J.; Feldman, Y. & Sa'ar. A. (2002) Dielectric relaxation and transport in porous silicon *Physical Review* B, Vol. 65, №16, (April 2002) , p.p. 165429(1-7) ,ISSN 0163-1829

[5] Azzam, R.M.A. & Bashara N.M. (1977) *Ellipsometry and polarized light*, North-Holland publishing company, ISBN 1704050000 , Amsterdam, New York, Oxford

[6] Barsoukov, E. & Macdonald, J. R. (2005) *Impedance Spectroscopy. Theory, Experiment, and Applications*, Second Edition, A John Wiley & Sons, Inc., Publication, New Jersey, USA and Canada ISBN: 0-471-64749-7

[7] J. L. Barton, (1966). *Verres Réfract.* Vol. 20, p.p. 328 (1966).

[8] Ben-Chorin, M.; Möller, F.; Koch, F.; Schirmacher, W. & Eberhard, M. (1995) Hopping transport on a fractal: ac conductivity of porous silicon *Physical Review* B, Vol. 51, №4, (January 1995) , p.p. 2199(1-15) , ISSN 0163-1829

[9] Bianchi, R.F.; Leal Ferreira, G.F.; Lepienski, C.M. & Faria, R.M. (1999). Alternating electrical conductivity of polyaniline. *Journal of Chemical Physics*, Vol.110,№9, (March 1999), p.p. 4602-4607, ISSN 0021-9606

[10] D.A.G. Bruggeman. (1935). Berechung verschiedener physikalisher konstanten von heterogenen substanzen. *Annalen der Physik* (Leipzig), Vol. 24, (1935), pp. 636-664

[11] Brus, L.E.; Szajowski, P.F.; Wilson, W. L.; Harris, T. D.; Schuppler, S. & Citrin. P. H. (1995) Electronic Spectroscopy and Photophysics of Si Nanocrystals: Relationship to Bulk c-Si and Porous Si. *Journal of American Chemical Society*, Vol. 117, (1995) pp. 2915-2922

[12] Campbell, I.H. & Faushet P.M. (1986) The effect of microcrystalline size and shape on the one phonon Raman spectra of crystalline semiconductors. Solid State Communications. Vol. 58, №10 (February 1986), p.p.739-741, ISSN 0038-1098

[13] Cao, Chao; He, Yao; Torras, J.; Deumens, E.; Trickey, S. B. & Cheng, Hai-Ping. (2007) Fracture, water dissociation, and proton conduction in SiO_2 nanochains. *Journal of Chemical Physics*, Vol. 126, №21, (June 2007) p.p. 211101(1-3), ISSN 0021-9606

[14] Cole, K. S. & Cole, R. H. (1941). Dispersion and Absorption in Dielectrics. *Journal of Chemical Physics*, Vol. 9, №, (February 1941) p.p. 341-351, ISSN 0021-9606

[15] Conte, G.; Feliciangeli, M. C.; & Rossi, M. C. (2006) Impedance of nanometer sized silicon structures. *Applied Physics Letters*, Vol. 89, № 2, (July 2006), p.p.022118(1-3), ISSN0003-6951

[16] Dorofeev, S. G.; Kononov, N. N.; Ishchenko, A. A.; Vasil'ev, R. B.; Goldschtrakh, M. A.; Zaitseva, K. V.; Koltashev, V. V.; Plotnichenko, V. G. and O. V. Tikhonevich. (2009). Optical and Structural Properties of Thin Films Precipitated from the Sol of Silicon Nanoparticles. *Rus. Semiconductors*, Vol. 43, No. 11, (April 2009), pp. 1420–1427. ISSN 1063-7826.

[17] Du, Mao-Hua; Kolchin, A. & Chenga, Hai-Ping. (2003). Water–silica surface interactions: A combined quantum-classical molecular dynamic study of energetics and reaction pathways. *Journal of Chemical Physics*, Vol. 119, №131, p.p.6418 – 6422 (2003) ISSN 0021-9606.

[18] Dyre, J.C. & Schrøder, T. B. (2000). Universality of ac conduction in disordered solids. *Reviews of Modern Physics*, Vol.72, №3, (July 2000), p.p. 873 – 892, ISSN 0034-6861.

[19] Ehbrecht, M.; Ferkel, H.; Huisken, F.; Holz, L.; Polivanov, Yu.N.; Smirnov, V.V.; Stelmakh, O.M. & Schmidt R. (1995) Deposition and analysis of silicon clusters generated by laser-induced gas phase reaction *Journal of Applied Physics*, Vol.78, № 9, (November 1995), p.p. 5302-5305, ISSN 0021-8979

[20] Glasser, L. (1975) Proton Conduction and Injection in Solids. *Chemical Reviews*, Vol.75, №1, (January 1974), p.p. 21 – 65.

[21] Goswami, A. & Goswami, A.P. (1973). Dielectric and optical properties of ZnS films. *Thin Solid Films* Vol. 16, (January 1973), p.p. 175 – 185

[22] Hubner, K. (1980). Chemical Bond and Related Properties of SiO_2. *Physica Status Solidi* (a) Vol. 61, (May 1980), p.p. 665-673

[23] Hunt, A.G. (2001). Ac hopping conduction: perspective from percolation theory. *Philosophical Magazine* B Vol. 64, №9, (2001), p.p. 875-913, ISSN 1364-2812 print/ISSN 1463-6417 online

[24] Isichenko, M. B. (1992) Percolation, statistical topography, and transport in random media. *Reviews of Modern Physics*, Vol. 64, №4, (October 1992), p.p.961 – 1043, ISSN 0034-6861.

[25] Jurbergs, D.; Rogojina, E.; Mongolini, L. & Kortshagen, U. (2006). Silicon nanocrystalls with ensemble quantum yields exceeding 60%. *Applied Physycs Letters*, Vol. 88, №23, (June 2006) 233116(1-3), ISSN 0003-6951.

[26] Kakinuma, H.; Mohri, M.; Sakamoto, M. & Tsuruoka T. (1991). Sructural properties of polycrystalline silicon films prepared at low temperature by chemical vapor deposition. *Journal of Applied Physics*, Vol.70, (December 1991), p.p. 7374 - 7381, ISSN 0021-8979

[27] Kononov, N. N.; Kuz'min, G. P.; Orlov, A. N.; Surkov, A. A. and Tikhonevich, O. V. (2005). Optical and Electrical Properties of Thin Wafers Fabricated from Nanocrystalline Silicon Powder. *Rus. Semiconductors*, Vol. 39, No. 7, (November 2005), pp. 835–839, ISSN 1063-7826

[28] Kononov, N. N.; Dorofeev, S. G.; Ishenko, A. A.; Mironov, R. A.; Plotnichenko, V. G. & E. M. Dianov. (2011). Dielectric and Transport Properties of Thin Films Precipitated from Sols with Silicon Nanoparticles *Rus. Semiconductors,* 2011, Vol. 45, No. 8, (August 2011), pp. 1038–1048. ISSN 1063-7826

[29] Kovalev, D.; Polisski, G.; Ben-Chorin, M.; Diener, J. & Koch F. (1996) The temperature dependence of the absorption coefficient of porous silicon *Journal of Applied Physics,* Vol.80, №10, (November 1996) , p.p.5978 -5983, ISSN 0021-8979

[30] Kousik, D. & De, S.K. (2007). Double dielectric relaxations in SnO_2 nanoparticles dispersed in conducting polymer. *Journal of Applied Physics,* Vol.102,.№ 8, (October 2007), p.p. 084110 -1,084110 -7, ISSN 0021-8979 print | 1089-7550 online

[31] Kuz'min, G.P.; Karasev, M.E.; Khokhlov, E.M. Kononov, N.N.; Korovin, S.B.; Plotnichenko, V.G.; Polyakov, S.N.; Pustovoy, V.I. & Tikhonevich O.V. (2000). Nanosize Silicon Powders: The Structure and Optical Properties. *Rus. Laser Physics* Vol.10, №4, (April 2000) p.p. 939-945

[32] Leite, V. B. P.; Cavalli, A. & Oliveira, O. N. Jr. (1998) Hydrogen-bond control of structure and conductivity of Langmuir films. *Physical. Review* E, Vol. 57, №6, (June 1998), p.p. 6835 – 6839, ISSN 1063-651X

[33] Luppi, M. & Ossicini S. (2005). Ab initio study on oxidized silicon clusters and silicon nanocrystals embedded in SiO_2: Beyond the quantum confinement effect. Phys. Rev. B V.71, (January 2005), p.p. 035340(1-15), ISSN 1098-0121

[34] Ma, Zhixun; Liao, Xianbo; Kong, Guanglin & Chu, Junhao (2000) Raman scattering of nanocrystalline silicon embedded in SiO_2. *Science in China* A, Vol.43, (2000), p.p. 414-420

[35] Min, R .B. & Wagner, S. (2002), Nanocrystalline silicon thin-film transistors with 50-nm-thick deposited channel layer, 10 cm² V^{-1}s^{-1} electron mobility and 10^8 on/off current ratio *Applied Physics A Materials Science &Processing,* Vol. 74, (August 2001), p.p.541–543 (DOI) 10.1007/s003390100927

[36] Moliton, A. (2007) *Applied Electromagnetism and Materials,* Ch.1, p.8. Springer Science and Business Media. ISBN-10: 0-387-38062-0

[37] Nakajima, T. (1971) *Annual Report, Conference on Electric Insulation and Dielectric Phenomena* (Washington, D.C., National Academy of Sciences, 1972), p. 168.

[38] Namikawa. H. (1975).Characterization of the diffusion process in oxide glasses based on the correlation between electric conduction and dielectric relaxation. Journal of Non-Crystalline Solids, Vol.18, №2, (September 1975), p.p.173-195, ISSN 0022-3093.

[39] Nogami, M. & Abe, Y. (1997). Evidence of water-cooperative proton conduction in silica glasses. *Physical Review* B, Vol. 55, №18, (May 1997), p.p.12108 - 12112, ISSN 0163-1829

[40] Nogami, M.; Nagao, R. & Wong, C. (1998). Proton Conduction in Porous Silica Glasses with High Water Content. *Journal of Physical Chemistry* B, Vol. 102, (July 1998), p.p. 5772-5775, S1089-5647

[41] Pickering, C.; Beale, M. I. J.; Robbins, D. J.; Pearson, P. J. & Greef. R. (1984). Optical studies of the structure of porous silicon films formed in p-type degenerate and non- degenerate silicon. *Journal Physics* C: *Solid State Physics,* Vol.17, (August 1984) p.p.6535-6552, ISSN 0022-3719

[42] Puzder, A.; Williamson, A.J.; Grossman, J.C. & Galli G. (2002). Surface Chemistry of Silicon Nanoclusters. *Physical Review Letters*, Vol. 88, (2002), p.p.097401(1-4), ISSN 0031-9007

[43] Richter, H.; Wang, Z. & Ley L. (1981). The one phonon Raman spectrum in microcrystalline silicon. *Solid State Communications*, Vol.39, №5, (May 1981), p.p.625-629, ISSN 0038-1098

[44] Saadane, O.; Lebib, S.; Kharchenko, A.V.; Longeaud, C. & R. Roca I Cabarrocas. (2003) Structural, opical and electronic properties of hydrogenated polymorphous silicon films deposited from silane-hydrogen and silane-helium mixtures. *Journal of Applied Physics*, Vol. 93, №11, (June 2003), p.p.9371 – 9379, ISSN 0021-8979

[45] Schrøder, T. B. & Dyre. J. C. (2002). Computer simulations of the random barrier model. *Physical Chemistry and Chemical Physics*, Vol.4, (June 2002), p.p. 3173–3178

[46] Schrøder, T. B. & Dyre J. C.(2008). AC Hopping Conduction at Extreme Disorder Takes Place on the Percolating Cluster *Physical Review Letters*, Vol.101, (July 2008) p.p.025901(1-4), ISSN 0031-9007

[47] Schuppler, S.; Friedman, S.L.; Marcus, M.A.; Adler, D.L.; Xie, Y. H.; Ross, F.M.; Chabal, Y.J.; Harris, T.D.; Brus, L.E.; Brown, W.L.; Chaban, E.E.; Szajowski, P.E.; Christman, S.B. & Citrin, P.H. (1995). Size, shape and composition of luminescent species in oxidized Si nanocrystals and H-passivated porous Si. *Physical Review* B, Vol.52, p.p. 4910-4925, ISSN 1098-0121

[48] Stuart B. (2004). *Infrared Spectroscopy: Fundamentals and Applications*, John Wiley & Sons, Ltd., ISBN 0-470-85427-8

[49] Eds. by Tompkins, G. H. & Irene, E.A. (2005). *Handbook of ellipsometry*. p.p. 282-289. William Andrew publishing, Springer-Verlag GmbH & Co. KG. New York, Heidelberg. ISBN: 0-8155-1499-9 (William Andrew, Inc.), ISBN: 3-540-22293-6 (Springer-Verlag GmbH & Co. KG)

[50] De la Torre, J.; Bremond, G.; Lemiti, M.; Guillot , G.; Mur, P. & Buffet. N. (2006) Silicon nanostructured layers for improvement of silicon solar cells efficiency: A promising perspective. Materials Science and Engineering C Vol. 26, (January 2006), p.p. 427–430, ISSN 0928-4931

[51] Tsang, J.C.; Tischler, M.A. & Collins R.T. (1992). Raman scattering from H and O terminated porous Si. *Applied Physics Letters*, Vol.60, №18, (May 1992) p.p. 2279-2281, ISSN 0003-6951

[52] Tsu, R.; Shen, H. & Dutta M. (1992). Correlation of Raman and photoluminescence spectra of porous silicon. *Applied Physics Letters*, Vol.60, №1, (January1992), p.p. 112-114, ISSN 0003-6951

[53] Tsu, R.; Babić, D. & Ioriatti, L. Jr. (1997). Simple model for the dielectric constant of nanoscale silicon particle. *Journal Applied Physics*, Vol. 82, № 3, (August 1997), p.p.1327 – 1329, ISSN 0021-8979

[54] Tsu, R. (2000) Phenomena in silicon nanostructure devices. *Applied Physics* A Vol.71, (September 2000) p.p.391–402, DOI 10.1007/s003390000552

[55] Urbach, B.; Axelrod, E. & Sa'ar A. (2007). Correlation between transport, dielectric, and optical properties of oxidized and nonoxidized porous silicon *Physical Review* B, Vol. 75, №20, (May 2007) , p.p. 205330(11) , ISSN 1098-0121

[56] Voutsas, A.T.; Hatalis, M.K.; Boyce, J. & Chiang A. (1995). Raman spectroscopy of amorphous and microcrystalline silicon films deposited by low-pressure chemical

vapor deposition *Journal of Applied Physics,* Vol.78, №12, (December 1995), p.p. 6999-7005, ISSN 0021-8979

[57] Wang, K.; Chen, H. & Shen, W.Z. (2003). AC electrical properties of nanocrystalline silicon thin films. *Physica* B, Vol. 336, (April 2003), p.p. 369-378, ISSN 0921-4526

[58] Wang, W. & MacDiarmid, A. (2002), New synthesis of phenil/phenil end-capped tetraaniline in the leucoemeraldine oxidation states. *Synthetic metals,* Vol. (2002) 129, p.p. 199-205, ISSN

[59] Wovchko, E. A.; Camp, J. C.; Glass, J. A. Jr. & Yates, J. T. Jr. (1995). Active sites on SiO_2: role in CH_3OH decomposition. Langmuir Vol.11, №7, (1995), p.p.2592-2599, ISSN 0743-7463

[60] G. Zuter.(1980) Dielectric and Optical Properties of SiO_x. *Physica Status Solidi* (a) Vol. 59, (April 1980), p.p. K109 –K113

8

View on the Magnetic Properties of Nanoparticles Co_m (m=6,8,10,12,14) and Co_6O_n (n=1-9)

Jelena Tamulienė[1], Rimas Vaišnoras[2],
Goncal Badenes[3] and Mindaugas L. Balevičius[4]
[1]Vilnius Uinversity, Institute of Theoretical Physics and Astronomy, Vilnius,
[2]Vilnius Pedagogical University, Vilnius,
[3]Institut de Ciences Fotoniques ICFO, Barcelona,
[4]Vilnius University, Vilnius,
[1,2,4]Lithuania
[3]Spain

1. Introduction

Currently there are several potential applications for magnetic nanomaterials in medicine including magnetic resonance imaging contrast agents, magnetic-field-directed drug delivery systems, bio-toxin removal, gene therapy, and magnetic fluid hyperthermia. Cobalt nanoparticles are is one the most promising material for both technological applications and academic studies as model system how effects the nanoparticle size, shape, structure, and surface anisotropy on macroscopic magnetic response. The magnetic behaviour of Co nanoparticles reveals how the magnetic metal nanoparticles can be used to enhance the signal due to their magnetic resonance imaging.

Today it is very well known that in a paramagnetic material there are unpaired electrons, that are free to align their magnetic moment in any direction, while paired electrons by the Pauli Exclusion Principle are to have their intrinsic ('spin') magnetic moments in to opposite directions, causing their magnetic fields to cancel out. It implies, that in many cases, the magnetic properties of the Co nanoparticles are explained by the presence of unpaired electrons because the particles consist of an odd number of cobalt atoms. However, in experimental studies the number of atoms in the particle has never been mentioned only the description of their size and main structure along with their magnetic properties have been provided. It is not a surprise, because a magnetic behaviour of materials depends on their electron configuration that is strongly related with a geometrical structure, and on temperature.

The dependence of magnetic anisotropy energy on crystal symmetry and atomic composition is observed in both ferromagnetic bulk materials and thin films. Even the structural parameters such as the shape of particles or the inter-atomic distances, in some cases, are affected by the above dependence. The importance of the electronic structure of particles exhibits the dependence of magnetic anisotropy energy on a single-atom coordination. Current, experiments exhibited that the coercivity of some particles at 10 K increased from 640 to 1250 Oe while the particle size increased from 1.8 to 4.4 nm. The saturation magnetization increases with decreasing of particle size. Pure CoO nanoparticles in the 4.5-18 nm exhibit a super-paramagnetic behaviour at room temperature, and a large orbital contribution to the magnetic moment at low temperatures was also observed.

It was mentioned, that an electronic structure of both the materials and particles is strongly related with the geometrical structure. However, there are some difficulties to identify the structure of a cobalt nanoparticle. The crystallinity was evidenced by the transmission electron microscope (TEM) indicating that Co particles sized around 4.7 nm are a well-crystallized FCC. While the particles with the average diameter smaller than 4,7 nm are almost perfectly spherical. The lattice of Co nanoparticles with inter-planar distance of around 0.23 nm was obtained and explained that such crystalline structure could originate either from BCC cobalt particles observed along the [001] direction or due to Co-FCC particles since the lattice would be formed by two [002] perpendicular planes. Both a high-resolution TEM and powder x-ray diffraction profiles reveal the presence of 8-15 nm diameter crystallites that are identified as hcp-Co, FCC-Co nanocrystals. S. Ram reports two crystalline phases of cobalt FCC and BCC structures, while S. P. Gubin and et al. report that hcp and FCC structures or their combination can be realized in Co nanoparticles. C. G. Zimmermann and et al. investigate Co nanoparticles the diameter of which is 13 nm and the variance of 4 nm; the first four FCC rings were visible in the diffraction pattern. Hence, there is no evidence what a crystalline phase of cobalt is more preferable and it is difficult to define which structure type of Co is realized in nanoparticles. Theoretical investigations of the Co clusters are not complete. J. Guevara and et al. calculated those Co clusters that are part of FCC or BCC block without distortion of the initial geometry structure. In other works, the structural distortion of the above clusters was performed by moving one or several atoms along the main axis of the clusters, i.e. this operation does not change the symmetry if the configuration of the cluster belongs to a point group with a single main axis. Hence, we begin at the results of the investigation of the structure of the Co nanoparticles aiming to recognize the most important structure features influencing the magnetic properties of the Co nanoparticles.

Other very important results obtained are that the nanoparticle behaviour is influenced by the proximity of neighbouring particles, i.e. dipolar inter-particle interactions lead to the appearance of collective behaviour. Such a collective behaviour due to dipolar interactions has been observed in the low susceptibility measurements corresponding to a highly ordered fine particles system. Puntes and et al. observe that when the density of particles per unit area is higher than a determined threshold, the two-dimensional self-assemblies behave as a continuous ferromagnetic thin film. A weak interaction among the assemblies of the Co nanoparticle is obtained by Park and et al. and this assembly leads to hysteresis disappearance. We shortly explain why the assembly of Co nanoparticles leads to loosing of the particle magnetic properties and make predictions how to avoid the loosing.

One of the reasons of the above assembly of the particles is their stability that is an important factor for the particle application in technology. Small cobalt nanoparticles not only self-assemble, but also easily oxidize in the air and, as a consequence, loose their magnetic properties. Thus, Co nanoparticles need to be coated with organic surfactants aiming to prevent them from both irreversible aggregation and loosing of magnetic properties.

For coating of Co nanoparticle different materials such as graphite, nanoroads, nanocapsules and oxygen are used. The core-shell nanoparticles (Co-CoO) are examined and, it is established, that the magnetic properties of these particle strongly depend on the plane coverage. The results reported demonstrate the essential role played by shells in stabilizing the magnetism of Co-CoO nanoparticles. Few reports on the preparation and properties of pure CoO in bulk are due to difficulties to obtain the materials in pure form by simple methods. The particles are often contaminated with Co_3O_4 or Co metal. The greater stability of Co_3O_4 than CoO is also established.

Herein, we report on the several very important issues related to magnetic properties of Co nanoparticles such as:

1. What are electronic and geometric structure properties of pure and oxidized Co nanoparticles and how these properties change with the increase of the size of particle ;
2. Could Co nanoparticles consisting of the even number of atoms exhibit magnetic properties because their electronic structure is such that an uncompensated electron-magnetic-moment appears? What are the main reasons of the above appearance?
3. Some Co oxide particles exhibited magnetic properties and have large perspective to be used in electronics.

2. Description of method

The structural origin of clusters has been studied by using the generalized gradient approximation for the exchange-correlation potential in the density functional theory (DFT) as it is described by Becke's three-parameter hybrid functional, using the non-local correlation provided by Lee, Yang, and Parr. The DFT method is commonly referred to as B3LYP, - a representative standard DFT method. The 6-31G basis set has been used as well. The basis set was chosen keeping in mind relatively minimum computational costs. The structures of the investigated nanoparticles have been optimized globally without any symmetry constraint and by starting from various initial geometries which have been constructed according to a certain symmetry in order to determine the lowest energy structures of each cluster. The GAMESS and Gaussian program suites were used for all simulations here.

It is necessary to mention that there are different ways to theoretically investigate the magnetic properties of the materials. Aiming to exhibit why closed shell particles could be paramagnetic, we have chosen the most simple method to investigate magnetic properties of the Co nanoparticles. Hence, magnetizability (commonly known as susceptibility) was investigated. The magnetizability is the second-order response to an external magnetic field:

$$\xi = \frac{-\delta^2 E(B)}{\delta B^2}\Big|_{B=0}$$

Where E is energy, B is an external magnetic field.

When $\xi < 0$, the induced magnetic moment is opposite to the applied field, i.e. the investigated materials are diamagnetic; while for paramagnetic materials the magnetizability is larger than zero ($\xi > 0$) in this case the induced magnetic moment enforces the magnetic field. Experimentally, magnetizability is often poorly determined or it is only known in the liquid or solid state, thus it is difficult comparisons between calculated and experimental results, while rotational g-tensors are known as precisely determined. However, a rotational g-tensor behaves in the same manner as magnetizability, with a near cancellation of large nuclear and electronic contributions in a large system.

A calculation of rotational g tensors is closely related to that of magnetizabilities via:

$$g = -4m_p \left(\xi^{LAO} - \xi_{cm}^{dia} \right) I_{nuc}^{-1} + \frac{1}{2\mu_N} \sum Z_k \left(R_K^T R_K I_3 - R_K R_K^T \right) I_{nuc}^{-1}$$

where m_p is the proton mass, ξ^{LAO} is the magnetizability tensor calculated with London orbitals, ξ_{cm}^{dia} is the diamagnetic contribution to the magnetizability tensor calculated with conventional orbitals and the gauge origin at the centre of mass, and the sum of all nuclei with charges Z_K and positions R_K, while I_{nuc} is the moment-of-inertia tensor. Although not explored in a large number of studies, obtained theoretical results fit experimental. Hence, the above close relationship allows us to expect that our methods chosen that are well suited to the calculation of rotational g tensors should also be well suited to the calculations of magnetizabilities. Moreover, this simple enough method is suitable to describe general magnetic properties of the investigated particles and to explain the results obtained.

The isotropic magnetizability of the most stable clusters was calculated by adopting quantum mechanical response theory and London atomic orbital to ensure both gauge-origin independent results and fast basis set convergence by using Dalton program. The approach used allows us to calculate accurate magnetizability even for quite large molecules at a moderate cost of computing time. In this case, the B3LYP method with Ahlrichs-pVDZ basis set was used. These basis sets were obtained by optimizing the exponents and contraction coefficients in the ground state ROHF calculations. There are total 241 contracted functions in the basis mentioned. It is showed, that the isotropic magnezitability and its anisotropy are remarkably constant with respect to the basis set and close to the experiment. So, the performances obtained allow us to foresee how magnetic properties of the particles depend on their structures.

3. Structure, stability and magnetic properties of Co_m (m=6, 8, 10,12, 14) nanoparticles

3.1 Structure and stability

Let us remember that magnetic properties of the materials are related with an electronic structure. The electronic structure is mostly geometrical-structure-depended. On the other, hand when the geometrical structure of a compound is know, it is possible to predict some properties of the compound. Thus, the first step to understand the nature of the magnetizability of the Co nanoparticles and why this property is shape- and size-depended is to investigate the geometrical structure.

Let us remember that, there is an infinite number of possible surfaces which can be exposed for every crystal system. In practice, only a limited number of planes are found to exist in any significant amount. Thus, the attention was concentrated on the above surfaces, because it is possible to predict the ideal atomic arrangement for a given surface of a particular metal by considering how the bulk structure is intersected by the surface. It is necessary to remember that investigated nanoparticles consist of a small number of atoms, thus, it is not possible to obtain a very strict crystalline structure; the crystalline structure that is expected to be in the investigated Co nanoparticles was obtained on the basis of the symmetry of a bond and atom location in the planes.

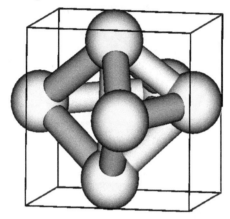

Fig. 1. Co_6 nanoparticle located in the cubic cell.

It is necessary to mention that the structure of the Co_4 particle was found, too. The results obtained indicate that Co_4 is planar and a nice equilateral is formed. In the case of Co_6 nanoparticle, we have the three-dimensional structure with a slightly disordered cubic symmetry. The structure was obtained after global optimization of the D_{4h} isomer of a Co_6 particle. It is important that each atom of the Co_6 nanoparticle is possible to approximately be located in the centre of the plane of the cubic cell (Fig. 1). The three surfaces are obtained. So, this nanoparticle is the element of a FCC structure.

It is possible to see two planes of the Co_8 nanoparticle (Fig. 2). The location of atoms on these planes as well as the symmetry of bonds allows us to predict that the element of FCC structure has been formed, too. This assumption is supported by following: i) each Co atom is four-fold coordinated; ii) the structural element of the Co_6 particle is obtained (see the structure that form atoms 5, 6, 7, 8 or 1, 2, 3, 4 in Fig. 2). So, the element of FCC structure has also been obtained in the Co_8 nanoparticle. The conformation of Co_8 nanoparticle has proved to be the most stable.

One of more interesting situations arises in case of Co_{10} nanoparticle. In this case, we have the two-dimensional disordered symmetry structure consisting of two planes and two atoms in the middle of each plane (Fig.3). The atoms mentioned join these planes. Roughly speaking, the structure of the Co_{10} nanoparticle is formed when the planes of the Co_8 nanoparticle are rotated in respect of each other when two Co atoms are added and a nice cubic structure is formed. This has also been confirmed by bond order investigations. On the other hand, each Co atom is four-fold-coordinated and a structural element of the Co_6 particle could also be foreseen. In case of Co_{12} and Co_{14} particles there are three planes where the location of atoms is as in the FCC structure: the atoms lie on the corners of the cube with additional atoms in the center of each of four cube of faces. The structure of Co_6 particle is also obtained. The element of BCC structure is also present because some atoms are out of the cube face. The most important for us is that the structure of a Co_6 particle was also obtained.

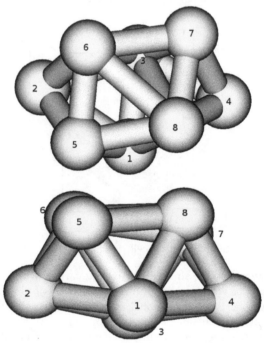

Fig. 2. The view of two planes of Co_8 particle from two different sides take places. T

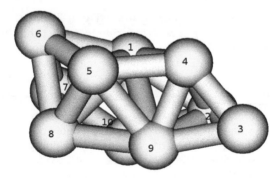

Fig. 3. The views of two formed by atoms 4,5,8,9 and 1,2, 7, 10 planes where atoms 3 and 6 is between the planes in Co_{10} particle.

Let us describe the structure of this Co$_{16}$ particle on the basis of the Co$_{14}$ particle. Firstly, it is necessary to mention that the additional atoms are joined with three-fold coordinated atoms placed above the centre of the cube face. The joining leads to the deformation of the cube because the above atoms of the Co$_{14}$ particle are pushed to the centre of the cube face. On the other hand, several structures of the Co$_6$ particle are possible to be seen. Hence, in this case, the deformed FCC structure also takes place, but the BCC structure element tends to disappear in the inner part of the particles although the exterior part the element remains unchanged. So, the tendency to form FCC is possible to predict. It allows us to speculate, that in large particles (particles with the diameter more than 10 Å) the main structure could be FCC, while in the external part the BCC structure could be present.

When considering the electronic properties of the above Co$_n$ particles, a singlet state is a ground one. The triplet state of these particles lies higher in total energy. These results disagree with the results presented by H.J. Fan and et al. It is necessary to mention that in the paper of H.J. Fan and et al. only high spin multiplicity particles were investigated applying Amsterdam density functional method with STO basis set with no report on how the geometry of the most stable compound was obtained. The calculated binding energies (per atom) of the Co nanoparticles, as a function of the number of these atoms in the particle, indicate that the Co$_{14}$ particle with the primitive cell of FCC structure is one of the most stable species among those presented in this section (Table 1). We also received, that Co$_6$ and Co$_{12}$ particles are more stable among the investigated by us particles that consist of less than 12 atoms and this result coincides with that presented by Q. M. Ma and et al. very well.

Number of atom	Binding energy per atom, eV	HOMO-LUMO gap, eV
6	0.45	1.47
8	0.20	1.32
10	0.36	1.18
12	0.48	1.53
14	0.69	1.64
16	0.78	1.24

Table 1. The dependence of calculated binding energy per atom and HOMO-LUMO gap on the atom number in the particle.

The above results confirm the investigation of HOMO-LUMO gap.

Hence, the main important observations on the geometrical structure of the pure Co nanoparticle are the following:

- The Co$_8$, Co$_{10}$, Co$_{12}$, Co$_{14}$, Co$_{16}$ particles consist of Co$_6$, thus this particle can be regarded to as the key element of the large Co nanoparticles.
- The face centered cubic structure which is slightly less close packed occurred in the Co$_{14}$ nanoparticle, while the other particles described are the elements of the FCC structure. When increasing the number of Co atoms in the particle, the atoms that are above the centre of the cube face are pushed both to the cube face centre and the inner part of the particle. Hence, in the inner part of the particle there is a FCC structure while the BCC structure element is obtained in the exterior part the particle. Thus, the obtained results allow us to speculate that in a large cobalt nanoparticle the FCC structure should be

clearly seen, while in a smaller one the FCC structure with the element of BCC structure should be obtained.

It is necessary to mention that the bond length and the bond order were also investigated. The obtained results are summarized in Table 2.

Atom number in a particle	Single bond length, Å	Double bond length, Å	Coordinated bond length
6	2.2	2.0	2.3
8	2.1 - 2.2	2.0	2.3
10	2.1 - 2.2		2.4
12	2.27		2.3
14	2.15- 2.27		2.3
16	2.15- 2.27		2.3

Table 2. The bond lengths obtained in the investigated particles

It is emphasized, that EXAFS MFT provides a Co-Co inter-atomic distance in the nanoparticle as equal to 2.561±0.015 Å. The comparison of the theoretical investigation of the Co particles with the corresponding experimental data is rather complicated quantitatively. The use of a restricted basis set of Co which can limit the quantitative analysis in the theoretical calculations should be taken into account. Hence, the obtained bond length fits well enough into the above-mentioned results. That obtained structures of the Co nanoparticle are to be mentioned as not fully spherical. The results obtained fit the results of the high magnification TEM image perfectly. In any case it is possible to see that double bonds are ruptured when the number of Co atoms is increased while coordinate bonds remain. The Co-Co bond elongation within the increasing atom number in the particles is not possible to explicitly be exhibited, while the presence of the coordinated bond allow us to foresee that the total electron density in this Co–Co bond is smaller than that in the other bonds what leads to non-compensation of the electron spins. On the other hand, the results exhibit that investigated Co nanoparticles are not homogeneous systems, i.e. the systems consist of different-fold-coordinated atoms. The obtained results indicate that a Co atom is three-to-seven-fold coordinated in the most stable nanoparticles. The presence of a coordinated bond, that is a kind of 2-centre, 2-electron covalent bond in which the two electrons derive from the same atom, prove the above results, too. Additionally performed analysis of the molecular orbital nature indicates that in the Co derivatives the number of bonding molecular orbitals, that may be occupied, is insufficient to locate all electrons of the system. It implies that some electrons are displaced on the anti-bonding orbitals, the energy of which is higher than that of the bonding orbitals. This electronic non-uniformity (a different oxidation state of an atom in the particle) of Co atoms, as we will prove below, and the electron displacement on the anti-bonding orbitals are important for the magnetic properties of the Co nanoparticles consisting of the even number of atoms.

3.1.1 Magnetic properties of the pure Co nanoparticles

Aiming to explain the results on magnetic properties of the particles investigated as well as particle dependence on the size, we paid our attention to the nature of molecular orbitals and their placements, because the studies on the Co_2O_m (m=0-7) compound indicate that both the increase of the number of oxygen atoms in the compound and the changeability of the oxidation state of the Co atoms led to the increase of the Co–Co bond length and weakening of the Co–Co bonds. The weakening of these bonds is important for the magnetic properties of these compounds. The results obtained indicate that the displacement of the two electrons on dz^2 orbitals of Co atoms creates Co–Co bonds. The energy of these orbitals is similar to that of other ones. Thus, the repulsion between the electrons on the dz^2 orbitals is larger than in other cases investigated, therefore these electrons tend to be as far as possible from each other and the correlation between them is weakened, resulting in the elongation of Co–Co bonds and, as a consequence, presence of an unpaired spin.

Let us remember that in the Co derivatives the number of bonding molecular orbitals that may be occupied is insufficient to locate all electrons of the system. As example, in Co$_6$ compounds all bonding orbitals are occupied and, as it has already been mentioned, some electrons are displaced on the anti-bonding orbitals, the energy of which is higher than that of the bonding orbitals.

It should be mentioned that the increased number of Co atoms in the compound leads to weakening of Co-Co bonds what, as we think, is important for the magnetic properties of these compounds, because magnetic properties depend on the bonds' nature and the number of bonds as well as on the charge distribution. Thus, aiming to explain the magnetic properties of the investigated particles, the attention is paid to the bonds' nature (what orbitals consist of bonds), the dipole moment and its components as well as on the isotropic g-tensor which depends on a spin angular moment.

In Table 3 the data on magnetizability, dipole moment, isotropic g tensor and the number of bonds consisting of anti-bonding orbitals are presented. The analysis of the most important orbitals (HOMO) of the described particles has been performed (Figs. 4-9). Fig.6 represents a full view of the HOMO of the Co$_6$ particle and the additional schematic presentation of the bond places in the particle is given to better illustrate the results presented.

Compound	Magnetizability, a.u.	Isotropic g-tensor	Dipole moment, a.u.	Number of bonds consist of anti-bonding character
Co$_6$	58.77	-0.513	0.097	5
Co$_8$	25.79	-0.289	0.468	3
Co$_{10}$	-26.13	-0.071	0.084	5
Co$_{12}$	-35.11	-0.038	0.213	20
Co$_{14}$	-39.27	-0.046	0.157	8
Co$_{16}$	69.84	-0.163	0.331	3

Table 3. The data on magnetizability, dipole moment, g tensor and the number of bonds consisting of anti-bonding orbitals.

Firstly, it is necessary to mention, that in the g-tensor of a molecule, there is a nuclear contribution and an electronic sum-over-states contribution. The electronic contribution represents an isotropic g-tensor. Here, it should be mentioned, that an isotropic g-tensor along with magnetizability is calculated to recognize what contributions (nuclear or electronic) are more important for the magnetic properties of the particle investigated.

It is possible to see that only Co_6, Co_8 and Co_{16} particles exhibit paramagnetic properties although the bonds that are of anti-bonding character are present in all the particles investigated. The different number of bonds formed of anti-bonding orbitals is present in the Co_6, Co_8 and Co_{16} particles. The view of the particles and location of the above bonds are presented in Figs. 4-9. The conclusion on the character of bonds was made on the basis of the analysis of the most important atomic orbitals on atoms, bond lengths, bond order and views of the orbitals.

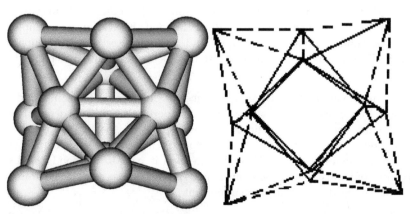

Fig. 4. The views of Co_{12} particle on the left and the view (on the right) when the bonds form of anti-bonding orbitals are marked by dash lines.

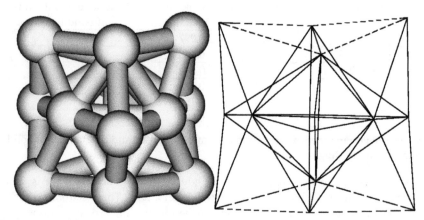

Fig. 5. The view of Co_{14} particle on the left and the view (on the right) when the bonds form of anti-bonding orbitals are marked by dash lines.

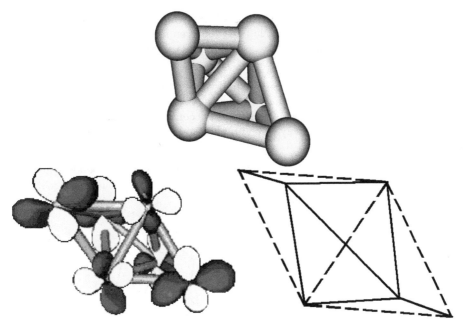

Fig. 6. The views of Co$_6$ particle (on the left) and their most important orbital (HOMO) (in the centre). The same view (on the right) is given when the bonds form of anti-bonding orbitals are marked by dash lines for simple guidance.

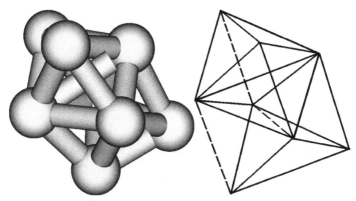

Fig. 7. the views of Co$_8$ particle (on the left) and the same view are given when the bonds consisting of anti-bonding orbitals are marked by dash lines.

Hence, it is possible to see that Co$_{12}$ and Co$_{14}$ are diamagnetic because in these particles there are 18 and 4 respectively symmetrically placed bonds with weakly interacting electrons what leads to the disappearance of non-compensate spins. These non-compensated spins quench each other what indicates the isotropic g-tensor value being equal to 0.038 and 0.046 in comparison to the value 2.00 for a free electron and indicates the absence of free electrons or a non-compensate spin.

In case of Co_{12}, the oxidation state of Co atoms is even. So, a non-compensate spin can not appear because the atoms of this particle loose the even number of electrons (below, it is exhibited that the oxidation state of atoms is also important to the explanation of Co particle magnetic properties).

The electronic properties of the Co_{14} particle fit described properties of the bond nature and oxidation state of the atoms very well. In the case the even number of bonds that are of anti-bonding character is found. Hence, electron spins are compensated and this particle exhibits diamagnetic properties. Additionally, even number (four) of atoms with oxidation state +3 are present

In case of Co_6, Co_8, Co_{10}, Co_{16} there are non - symmetrically placed bonds with weakly interacting electrons. Thus, we may suspect that these particles could be paramagnetic.

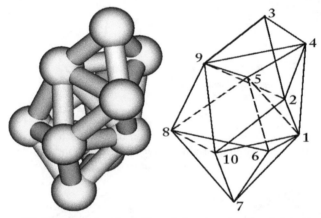

Fig. 8. The views of Co_{10} particle (on the left) and the view (on the right) when the bonds form of anti-bonding orbitals are marked by dash lines.

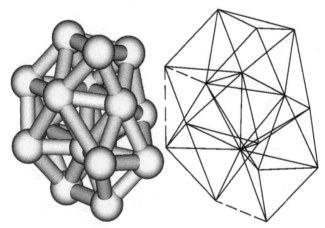

Fig. 9. The views of Co_{16} particle (on the left) and the view (on the right) when the bonds form of anti-bonding orbitals are marked by dash lines.

The magnetizability and g-tensor of the Co$_6$ particle are approximately twice larger than those of the Co$_8$ particle. In the Co$_6$ particle the number of bonds of anti-bonding character is five and these bonds are non-parallel. The dipole moment of the particle is approximately zero. It allows us to conclude that this particle is paramagnetic due to the electronic contribution, i.e. the repulsion between the electrons located on the anti-bonding orbital is large, therefore, they tend to be as far as possible from each other and become non-strongly correlated. Thus the spins of the electrons are not compensated, while the unparalleled displacement of the bonds leads to that that spins of all non-strongly-correlated-electrons are not compensated. It implies, that magnetic properties of the Co$_6$ particle are related with an electronic contribution.

A similar situation is obtained in case of Co$_8$ particle. Approximately twice smaller magnetizability of this particle than that of Co$_6$ is present because in the particle the number of non-strongly- correlated electrons is smaller than that in Co$_6$.

The largest magnetizability is the Co$_{16}$ particle, although, its isotropic g-tensor is approximately twice smaller than that of Co$_6$ particle. To explain the above mentioned contradictions, we investigated a dipole moment of these particles. The dipole moment indicates electron concentration places in a particle. On the other hand, the components of these dipole moments allow us to foresee the distribution of the above places. Both the concentration of electrons and their distribution helps us to find the appearance of the additional spins due to the different oxidation states of the Co atoms, i.e. if the even number of atoms loose the odd number of electrons and the particle possesses a dipole moment, we may suspect the presence of the localization of electrons and non-compensation of their spins. We named the above spin an ion one to simplify the discussion.

The dipole moment components of the particles are presented in Table 4 and indicate the electron charge delocalization in the Co$_8$ and Co$_{16}$ particles, while in case of the Co$_{10}$ and Co$_{12}$ particles, the charge localization occurs (see the component of dipole moment). It is necessary to add, that in Co$_8$ and Co$_{12}$ cases, the oxidation states of Co atoms are even. It allows us to predict, that electron spins occurring when the atoms loose an electron are compensated.

In case of Co$_{16}$ particle, the dipole moment components indicate charge delocalization, while the isotropic g- tensor value is smaller than that of Co$_6$ and Co$_8$. It allows us to conclude that the magnetic properties of this particle are mostly related with nuclear contribution. However, it is not explicitly possible to recognize the folding of atoms such as 3.49 or 3.51 on the results of these calculations. Thus, it is only speculation based on the comparison of the magnetizability of the investigated results that, in case of Co$_{16}$ particle, the ion spins and electron spins are not compensated.

The Co$_6$, Co$_8$ and Co$_{16}$ particles are paramagnetic, while Co$_{10}$, that possesses the odd number of anti-bonding character bonds as the particles mentioned, indicates diamagnetic properties. In case of Co$_{10}$ particle, the oxidation state of the Co1 atom is +5 (Fig. 8). The four bonds with anti-bonding character are displaced like in case of Co$_6$, however, one bond is in the same direction of the largest component of the dipole moment. Thus, it is possible to suspect, that in this case a weakly interacting electron spin is quenched by the ion spin. It may be concluded that paramagnetic behaviour is dominating when the uncompensated spin is present due to the presence of a weakly interacting electron on the anti-bonding orbital and this spin is not quenched by the ion spins.

Compounds	Dipole moment, a.u.	Dipole moment components, a.u.		
		x	y	z
Co_6	0.097	-0.09	-0.01	-0.01
Co_8	0.468	0.147	0.298	0.329
Co_{10}	0.084	0.076	0.031	0.021
Co_{12}	0.213	0.198	-0.051	-0.061
Co_{14}	0.157	-0.077	0.094	-0.098
Co_{16}	0.331	-0.107	0.167	-0.264

Table 4. Dipole moments and their components of the investigated particles that are paramagnetic or weakly diamagnetic.

It is possible to see that the investigated systems are very flexible and it is possible to predict that any dipole interaction or Co particle agglomeration could change their magnetic properties. To confirm the above prediction, the magnetic properties of the Co_6 and Co_6 as well as those of Co_6 and Co_{12} derivatives have also been investigated.

The structure of Co_6 and Co_6 particles was found after global optimization. The results obtained indicate possible agglomeration of these particles, i.e. the Co_{12} particle should formed. The magnetizability of this compound is -12.55 a.u., what indicates diamagnetic properties.

In case of the Co_6 and Co_{12} compound, we did not perform any geometry optimization to avoid agglomeration of particles because the changes of geometrical structure lead to dramatical changes of the electronic structure and consequential changes of magnetic properties. The investigated particles were placed randomly. Indeed, a compound consisting of Co_{12} and Co_6 particles is paramagnetic and its magnetizability is equal to 24.65 a.u. The results clearly indicate that dipole interaction and particle agglomeration change magnetic properties of the Co nanoparticle.

4. Structure, stability and magnetic properties of $Co_6 O_m$ nanoparticles

4.1 Structure and stability of Co_6O_m particles

As exhibited above, the stability of small Co nanoparticles is not very high. On the basis of our previous investigations it was speculated that those nanoparticles were non-rigid structures. It implies that the geometrical structure of these particles could change very quickly due to the tunnelling effect. Let us remember that Co_6 particle is found as the most stable one and it is the key element of other particles investigated. So, Co_6O_n (n=0-9) derivatives were also investigated to establish how the magnetic properties of the Co particles may change due to oxidation. On the other hand, these investigations allow us to foresee the conditions under which the metal Co-Co bond is broken. It is also proved our prediction that the particles consisting of the even number of atoms possess magnetic properties due to the weakly interacting electrons on the anti-bonding orbital and this spin is not quenched by additional spins that occurred because some atoms of the nanoparticle loose the odd number of electrons.

The most stable structures of the Co_6O_n derivatives are presented in Fig.10.

Firstly, it is necessary to mention that oxygen stabilizes the Co nanoparticle and the increasing number of oxygen atoms increases the binding energy per atom up to n=7 (Table 5). Furthermore, when a certain limit is reached, oxygen atoms do not influence the stability of the Co_6O_n particles.

The Co_6O_{12} particle was investigated too. The binding energy per atom of this particle is equal to 3.26 eV what is similar to that of Co_6O_n (n=7, 8, 9). The difference of the binding energy of the above particles is too small (0.2 eV or less) to make the conclusion on the most stable particle.

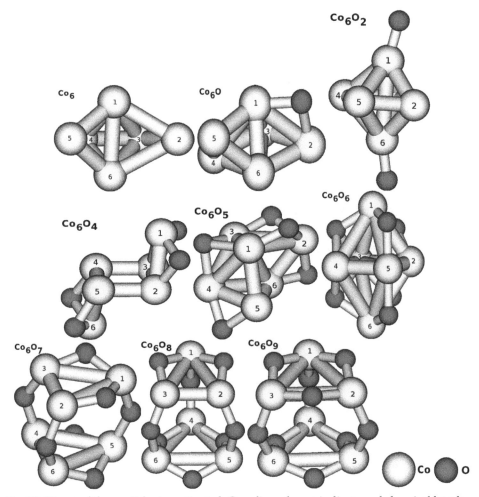

Fig. 10. Views of the particles investigated. Grey lines do not indicate real chemical bonds, but are implemented for the sake simple guidance.

Particle	Co_6	Co_6O	Co_6O_2	Co_6O_3	Co_6O_4	Co_6O_5	Co_6O_6	Co_6O_7	Co_6O_8	Co_6O_9
Binding energy per atom, eV	0.45	0.93	1.21	2.22	2.49	2.85	3.22	3.43	3.33	3.48

Table 5. Binding energy per atoms for the Co_6O_n (n=0-9) particles

The difference of the binding energy per atom for Co_6 and Co_6O is equal to 0.48 eV, while that between Co_6O_6 and Co_6O_7 is only 0.21 eV, i.e. twice less. On the other hand, the changing of the number of oxygen atoms from 2 to 3 leads to the largest increase of the binding energy per atom (1.01eV), while the binding energy per atom increase only up to 0.27 eV when the oxygen atom number in a particle increases from 3 to 4. Thus, the results of our investigations allow us to foresee that starting with n=6 (n is the number of oxygen atoms) the further increase of the number of oxygen atoms will not influence the stability of these particles very strongly and the main structure (the key-element) is not considerably changed (Fig. 10). The binding energy per atom of the Co_6O_6, Co_6O_7, Co_6O_8 and Co_6O_9 is approximately equal and proves these particles to be the most stable. These results coincide with the experimental measurements that indicate the presence of CoO and Co_3O_4; CoO_2, Co_2O_3 and Co_6O_7 particles should also be found among them what was proved by the results we obtained.

Such a changeability of the binding energy per atom in some cases could be explained by changes in geometrical structure of Co particle. In case when the additional oxygen atom does not significantly increase the binding energy per atom, the main part of the energy of this atom is used to deform the structure of the key element (Co_6). Thus, the binding energies per atom of Co_6O_3 and Co_6O_4 or Co_6O_6, and Co_6O_7 are approximately equal.

The key element of the Co_6 is also present in the Co_6O_n (n=0-9) derivatives. However, this key element is slightly deformed. The changeability of the initial form is oxygen atom depended. The largest deformation is obtained in Co_6O_7, when the distance between the planes (formed of atoms 1, 2, 3 and of 4, 5, 6) is increased and one plane is rotated in respect of the other one by angle of $\pi/4$. Actually, one more structure of the Co_6O_7 which looks like Co_6O_6 was also obtained, but the energy of this formation of the particle is 1.23 eV higher than that of the particle, the structure of which was described above.

In the Co_6O_4 particle the key element (Co_6) is deformed twice: 1. firstly, when the distances between the atoms Co2-Co5 decrease; 2. Secondly, when Co1 and Co6 positions in respect of the plane that is formed of atoms 2,3,4,5 is changed. Here, it should be emphasized, that this structure of the particle has been obtained after global geometry optimization starting with several completely different initial geometries. Thus, the geometrical structure of the Co_6O_4 particle is confirmed.

Hence, the largest deformations of the Co_6 particle are obtained when the number of oxygen atoms is changed from 3 to 4 and from 6 to 7. In these cases the stabilization energy per atom is smaller than in other cases investigated. Thus, the main part of Oxygen energy is used to deform the key structure of Co_6.

It is necessary to mention, when the number of oxygen atom is 2 and 6, the structure of the Co_6O_m particle looks like the octahedron, while in case of odd numbers of oxygen the octahedron form is strongly deformed (except the results for Co_6O_4). It is interesting to note that the most stable structure of Co_6O_8 (prototype of Co_3O_4) has a deformed spinel structure. Thus, it is not surprising that a large effective magnetic moment estimated from the inverse susceptibility has not been explained properly.

Compound	Co-Co bond length, Å								
	1-2	1-3	1-4	1-6	2-3	2-6	4-5	4-6	3-4
Co_6	2.15	2.33	2.15	2.24	2.04	2.23	2.04	2.31	2.33
Co_6O	2.54								
Co_6O_2				3.01					
Co_6O_3				2.18	2.37	2.33	2.14		
Co_6O_4	2.27	2.27	2.27	4.72				2.27	
Co_6O_5	2.25				2.61	2.25			2.39
Co_6O_6	2.83	2.67	2.89		2.32	2.87			2.44
Co_6O_7	2.93			3.11	2.93		2.93	2.93	

	1-2	1-3	1-4	2-3	2-5	3.6	4-5	4-6	5-6
Co_6O_7	2.92	2.93	3.21	2.21*	3.07	3.11	2.92	2.91	2.22*
Co_6O_8	2.88	2.90	3.21	2.24*	3.18	3.16	3.03	3.03	2.93
Co_6O_9	3.04	3.04	3.15	3.04	3.14	3.15	3.04	3.05	3.05

* the Co-Co bond is present.

Table 6. The distance between the Co atoms which are connected with the O atom.

According to the results of our investigations, the Co-Co bond length of the single bond is longer (2.2 Å) than the bond length of a double bond (2.0 Å) in a Co_6 particle. On the other hand, three bonds were obtained where the length is equal to 2.3 Å. The bond order of the largest bond is twice smaller than that of a single bond. Here, we the commonly observed that the Co-Co bond lengths are marginally changed only between the atoms that are connected with the oxygen atom (Table 6) and, as a consequence, the bond enlargement leads to Co-Co bond dissolving. For example: in the Co_6 particle the bond order between Co1-Co5 is equal to 1.018, while that in Co_6O_4 is approximately twice smaller and equals to 0.55. Additionally, the two, one and zero Co-Co bonds are respectively found in the Co_6O_7, Co_6O_8 and Co_6O_9 nanoparticles. To shed some light on the present observation, the analyzes of the most important orbitals of the Co_6 particles have been investigated. HOMO (the highest occupied orbital)- LUMO (the lowest unoccupied orbital) gap dependence on the number of oxygen atoms is represented in Fig. 11. The HOMO-LUMO gap indicates that chemical stability of Co_6, Co_6O_4 and Co_6O_6 is very low, i.e. they tend to form new chemical bonds. These results coincide well with the results of binding energy per atoms.

However, the electronic structure of the investigated particles is quite different because the oxidation state of Co atoms exchanges when the number of oxygen atoms in the particle is increased. For example, in Co_6O particle oxidation state of Co atoms is +3 and +4; in the Co_6O_2 particle the oxidation state of these atoms is +4 and +5 and in Co_6O_4 it is +1 and +3. We have not observed any relationship between number of oxygen atom in the particle and the oxidation state of Co atoms.

Let us remember that in the Co derivatives the number of bonding molecular orbitals, that may be occupied, is insufficient to locate all the electrons of the system. This leads to the presence of electrons on the anti-bonding orbital and, as a consequence to, the dissolution of Co-Co bonds.

On the other hand, the electronic configuration of cobalt for the ground state neutral gaseous atom is $[Ar].3d^7.4\ s^2$, while that of oxygen is $[He].2s^2.2p^4$. The configuration, associated with Cobalt in its compounds, is not necessarily the same, but it could be used to explain formally obtained results.

As it was mentioned above in Co_6 compounds some electrons are displaced on the anti-bonding orbitals, the energy of which is higher than that of the bonding orbitals. Therefore, the stability of the pure cobalt nanoparticle is low. When the Co_6 nanoparticle is joined to one or two oxygen atoms, the number of electrons that occupy anti-bonding orbitals, decreases because these electrons occupy the oxygen orbitals (Fig.12)

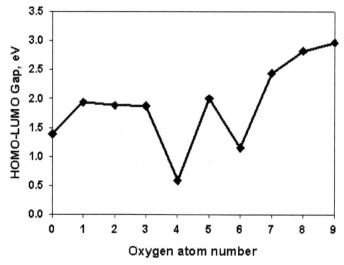

Fig. 11. The HOMO-LUMO gap of the Co_6O_n (m = 1 - 9)

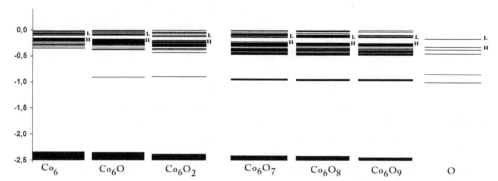

Fig. 12. Displacement of orbitals of several Co_6O_m (m=0, 1, 2, 7, 8, 9) and oxygen atoms in respect of each other. Here, H and L indicate HOMO and LUMO respectively. Additionally, the ground state (triplet) of oxygen atoms are calculated. It is possible to see that with the

increasing number of the oxygen atoms in the particle, the number of occupied orbital also increases. i.e. the number of bonds of anti-boding nature decrease.

Oxygen atoms in the Co$_6$O$_4$, Co$_6$O$_3$, Co$_6$O$_2$, and Co$_6$O particles are joined to atoms between which the anti-bonding orbitals occur. Having in mind that the joining of oxygen atoms leads to the increase of the bond length and dissolution of Co-Co bonds, what confirms the above mentioned prediction. In case of the Co$_6$O$_3$ particle, one O atom is joined to Co4-Co5 atoms (Fig.10). The anti-bonding nature of the bonds has not been observed between those atoms. In this case, a steric effect is more preferable because other positions of the oxygen atom should complicate Co1-Co2 and Co2-Co6 elongation or leads to the destruction of this particle. Hence, oxygen atoms stabilize Co$_6$ particles due to dissolving of Co-Co bonds that possess anti-bonding character.

It is very well known, that a semiconductor must have at least two characteristics: 1. the bonding and anti-bonding orbitals must form a delocalized band; 2. the HOMO-LUMO gap in molecular species should be generally of the order of 0.5eV to 3.5 eV. HOMO-LUMO gaps of the investigated derivatives belong to the above range. However, the number of anti-bonding orbitals decreases with increasing of the number of oxygen atoms. The results allow us to predict that Co$_6$O$_m$ are semiconductors but the particles should loose their semiconductor properties if the number of oxygen increases.

4.1.1 Magnetic properties of Co$_6$O$_m$ particles

In the above chapter we proved that oxygen atoms stabilize cobalt nanoparticles, although, in some cases, the structure of particles changes insufficiently while the electronic structure is dramatically changed because the increasing number of oxygen atoms decreases the difference between the number of electrons and the number of atomic orbitals that they may occupy. Hence, the bonds of anti-bonding nature as well as uncorrelated spins disappear. So, we may suspect, that all Co oxide particles could be diamagnetics. Let us analyze the results presented in Table 7.

Compound	Magnetizability, a.u.	Isotropic g tenzor a.u	Co-Co bond number
Co$_6$	58,77	-0,51	5
Co$_6$O	-11,24	-0,14	6
Co$_6$O$_2$	-15,47	-0,11	6
Co$_6$O$_3$	-3,26	-0,16	5
Co$_6$O$_4$	-9,62	-0,12	4
Co$_6$O$_5$	-15,97	-0,11	2
Co$_6$O$_6$	-2,11	-0,16	1
Co$_6$O$_7$	-25,94	-0,07	2
Co$_6$O$_8$	25,42	-0,18	1
Co$_6$O$_9$	-24,28	-0,05	0
Co$_6$O$_{12}$	-24,25	-0,07	0

Table 7. Data on magnetizability, isotropic g tensor, and Co-Co bonds number and the number of Co-Co bonds that was found based on the electron density investigation results. Only those bonds with unpaired spin electrons are mentioned.

It is obvious to see, that the particles with odd number of Co-Co bonds are paramagnetic or lightly diamagnetic. On the other hand, the isotropic g-tensor value of the cobalt oxide particles is not large, thus we may suspect that a ion spin in these cases is very important.

It is necessary to mention that based on the results described above, we may divide the described particles into the following groups:

1. The particles that posses shape of Co_6: Co_6, Co_6O, Co_6O_2, Co_6O_3, Co_6O_5, Co_6O_6 (A group)
2. The particles Co_6O_7, C_6O_8, Co_6O_9 in which the distance between the planes (formed of atoms 1, 2, 3 and of 4, 5, 6) is increased and one plane is rotated in respect the other one by the $\pi/4$ angle (B group).
3. The rest (Co_6O_4)

It has to be pointed out, that a lot of reports concluded that magnetic properties of the nanoparticles depend on their shape. So, we suspected that magnetizability of the particles belonging to one group should be the same. However, the results of our investigations do not prove the above prediction (Table 7).

According to our investigations, the Co_6 nanoparticle is a strong paramagnetic, while other particles, belonging to group A, are diamagnetics. The same phenomenon is obtained in case of B group. In this case, the Co_6O_8 particle is paramagnetic, while other particles are diamagnetics. Moreover, the diamagnetic properties of the similarly shaped particles are quite the same only in the following cases: Co_6O_7, Co_6O_{12}, Co_6O_9; Co_6O_3, Co_6O_6; Co_6O_2, Co_6O_5. It implies that the shape of the particle has no influence on the magnetic properties of the nanoparticles. To confirm this conclusion, we have calculated magnetizability of several isomers of Co_6O_8 particles (Fig. 13). It is possible to see, that the shapes of isomers II and III are similar, but the shape of isomer I differs. However, the magnetizability of isomers II and I with different shapes is approximately alike, while the magnetizability of isomer III is smaller than that of isomer II with the same shape (Table 8).

Isomers	I	II	III
Magnetizability, a.u.	25.42	24.76	14.24

Table 8. The Magnetizability of different isomers of Co_6O_8 particle.

Hence, the magnetic properties of these particles does not depend on their shape.

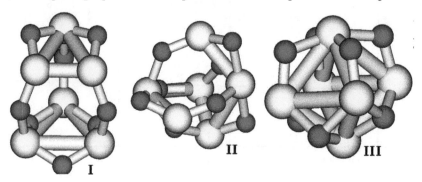

Fig. 13. The view of several isomers of Co_6O_8.

Let us remember, that the nanoparticles could be paramagnetic due to several reasons: 1) the unpaired electron location on the Co-Co bonds; 2) the small total electron charge density between Co atoms which appears due to overlapping of p orbitals of oxygen atoms; 3) the significant contribution of atoms that loose odd number of electrons. The second reason mentioned could not be realized in case of the Co$_6$O$_n$ particles due to their relatively large size and small number of oxygen atoms. The first and third reasons could be realized thus supporting the previously found results. It is necessary to mention, that non-compensation of spin for CoO/SiO2 multilayers was also observed.

Now, we shall describe the particles of group B in detail. Firstly, it is necessary to mention, that the particles of this group have the different number of Co-Co bonds: Co$_6$O$_7$, has two, Co$_6$O$_8$ has one, and Co$_6$O$_9$ has zero. Only Co$_6$O$_8$ particle exhibits paramagnetic properties.

Let us remember that in the Co derivatives the number of bonding molecular orbitals, that may be occupied, is insufficient to locate all the electrons of the system. This causes the presence of electrons on the anti-bonding orbital and, as a consequence, a weaker correlation of these electrons. Similar states are obtained in biradicals where the number of atomic orbitals, that may be occupied, is smaller than that of electrons. That leads to the appearance of electrons on the anti-bonding orbitals and serves predicts a large orbital contribution to the magnetic moment of a small Co$_m$O$_n$ particle (Fig.14).

Fig. 14. HOMO orbital antibonding character of the Co$_6$O$_8$ particle.

It implies, that a non-compensate electron spin should be obtained. This situation is realized in the Co$_6$O$_7$ and Co$_6$O$_8$ particles. However, in the Co$_6$O$_7$ particle two pairs of weakly correlated electrons are present what leads to the disappearance of non-compensate spins. This is indicated by the isotropic g-tensor value which equals to 0.007. However, in case of the Co$_6$O$_8$ particle, only one Co-Co bond is present and only one pair of weakly correlated electrons should be found. This weak correlation indicates the nature of HOMO orbital that consists of anti-bonding dz^2 type orbitals (Fig.14). Hence, the total spin of electrons is not compensated and, as a consequence, the particle exhibits paramagnetic features. This presumption is also confirmed by the isotropic g-tensor value, that is one of the largest between the particles described (Table 7). The small value of the isotropic g-tensor indicates that the electronic contribution to the magnetic properties of the particle is not very large, but it is essential.

Additionally, we may supplement the proposition, that Co_6O_n particles should be paramagnetics when the number of Co-Co bonds on which the unpaired electrons are located might be odd (Table 7). It should be emphasized, that the number of Co-Co bonds was found on the basis of the electron density investigation results and only the bonds, where unpaired spin electrons could be presented, are mentioned. Indeed, the investigated particles with the odd number of Co-Co bonds exhibit paramagnetic or weak diamagnetic properties. However, it is not clear why the magnetic properties are different, i.e., formally, some, different features should appear.

Aiming to explain the above mentioned discrepancy, we investigated a dipole moment of these particles. The dipole moment indicates electron concentration places in the particle. On the other hand, the components of these dipole moments allow us to foresee the distribution of the above places. Both the concentration of electrons and their distribution helps us find additional spins that appeared due to the different oxidation state of the Co atoms (formally, we call the above spin as an ion one). The exception concerns Co_6.

The components of the dipole moment of the particles that are paramagnetics or weak diamagnetics are shown in Table 9.

Compounds	Dipole moment, a.u.	Dipole moment components, a.u.		
		x	y	z
Co_6	0.096	-0.09	-0.01	-0.01
Co_6O_3	1.689	0.55	0.23	-1.58
Co_6O_6	1.639	-1.06	-1.16	0.44
Co_6O_8 (I isomer)	2.652	2.60	0.45	0.23
Co_6O_8 (II isomer)	2.059	-1.08	-0.01	-1.75
Co_6O_8 (III isomer)	1.372	1.37	-0.06	-0.03

Table 9. Dipole moments and their components of the investigated paramagnetic or weak diamagnetical particles.

The Co_6O_8 particle is a paramagnetic due to the presence of non-compensate spin what indicates the value of the isotropic g-tensor of 0, 51 (a free electron g-value is 2.00) because of the appearance of electrons on the anti-bonding orbitals.

So, as it was mentioned, the following different types of magnetic interactions could be obtained in the Co_6O_m nanoparticles: 1. an uncompensated spin of weakly interacting electrons on the anti-bonding orbital; 2) the presence of Co ions that looses the odd number of electrons (Co^{+3} and the like) leads to the emergence of the additional non-compensated spin.

The results obtained exhibit that the magnetic properties of nanoparticles could depend on the above interactions. The paramagnetic behaviour dominates when the non-compensated spin is present due to weakly interacting electrons on the anti-bonding orbital and this spin is not quenched by the ion spins. Let us remember, that Co_6O_3 and Co_6O_6 particles are weak diamagnetics, thought the isotropic g-tensor is not smaller than that of the Co_6O_8 particle. In these particles the ion spin is also presented what indicates a high dipole moment. The number of the Co^{+3} ions is 2 and 4 respectively in the Co$_6$O$_3$ and Co$_6$O$_6$ particles. However, the components of the dipole moment indicate that the ion spins are delocalized. The interaction between these spins leads to the quench of an electron spin, i.e. both spins (ion and non-compensated spin of electrons located on the anti-bonding orbital of Co-Co bond) are oriented so that the total spin equals to zero.

The opposite situation is realized in the Co$_6$O$_8$ particle: an ion spin is localized and one Co-Co bond is present. In this case, the spins are oriented so that they are relatively parallel to each other. This prediction is supported by additional investigations of the isomers of the Co$_6$O$_8$ particle. It is necessary to mention, that one Co-Co bond is present in isomer II and a detailed investigation of the dipole moment indicates that it lies approximately in parallel to the Co-Co bonds. Therefore, the unpaired spins of a different nature support each other. Thus, the magnetizability of the I and II isomers of the Co$_6$O$_8$ particle is the same. In case of isomer III, all Co-Co bonds are dissolved, but an ion non-compensated spin is present. It implies that magnetic properties of the particle are determined by the localized ion spin only. Thus, the magnetizability of isomer III is lower than that of the other isomers investigated.

Hence, the paramagnetic behaviour of the cobalt oxide particle is dominating when the non-compensated spin is present due to weakly interacting electrons on the anti-bonding orbital and this spin is not quenched by the ion spins.

It is necessary to pay attention to other important observations. As it was earlier mentioned, the cobalt oxide particles are semiconductors and Co$_6$O$_8$ exhibits magnetic properties. It implies that this Co$_6$O$_8$ particle could be magnetic superconductor and could be implemented in electronic devices to provide a new type of the control of conduction, i.e. of the charge carrier and quantum spin state. Hence, this particle could be used in quantum computing.

5. Absorption spectra of the C$_6$O$_m$ (m=1-9) nanoparticles

It is known that the growing metallic particles are stabilized by the absorption of the polymer chains on the surface of the growing metal fragments, lowering their surface energy and creating a barrier to further aggregation. On the other hand, the organic coating of a particle prevents the surface from oxidation, rendering the particle stable over a long period. So, it is necessary to have a tool to investigate and control the process of stabilization of nanoparticles because the stabilization could be related to the oxidation of a metal particle and, as a consequence, it looses its magnetic properties. We believe that the knowledge concerning the shape and the nature of the absorption spectra of the Co$_6$O$_m$ particles in the Vis and UV region could be a good tool for the investigation of the oxidation processes of Co nanoparticles. The above assumption is based on the results obtained that indicate the Co$_6$ particle as a key element of larger particles.

As it was to mention in 3.1.1 we may divide the described particles into groups based on the changes of the Co$_6$ structures. The groups are the following:

1. The particles that posses a shape of Co$_6$: Co$_6$, Co$_6$O, Co$_6$O$_2$, Co$_6$O$_3$, Co$_6$O$_5$, Co$_6$O$_6$ (group A).
2. The particles in which the distance between the planes (formed by atoms 1, 2, 3 and of 4, 5, and 6) is increased and one plane is rotated in respect of the other one by $\pi/4$ angle: Co$_6$O$_7$, C$_6$O$_8$, Co$_6$O$_9$ (group B).

3.   The rest (Co_6O_4).

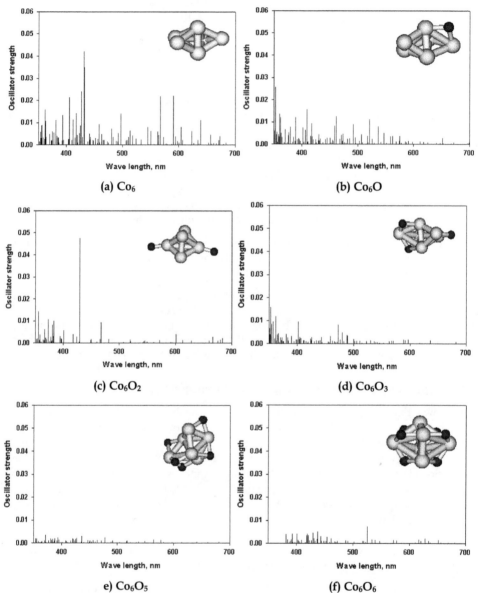

Fig. 15. Absorption spectra of group A. At (**a**) there are spectra of a Co_6 particle, while
(**b**) - those of Co_6O; (**c**) in the middle, on the left, there is a spectrum of Co_6O_2;
(**d**) on the right – Co_6O_3; (**e**) at the bottom, on the left, there are spectra of Co_6O_5;
(**f**) on the right – those of Co_6O_6. The black circle indicates oxygen atoms,
while the grey one – cobalt atoms.

The obtained absorption spectra of the particles making up groups A and B and their structures are presented in Fig. 15 and Fig. 16. The case of Co$_6$O$_4$ particle is different and should be investigated deeper, although the general tendency of absorption spectra changes described below are possible to foresee in the spectra of this particle, too.

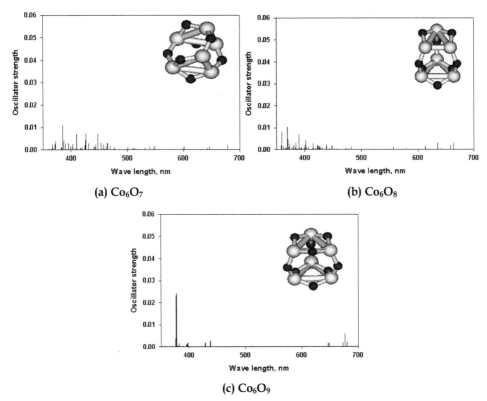

(a) Co$_6$O$_7$ (b) Co$_6$O$_8$

(c) Co$_6$O$_9$

Fig. 16. Absorption spectra of group B: **(a)** at the top on left there is a spectrum of Co$_6$O$_7$ particle, **(b)** on the right -those of Co$_6$O$_8$; **(c)** at the bottom, there are spectra of Co$_6$O$_9$. The black circle indicates oxygen atoms, while the grey one – those of cobalt.

Let us analyze the spectra of group A. It is obvious, that the intensity of absorption decreases especially in the [500;700] nm region with increasing of the oxygen number up till 5, and starts increasing again when the number of oxygen atoms is 6. The appearance of more intense absorption in the above region of the Co$_6$O$_6$ is related to the structure of this particle (Fig.15). The structure of the Co$_6$O$_6$ particle looks like the octahedron, while in the case of other particles investigated, the octahedron form is strongly deformed. The intensity of absorption in the region 300 to 400 nm increases when the number of oxygen atoms in the particle increases from 7 to 9 (Fig.16).

It is obvious, that with the increasing of the number of the oxygen atoms by one, the number of occupied orbitals in the [-1;0] a.u increase by three (Table 10). Moreover, the gap of the Co$_6$ particle between occupied orbitals in the]-2; -1[a.u. region is not filled what is explained by the displacement of the orbitals of both Co_6O_m (m=0-9) and an oxygen atom in respect of each other (Fig. 12). In case of Co_6 particle, only three orbitals (HOMO, HOMO -1 and LUMO) of oxygen interact with the occupied orbitals of the particle, while in case of Co_6O_m particles, the number of interacting orbitals increases. Starting with Co_6O, the additional occupied level occurred in the gap of the Co_6 particle between the occupied orbitals in the region of]-2; -1[a.u. However, the HOMO-LUMO gap increases. So, semiconductor properties of the Co_6O_m particles become stronger.

Naturally, that with the increasing number of oxygen atoms in the Co_6 particle, the mixing orbital (the molecular orbital consists of cobalt and oxygen atomic orbitals) increases due to the Co and O atomic orbital interaction. The analysis of the contribution of the atomic orbital to the molecular orbitals confirms the predicted interaction. Moreover, due to the above interaction, the orbital splits and several orbitals that are occupied in Co or Co oxide nanoparticles should become virtual and vice versa. Hence, the transitions in the spectra region of [350;700] nm are of Co3d → Co3d type and they are allowed in a pure Co particle or particles with the oxygen number of 1-2 because the above mixing is not very strong.

When the number of oxygen atoms in the Co particle is 3-7, the transitions in the spectra region of [350;700] nm are of Co3d→Co3dO2p orCo3dO2p→Co3d types. It is emphasized, that starting with the number of six of oxygen atoms, only occupied orbitals of nanoparticles interact with the occupied orbital of oxygen atom, i.e. the above mentioned interaction between LUMO of the oxygen atom and the occupied orbital of Co_6O_m nanoparticle does not occur. The analysis of the most important orbitals for excitation indicates, that in the spectra of Co_6O_m (m=0-5) the most intensive excitations correspond to Co3d→Co3d ones. Other partly allowed excitations correspond to Co3d→Co3dO2p ones. So, the number of Co3d→Co3dO2p excitations increases with the increased mixture of orbitals. Moreover, when the number of oxygen atoms is up to 7, the Co3dO2p→Co3dO2p excitations are more relevant. On the other hand, the symmetry of particles is different what leads to different number of the transitions allowed. It is very well known, that a part of the possible excitations is forbidden when the symmetry group of the particles is high, while all possible excitations are allowed when the symmetry group of the particle is the lowest (C_1).

Particles	Virtual orbitals	Occupied orbitals		
	[-1;0] a.u.	[-1;-0] a.u.	[-2;-3] a. u.	[-3;-4] a. u.
Co_6	11	27	18	6
Co_6O	13	30	18	6
Co_6O_2	13	33	15	5
Co_6O_3	12	36	17	6
Co_6O_4	15	39	14	6
Co_6O_5	13	42	18	4
Co_6O_6	13	45	18	6
Co_6O_7	15	48	18	6
Co_6O_8	15	51	18	6
Co_6O_9	16	54	18	6

Table 10. The number of states of the Co_6O_m (m=0-9) particles in the regions of different energy.

Particle	Co_6	Co_6O	Co_6O_2	Co_6O_3	Co_6O_5	Co_6O_6	Co_6O_7	Co_6O_8	Co_6O_9
Symmetry group	C_{2v}	C_1	C_{2v}	C_1	C_1	C_{2v}	C_{2v}	C_{1h}	C_{3v}

Table 11. Approximate symmetry of the particles investigated.

Hence, the particles with higher symmetry absorb certain wave lengths more intensively, while the absorbance of non-symmetrical particles is not intensive, but a very broad one (Table 11, Figs. 15, 16). It allows us to conclude, that the investigated spectra of the Co nanoparticles in the region of [300; 700] nm could explain the oxidation of the particles and, as a consequence, their structure changes what lead to changes of magnetic properties.

Basing on the results obtained, we speculate that the dependence on the place of excitation could be related with the particle oxidation when considering the excitation of large particles (approximately of 200 nm).

6. Conclusions

Herein, we report on the several important results related to magnetic properties of the Co nanoparticle.

The main important observations of the pure Co and oxidized nanoparticle are the following:

- The Co_8, Co_{10}, Co_{12}, Co_{14}, Co_{16} particles consist of Co_6, thus these particles could be regarded to as the key element of the large Co nanoparticles.
- The face centered cubic structure which is slightly less closely packed, occurred in the Co_{14} and Co_{16} nanoparticles, while the other particles described are the elements of the FCC structure in the sense of the above conclusions.
- The key element of the Co_6 is present in the Co_6O_n (n=0-9, 12) particles.
- The present investigations of the magnetic properties of Co and Co oxide particles resulted in the conclusion that a paramagnetic behaviour is dominating when the non-compensated spin is present due to the anti-bonding orbitals and such a spin is not quenched by the ion spins.
- The results of our investigations indicate that both a dipole interaction and particle agglomeration change magnetic properties of the Co nanoparticle.
- The intensity of absorption of Co_6O_m (m=0-9) particles should be decreased in the [500;700] nm region with increasing of the number of the oxygen atom up to 5, and should be increased again when number of oxygen atoms is 6.
- The spectra of investigated particles become linear when the number of oxygen atoms in the above particle is even, while the absorption lines in spectra should be difficult to observe with odd number of oxygen.
- It is obtained, that in the spectra of Co_6O_m (m=0-3) the most intensive excitations correspond to Co3d→Co3d excitations. The Co3dO2p→Co3dO2p excitations are more relevant in the spectra of the particles where the number of oxygen atoms is up to 7, while in the rest particles the Co3d→Co3dO2p or Co3dO2p→Co3d types of excitation are obtained.

7. References

Zalich M. A. and et al. (2006). Structural and Magnetic properties of oxidatively stable cobalt nanoparticles encapsulated in graphite shell. *Chemistry of Materials,* Vol. 18,p.p. 2648-2655, ISSN 0897-4756

Simeonidis K. and et al. (2008). Shape and composition oriented synthesis of Cobalt nanoparticles, *Physics and Advanced materials Winter school,* pp.1-8, Thessaloniki, Greece, January 14-18

Neamtu J and et al. (2005). Synthesis and Properties of Magnetic Nanoparticles with Potential Applications in Cancer Diagnostic. *Technical Proceedings of the 2005 NSTI Nanotechnology Conference and Trade Show,* Vol. 1, p.p. :222 - 224, ISBN:0-9767985-0-6, Anaheim, California, USA,May 8-12, 2005.

Sakurai M.I. and et al. (1998) Magic Numbers in Fe Clusters Produced by Laser Vaporization Source. *Journal of the Physical Society of Japan,* Vol. 67, No. 8, (August), pp. 2571-2573, ISSN 0031-9015

Gambardella P. and et al. (2003) Giant Magnetic Anisotropy of Single Cobalt Atoms and Nanoparticles. Science, Vol.300, No. 1130, (May), pp. *1130-1133, ISSN 0036-8075*

Chen J. P. and et al. (1994) Magnetic properties of nanophase cobalt particles synthesized in inversed micelles. *Journal of Applied Physics,* Vol. 76, No. 10, (November), pp. 6316 - 6318 ISSN 021-8979

Ghosh M.; Sampathkumaran E. V.; Rao C.N.R. (2005) Synthesis and Magnetic Properties of CoO Nanoparticles. *Chemistry Materials,* Vol. 17, No. 9, (March), pp.2348 -2352 ISSN 0897-4756

Zhao Y. W. and et al.(2003) A simple method to prepare Uniform Co nanoparticles. *IEEE Transactions on magnetics,* Vol.39, No. 5, (September) , pp. 2764-2766, ISSN 0018-9464

Sun X. Ch.; Reyes-Gasga J.; Dong X. L. (2002) Formation and microstructure of Carbo encapsulated superparamagnetic Co nanoparticles. *Molecular Physics,* Vol. 100, No 19, (March), pp. 314-315, *ISSN 0026-8976*

Ram S. (2001) Allotropic phase transformations in HCP, FCC and BCC metastable structures in Co- nanoparticles. *Material Science and Engineering A,* Vol. 304-306, (May), pp. 923-927, *ISSN 0921-5093*

Gubin S.P. and et al. (2003) Magnetic and structural properties of Co nanoparticles in a polymeric matrix. *Journal of Magnetism and Magnetic Materials,Vol.* 265, No. 2, (September) pp. 234-242 *ISSN 0304-8853*

Tsukamoto S.; Koguchi N. (2000) Magic numbers in Ga clusters on GaAs (0 0 1) surface. *Journal of Crystal Growth,* Vol. 209, No.1-2, (February), pp. 258-262, *ISSN 0022-0248*

Guevara J.and et al. (1997) Electronic properties of transition-metal clusters:Consideration of the spillover in a bulk parametrization. *Physical Review B,* Vol. 55, No. 19, (May), pp.13283-13287 *ISSN 1098-0121*

Ma Q. M.and et al. (2006) Structures, stabilities and magnetic properties of small Co clusters. *Physics Letter A,* Vol. 358, No. 4, (October), pp. 289-296, *ISSN 0375-9601*

Galvez N. and et al. (2002). Apoferritin-encapsulated Ni and Co superparamagnetic nanoparticles. *Erophysics. Letter ,* Vol. 76, No.5, (December), pp.142-148, *ISSN 0295-5075*

Park I.-W. and et al. (2003). Magnetic properties and microstructure of cobalt nanoparticles in a polymer film. *Solid State Communications,* Vol. 44, No.7 (May), pp. 385-389 *ISSN* 0038-1098

Yang H.T.and et al. (2004). Synthesis and magnetic properties of e-Co nanoparticles. *Surface and interface analysis,* Vol. 36, No. 2, (February), pp.155-160, *ISSN* 1096-9918

Becke A. D. (1993) Density-functional thermochemistry. iii. the role of exact exchange. *Journal of Chemical Physics,* Vol. 98, No 7, (April), pp.5648-5652, *ISSN* 0021-9606

Gordon M. S. and et al. (1982) Self-consistent molecular-orbital methods. 22. Small split-valence basis sets for second-row elements. *Journal of the American Chemical Society,* Vol. 104, No.10, (May), pp. 2797-2803, *ISSN* 0002-7863

Schmidt M.W. and et al. (2004) *Gaussian, Inc.,* Wallingford CT, *ISBN* 0963676938, Pittsburgh, USA

Lutnæs O. B. and et al. (2005) Benchmarking density-functional-theory calculations of rotational *g* tensors and magnetizabilities using accurate coupled-cluster calculations . *Journal of Chemical Physics,* Vol. 131, No.14, (September), pp.144104-144119, *ISSN 0021-9606*

Schafer, A.; Horn, H.; Ahlrichs, R. (1992) Fully optimized contracted Gaussian basis sets for atom Li to Kr. Journal of Chemical Physics, Vol. 97, No.4, (May), pp. 2571-2577, ISSN 0021-9606

Wu N. and et al. (2004). Interaction of Fatty Acid Monolayers with Cobalt Nanoparticles. Nano Letters, Vol. 4, No. 2, (January), pp. 383-386, ISSN 1530-6984

Fan, H. J.; Liu, Ch. W.; Liao, M. Sh. (1997). Geometry, electronic structure and magnetism of small Co$_n$ (n=2-8) clusters. Chemical Physics Letter, Vol. 273, No. 5-6, (July), pp. 353-359, ISSN 0009-2614.

Meldrum, A.; Boatner, L. A.; Sorge, K. (2003) Nuclear Instruments and Methods in Physics Research Section B: Beam Interactions with Materials and Atoms, Vol. 207,No.1, (May), pp. 36-44, ISSN 0168-583X

Ichiyanaga, Y.; Yamada, S. (2005) The size-depended magnetic properties of Co$_3$O$_4$ nanoparticles. Polyhedron, Vol.24, No. 16-17, (November), pp. 2813-2816, ISSN 0277-5387,

Graf, Ch. P.; Birringer, R.; Michels, A. (2006) Sythhesis and magnetic properties of cobalt nanocubes. Physical Review B, Vol. 73, No. 21, (April), pp. 212401 – 212404, ISSN 1550-235x

Resnick, D.A and et al. (2006) Magnetic properties of Co$_3$O$_4$ nanoparticles miniralized in Listeria innoucua Dps. Journal of Applied Physics. Vol. 99 (April), 08Q501-3, ISSN 1089-7550

Salavati-Niasari, M.; Afsaneh Khansari, A.; Davar, F. (2009) Synthesis and characterization of cobalt oxide nanoparticles by thermal treatment process. *Inorganica Chimica Acta,* Vol. 362, No. 14, (November), p.p. 4937-4942, *ISSN*: 0020-1693

Papis, E. and et al. (2009) Engineered cobalt oxide nanoparticles readily enter cells, *Toxicology letters,* Vol. 189 ,(June), p.p.253-259, *ISSN*: 0378-4274

Sakurai, K. and et al. (1998) Magic numbers in Fe clusters produced by laser vaporization source. *Journal of the Physical Society of Japan,* Vol. 8 : p.p. 2571-2573.

Gambardella, P. and et al. (2003) Giant magnetic anisotropy of Co atoms and nanoparticles. *Science,* Vol. 300, No. 5622, (May 16), p.p.1130-1133, *ISSN 1095-9203*

King, S.; Hyunh, K.; Tannenbaum, R. (2003) Kinetics of Nucleation, Growth, and stabilization of cobalt oxide nanoclusters. *Journal of Physical Chemistry B,* Vol. 107, No.44, (October 15), p.p. 10297-12104, *ISSN 1520-6106*

Nogues, J. and et al. (2006) Shell-driven magnetic stability in core-shell nanoparticles. Physical Review Letter, *Vol 97, No. 15, (October 13), p.p.1572031.1572034, ISSN · 0031-9007*

9

Organic Semiconductor Nanoparticle Film: Preparation and Application

Xinjun Xu and Lidong Li
School of Materials Science and Engineering,
University of Science and Technology Beijing,
Beijing,
P. R. China

1. Introduction

Organic semiconductors usually comprising π-conjugated structure in their molecules can exhibit excellent optical and electronic properties. They have advantages of simple fabrication and ease of tuning the chemical structure to give desired features. So they can serve as attractive candidates for applications in bio/chemical sensors and optoelectronic devices.[1,2] To meet the requirement of domains including information, energy and healthcare, nanoscale materials have emerged as new building blocks for optoelectronic devices, bioimaging agents, and drug delivery carriers in recent years.[3-5] These nanomaterials especially nanoparticles have already shown great potential to offer exciting opportunities in these areas.

Currently, most of the relevant works have been focused on inorganic semiconductor nanoparticles. Besides inorganic ones, organic semiconductor nanoparticles (OSNs) are desirable for a number of reasons. Their properties can be easily tuned for desired applications through the choice of functional molecules and surface modification. Additionally, their facile synthesis, good processability, high photoluminescence (PL) efficiency, high reaction activity, tunable properties, low toxicity and good biocompatibility further make them complementary to the inorganic nanomaterials and highly attractive in the material choice. As a result, OSNs have captured more and more interests. These OSNs can exhibit unique optical and electrical properties different from both the bulk solid samples and their molecular precursors. In comparison with molecule dispersed systems, OSNs are expected to show improved photostability and enhanced emission in various media.[6,7] These properties are essentially important in fluorescent labeling applications, such as fluorescence bioimaging and single molecular spectroscopy. For example, single molecules of most commercial dyes undergo photo-bleaching in a few milliseconds under typical excitation conditions under the radiation of a laser beam. On the contrary, because large numbers of chromophores are incorporated in single nanoparticles, they can show bright fluorescence even at a low excitation power. Thus, the fluorescent nanoparticles do not undergo rapid photo-bleaching and give less emission blinking which are generally observed in single molecule experiment.[8]

Up to now, most of the OSNs are used in aqueous solutions to serve as biological labels,[9,10] chemical sensors,[11] and photocatalysis materials.[12] To expand the application area of OSNs, there is an increasing effort to prepare OSNs as an active solid film in chemo/biosensors and optical and electronic devices.[13–16] Compared with bulk solid samples, nanoparticle films provide larger contact interface area, which is highly desired for chemical and biological sensing in sensors. So OSN based functional films tend to become a promising research area for applications in biosensing, energy conversion, photonic and optoelectronic devices.

In this review, after a brief introduction of organic semiconductor materials, we will summarize the methods for preparation OSN films. Then, its application in optical/electronic devices and chemo-/biosensors will be described. We hope this review can cast light on the advances and main problems in the research field of nanoparticle-based devices and sensors.

2. Organic semiconductor materials

Organic semiconductor materials are mainly classified into two categories. One is small molecules and the other is polymers made from repeated small conjugated monomer units.

2.1 Small molecule semiconductors

Polyphenyl derivatives (**1–3**),[17,18] fused aromatic rings (**4–8**),[19,20] porphyrin derivatives (**9,10**),[21,22] metal phthalocyanines (**11**),[23] fullerenes (**12**),[24] and some fluorescent dyes (**13,14**) have been made into nanoparticles.[25,26] Their molecular structures are illustrated in Figure 1. Small molecules are more easily packed to form crystals than polymers, so in some cases the nanoparticles of small-molecule semiconductors can transform to nanorods, nanotubes and nanoflakes.

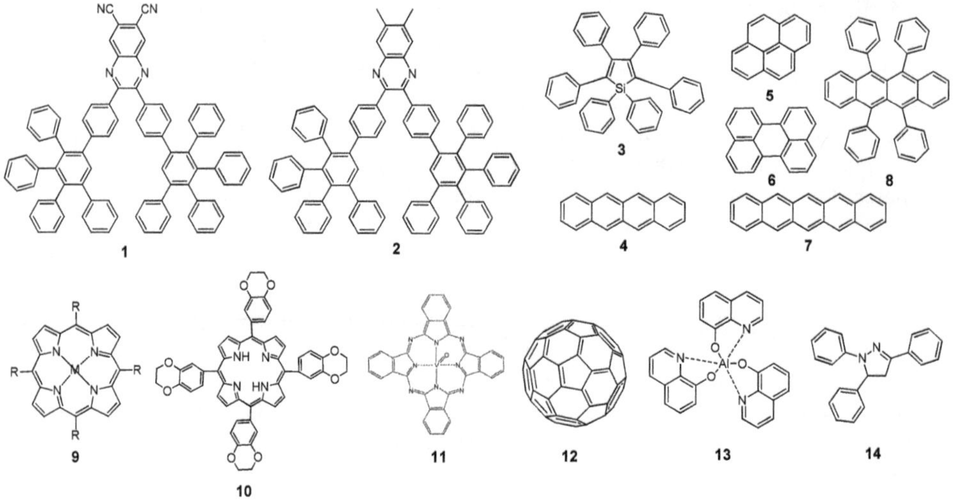

Fig. 1. Molecular structures of some small molecule semiconductors for synthesizing OSNs.

2.2 Polymer semiconductors

Poly(phenylene vinylene)s (**15**),[27] polyfluorenes (**16–18**),[28–30] polythiophenes (**19**),[31] ladder-type poly(para-phenylene)s (LPPP) (**20**),[32] poly(phenylene ethynylene)s (PPEs) (**21,22**),[33] polyanilines (PANIs) (**23**),[34] and some copolymers (**24,25**) have been utilized to prepare nanoparticles.[35,36] Their molecular structures are shown in Figure 2. Polymers have longer chains than small molecules, so it is possible for polymers to form nanoparticles even with a single molecule, which is advantageous for researches on single molecule behavior.

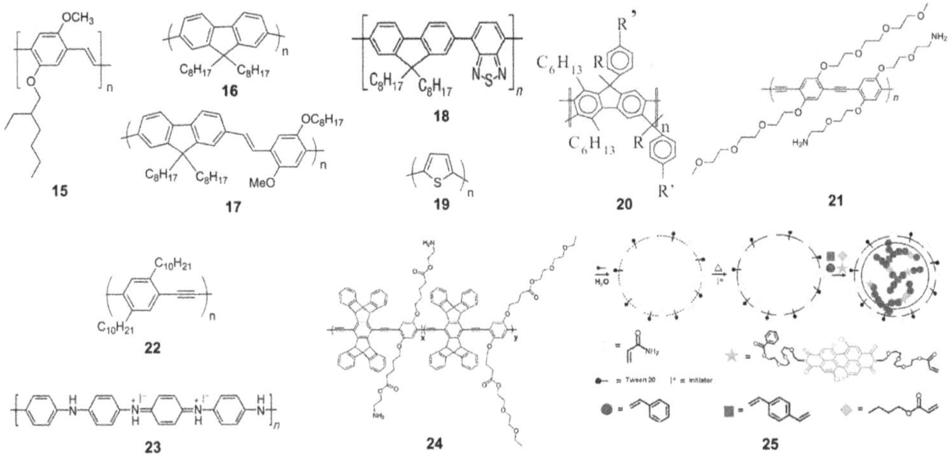

Fig. 2. Molecular structures of some polymer semiconductors for synthesizing OSNs.

3. Methods for synthesizing organic semiconductor nanoparticles

3.1 Reprecipitation

In 1992, Nakanishi and co-workers proposed the reprecipitation method and demonstrated the nanoparticles with the particle size less than 100 nm dispersed in water.[37] Since then, this method has been widely used in nanoparticle preparation for various kinds of molecules. In this method, a hydrophobic organic semiconductor material is dissolved in a good solvent (e.g., THF) for it and poured into a poor solvent (e.g., water), which is miscible with the good solvent. The resulting mixture is stirred vigorously using a magnetic stir bar or a sonicator to assist the formation of nanoparticles. After the nanoparticle formation the organic solvent is removed either by partial vacuum evaporation or by repeated dialysis process to leave behind water-dispersible nanoparticles. The main driving force for the formation of nanoparticles is the hydrophobic effect. When the solution of an organic semiconductor material in organic solvent is added to water, the compound molecules tend to avoid contacting with water. Consequently, in order to achieve minimum exposure they fold or packed into spherical shapes. The preparation does not involve the use of any additives such as surfactants and can be applied to a wide variety of organic semiconductors including both polymers and small molecules given that they are soluble in organic solvents. Moreover, using this method, it is possible to tune the size of nanoparticles by adjusting the concentration and the temperature of the solutions.

3.2 Miniemulsion

This is another commonly used method in the synthesis of OSNs. Using this method, Landfester and co-workers prepared nanoparticles from various polymers.[38] To prepare OSNs, the compound is dissolved in a water immiscible organic solvent and then the resulting solution is injected into an aqueous solution of an appropriate surfactant. The mixture is stirred vigorously by ultrasonicating to form stable miniemulsions containing small droplets of the polymer solution. The organic solvent is then evaporated to obtain a stable dispersion of polymer nanoparticles in water. The size of nanoparticles could vary from 30 nm to 500 nm depending on the concentration of the polymer solution. However, the droplets could also be destabilized by Ostwald ripening as well as the flocculation caused by the coalescence of droplets. To prevent flocculation appropriate surfactants are needed, while Ostwald ripening can be suppressed by addition of a hydrophobic agent (hydrophobe) to the dispersed phase. The hydrophobic agent promotes the formation of an osmotic pressure inside the droplets that counteracts the Laplace pressure (the pressure difference between the inside and the outside of a droplet) preventing diffusion from one droplet to the surrounding aqueous medium.

3.3 Pulsed-laser ablation

In this method, OSNs are formed by pulsed-laser ablation of large, several-micrometer-sized, organic crystals suspended in a liquid.[39,40] The powder of organic semiconductors was added to an aqueous solution containing surfactants such as sodium dodecyl sulfate (SDS). Then, the suspension was sonicated for a while. The mixture was put into a quartz cuvette, stirred vigorously with a magnetic stirrer, and then simultaneously exposed to the second harmonic of a nanosecond YAG laser. The spot area was approximately tens of mm[2], and the laser intensity was adjusted using a polarizer. The laser ablation mechanism for nanosecond laser ablation is based on photothermalization. The organic crystals in solutions absorb the laser light leading to a local increase in temperature and evaporation of a small amount of material from the crystal surface. The vaporized material is rapidly cooled by the surrounding liquid to form nanoparticles.

For nanosecond photothermal ablation in a solvent, rapid temperature elevation upon pulse excitation is compensated by a cooling process due to thermal diffusion to the solvent, and its balance gives the transient temperature determining the nanoparticle size. Higher fluence gives higher effective transient temperature, leading to efficient fragmentation to smaller particles. One advantage of the laser ablation method is its high controllability of size and phase of nanoparticles by tuning laser pulse width, wavelength, fluence, and shot number. However, this method is limited to fabricate OSNs based on small molecules only.

3.4 Direct condensation of organic vapor

Due to the fact that in the reprecipitation or miniemulsion process a solution of organic material (of typically about millimolar concentration) is added to a large excess of non-solvent, only very dilute particle dispersions can be obtained. That is one main disadvantage of these methods. The second one is that the reprecipitation or miniemulsion method is not applicable for organic materials that are poorly soluble in organic solvents (such as pentacene). As for laser ablation, more concentrated nanoparticle dispersions can be prepared compared to the repreciptation and miniemulsion method. However, as particle formation only occurs within the narrow laser beam, only small amounts of these nanoparticle dispersions can be prepared. Furthermore, the intense laser light may cause severe photochemical damage especially in the case of rather sensitive organic materials.

To overcome the above mentioned drawbacks, an approach for preparation of concentrated dispersions of organic nanoparticles by direct condensation the vapor of an organic semiconductor material into a liquid dispersion medium has been developed.[19] This approach combines elements from the physical vapor deposition (PVD) technique with cooling and condensation of the vapor directly inside a liquid. An illustration of the apparatus used in the direct condensation method is shown in Figure 3. The apparatus consists of four main parts: a tube furnace, a double-walled heated-vapor injection tube, a condensation and receiving vessel, and a vacuum pumping system. Temperatures in the different zones are adjusted according to the organic material to be evaporated, and were maintained such that no condensation of the organic materials occurred in the tubes. The evaporated organic material will be carried by the inert gas flow to the vapor-injection tube, which guides the organic material into a liquid condensation medium. The condensation liquid typically consists of an aqueous solution containing surfactants or polymeric stabilizers. It rapidly cools down the gas leading to condensation of the organic vapor and formation of nanoparticles. These nanoparticles are subsequently stabilized in situ by the surfactant or polymeric additive at the bubble/liquid interface to form a stable dispersion. The size of OSNs prepared by this method is in the range of 100–200 nm for fused aromatic hydrocarbons such as pentacene, rubrene, and tetracene.

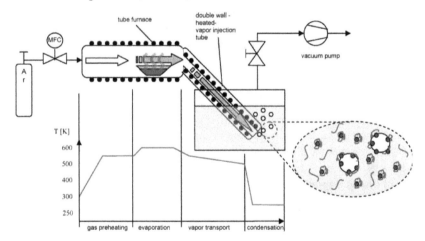

Fig. 3. Apparatus for direct condensation of organic vapor (Reproduced from Ref. 19, Copyright 2009 Wiley-VCH Verlag GmbH & Co.)

3.5 Template-based approaches

3.5.1 Soft templates

Micelles can be used as soft templates to conduct the polymerization in the aqueous heterophase system. By dispersing the appropriate monomers, surfactant, solvent, and catalysts in an aqueous medium, the Glaser coupling reaction can be carried out exclusively within the hydrophobic interior of surfactant micelles to produce the poly(arylene diethynylenes) nanoparticles.[41] Similarly, poly(p-phenyleneethynylene) nanoparticles can also be prepared by this method.[42]

The molecular structure of surfactant used in the aqueous heterophase system has a big influence on the shape of the formed nanoparticles. Using dodecylbenzene sulfonic acid as a surfactant and doping agent for poly(3,4-ethylenedioxythiophene) (PEDOT) yielded amorphous and polydisperse particles with diameters in the range of 35–100 nm.[43] Short chain alcohol ethoxylate surfactants yielded more spherical particles, but significant amounts of surfactant residue were trapped on the PEDOT latex, and secondary nucleation could not be completely suppressed.[44]

These examples show that the soft template approach has been a versatile method for preparing conjugated polymer nanoparticles. However, control over important parameters such as particle diameter and polydispersity by this method is often not easy. Many of these issues can be addressed by the use of a hard template.

3.5.2 Hard templates

Due to the shape persistence of hard templates, they typically offer a more reliable way of directing the shape of conjugated polymer nanostructures. Monodisperse nanoparticles such as silica and polystyrene particles can be used as a hard template for preparing core-shell structures. Conjugated polymers such as polypyrrole, PANI and PEDOT, highly fluorescent polymers such as PPE have also been attached to the surface of colloidal particles.[45] The conjugated polymers can either be polymerized in situ from monomers absorbed on the surface of the particle templates or be deposited from a layer-by-layer technique through electrostatic interactions.[46]

4. Methods for preparing nanoparticles film

Since OSNs are usually synthesized in solution with a low concentration, conventional thin film forming processes such as spin-coating or dip-casting are not appropriate for preparing OSN films. So other methods have been developed to prepare good OSN films.

4.1 Electrophoresis deposition

Electrophoretic deposition, which is based on the electrical collection of small, charged particles dispersed in dielectric liquids, is one of the most widely used coating methods capable of patterning. It has been reported that the phosphors for a cathode ray tube,[47] the oxide superconductors,[48] and the carbon nanotubes for a cold cathode have been successfully coated by electrophoretic deposition.[49] As for OSNs in solutions, they often carried charges on their surface. Such surface charges are generated according to Coehn's empirical rule. That is, the electrostatic charge separation may occur when two dielectrics are in intimate contact. The substance with the higher dielectric constant will receive the positive charge, while the other one will receive the negative charge. As illustrated in Figure 4, a DC voltage (usually hundreds of volts) was applied between two ITO-coated glass plates soaked in the nanoparticle suspension. Then nanoparticles will move towards the corresponding electrode under the driving of electric field force.[50] The films thus obtained were washed with clean solvent and dried in air.

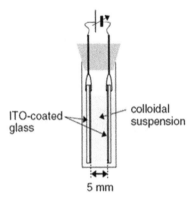

Fig. 4. Apparatus for electrophoretic deposition of OSNs

4.2 Rapid expansion of supercritical solution (RESS) technique

This technique for prepare OSN film is based on a rapid expansion process of supercritical solution (e.g. CO_2) which contains dissolved organic semiconductors.[51] By using an apparatus illustrated in Figure 5, organic semiconductors dispersed in supercritical CO^2 solution can be sprayed on the substrate through a long stainless steel capillary tube attached to the chamber. After the rapid evaporation of CO_2, OSNs are precipitated on the surface of substrates. Using process conditions of compressed-fluid precipitation and formulation, it appears possible to produce organic nanoparticles with tunable sizes and optical properties. This capability opens up avenues to create devices and functional films using organic nanoparticles as building blocks, which may be tailored for the application.

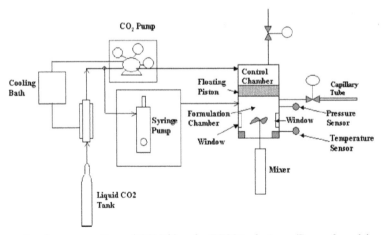

Fig. 5. Apparatus for preparation of OSN films by RESS technique (Reproduced from Ref. 51, Copyright 2006 Wiley-VCH Verlag GmbH & Co.)

Additionally, it is possible to mix building blocks of organic nanoparticles or combine different molecules within a building block. However, the limitation of this technique is that only small molecule based OSN film can be prepared while preparation of polymer based OSN films by this technique is not available. Also, surfactants such as the ammonium-exchanged Fluorolink 7004 ($Cl(CF_2CF(CF_3)O)_nCF_2COO-NH_4^+$) need to be introduced into the supercritical solution to adjust the size of ultimate nanoparticles.[51]

4.3 Solvent-evaporation induced self-assembly

The evaporation behavior during the drying process of a solution plays a vital role in controlling the film morphology and the distribution of solute in the final films. It is well known that when a liquid drop containing dispersed solids evaporates on a surface, it commonly leaves a dense, ring-like deposit along the perimeter. The reason is that the contact line is pinned during the drying process, leading to a fixed contact area on the substrate. Therefore, a capillary flow of the solvent occurs from the center of the drop to the contact line to replenish the evaporation loss, and this flow transports the solutes to its periphery.[52] As far as the OSNs solution is concerned, such phenomenon will result in an undesirably uneven distribution of nanoparticles across the deposited films. However, if another flow which has an opposite direction to the capillary flow is introduced into the OSNs solutions during the drying process, the transportation of nanoparticles towards the contact line by the capillary flow is expected to be counteracted. Marangoni effect is usually observed in a solution containing two kinds of solvents with different surface tensions and boiling points, and a flow is induced by the surface tension gradient existed in the solution caused by solvent evaporation. Such a flow is named as the Marangoni flow, and its direction can be controlled to be the same as the spreading of a drop on a solid surface (outward) or opposite to the spreading (inward), depending on the boiling points and surface tensions of the two solvents to be mixed. Consequently, by proper introduction of a second solvent into the solution, a Marangoni flow with an opposite direction to the capillary flow can be achieved.

The solvent-evaporation induced self-assembly method for preparing the thin nanoparticles films from their OSNs solutions is illustrated in Figure 6.[53] By using ethylene glycol (EG) as the second solvent with a high boiling point but a low surface tension, the capillary flow in the solution can be counterbalanced by the Marangoni flow. The self-assembly of nanoparticles on the substrate can thus be achieved through the nanoparticle-substrate and nanoparticle-nanoparticle van der Waals interactions.

4.4 Vapor-driven self-assembly

The vapor-driven self-assembly process is based on the selective phase demixing and self-assembled aggregate formation. Such behaviors occur from a molecularly dispersed solid solution of specific fluorescent molecules in a polymer matrix when it is exposed to volatile organic solvent vapors.[54] After solvent exposure, the supramolecular self-assembly of organic semiconductor materials leads to the formation of spherical nanoparticles (see Figure 7). The advantage of this method is to form nanoparticles films *in situ* on the substrate. Nevertheless, this kind of method is only appropriate for small-molecule

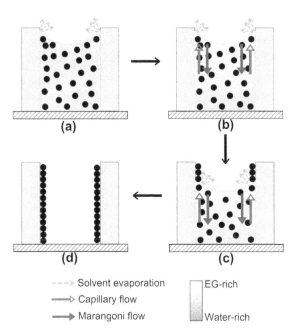

Fig. 6. Scheme for the solvent-evaporation-induced self-assembly of OSNs on the substrates to form films. (Reproduced from Ref. 53, Copyright 2010 The American Chemical Society)

compounds with certain structures and is not a universal method for most polymer semiconducting materials. In addition, OSNs formed by this method are discrete and continuous OSNs films can not be obtained.

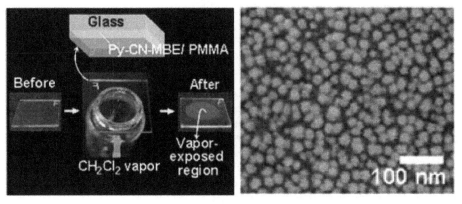

Fig. 7. Photograph of the 1-cyano-trans-1-(4′-methylbiphenyl)-2-[4′-(2′-pyridyl) phenyl]ethylene (Py-CN-MBE)/poly(methyl methacrylate) (PMMA) film before and after exposure to dichloromethane vapor (Left panel) and the SEM image of the Py-CN-MBE nanoparticles formed by the vapor-driven self-assembly process (Right panel). (Reproduced from Ref. 54, Copyright 2007 Wiley-VCH Verlag GmbH & Co.)

4.5 Inkjet printing

As mentioned above, when a droplet of OSN solution is dripped on the surface of a substrate, the OSNs tend to form coffee-stains after the evaporation of the solvent. So direct inkjet printing of OSN solutions can not provide good film morphology. To avoid this drawback, an aqueous dispersion of semiconducting polymer nanospheres is deposited by inkjet printing onto a polymer surface patterned by soft embossing.[55] By interaction between the spheres and the undulated surface a self assembly process is triggered, resulting in the formation of OSN nanostructures determined by the template.

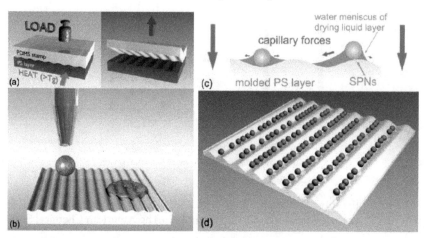

Fig. 8. The fabrication process for functional nanostructures from inkjet printing. Reprodcued from Ref. 55 (Copyright 2008 The Royal Society of Chemistry).

As shown in Figure 8, after a droplet of the OSN solution was printed on the surface of the structured polymeric template layer, OSNs assemble in the grooves of the embossed surface. This method relies on the application of a polymer template layer, so that the patterned structure that is formed with the OSNs can be incorporated into a device.

4.6 Spin-coating

Due to the very low concentration of OSN solutions, good film can hardly be formed by spin-coating or dip-coating method without any additives. As a result, auxiliary underlayers or additives such as surfactants or polymer matrix have to be introduced to assist the deposition of nanoparticle films. As mentioned above, nanoparticles usually carry charges on their surfaces when they are dispersed in solutions. Therefore, negatively charged nanoparticles can be formed on polycationic films with the help of electrostatic interactions via spin-coating and vice versa. Layers of LPPP nanoparticles were spin-coated on poly(allylamine hydrochloride) (PAH) can exhibit a homogeneous fluorescence over large areas.[38] Similarly, conjugated polymer nanoparticles such as polyfluorene derivatives and LPPP spin-coated on poly(3,4-ethylenedioxythiophene):poly(4-styrenesulfonate) (PEDOT:PSS) film can also exhibit a good film morphology.[32,56] Besides auxiliary underlayers, polymer matrix can also been utilized to act as a binder to improve the film quality deposited from OSN solutions. This kind of binder for the nanoparticles can also assist in the reduction of electric field singularities around the particles that may result in regions of pinhole formation in electronic devices. Poly(vinyl alcohol),[57] hexadecyl-modified poly(ethylene oxide) (PEO),[55] and PEDOT:PSS [58-60] are polymer matrixes used for this purpose. These binders have virtually no effect on the color characteristics of the electroluminescence spectrum since PVA, PEO, and PEDOT:PSS have a negligible absorption in the luminance regime of OSNs. Although by this means the film quality is improved, the additives remained in the nanoparticles films will be disadvantages to the optical and electrical properties of OSNs. For example, when using PEDOT:PSS as an additive to the OSNs aqueous solution for preparing thin nanoparticles films by spin-coating, the acidity of PEDOT:PSS will deteriorate the luminescent properties of the conjugated compounds largely.

5. Applications of OSN film in optical and electronic devices

5.1 Organic light-emitting diodes

LPPP, poly(9-vinylcarbazole) (PVK), 2-(4-tert-butylphenyl)-5-(4-biphenylyl)-1,3,4-oxadiazole (tBu-PBD), coumarins, nile red, nanoparticles prepared by miniemulsion method, core-shell nanoparticles with perylene as the core and poly[methyl methacrylate-co-vinylcarbazole-co-2-(3'-nitrophenyl)-5-(4'-acryloylphenyl)-1,3,4-oxadiazole] as the shell formed by emulsion copolymerization, poly(3-octadecylthiophene) nanoparticles prepared by reprecipitation method, multi-component nanoparticles prepared by RESS process or miniemulsion method have been reported to serve as an active layer in organic light-emitting diodes.[61] For nanoparticles synthesized by both miniemulsion and RESS methods, surfactants, stabilizing agents or hydrophobes are necessary and can hardly be removed. Such additives will be disadvantageous to the native optoelectronic properties of OSNs in devices. It would be still interesting to fabricate optoelectronic devices from OSNs prepared via the reprecipitation method in which there would be no additives including surfactants, stabilizing agents and hydrophobes. Electrophoretic deposition of OSN films from reprecipitation-processed nanoparticle solutions has been employed in fabricating OLEDs.[62] Although an electroluminescent emission from the device could be observed, the emission is not uniform because the nanoporosity of the OSN film prepared by electrophoretic deposition probably causes fatal pin-holes. As a result, an approach for preparing high-quality OSN films from reprecipitation-processed nanoparticle solutions is highly desired. Fortunately, the solvent-evaporation induced self-assembly method introduced above can meet this requirement. Currently, the main drawback of this method for fabricating OLEDs is that the solvent evaporation period is time-consuming. If there are some ways are found to overcome this drawback, this method is very promising for fabricating OSN based OLEDs.

5.2 Organic field-effect transistors

Organic field-effect transistors (OFETs) fabricated using solution-deposition techqiques are particularly well-suited for large-area electronic devices. For meaningful practical applications, the organic semiconductors need to provide FET mobilities close to that of amorphous silicon. This will necessitate establishment of proper molecular order in the semiconductors to achieve high mobilities, since charge-carrier transport in organic semiconductors is dominated by hopping and disordered materials are not efficient charge-transporting media. In nanoparticles, molecules are closely packed and they are usually highly ordered. For example, both poy(3-hexylthiophene) (P3HT) and poly(9,9'-dioctylfluorene) (PFO) nanoparticles can exhibit highly ordered structures and can be distinguished from the UV absorption spectrum with occurrence of a new peak.[31,63] Poly(3,3'''-dialkylquarterthiophene)s (PQTs) nanoparticles have been explored for using in OFETs.[64] As expected, the presence of lamellar π–stacking order in the nanoparticles can be verified by XRD and UV spectrum data. OFETs based on PQT nanoparticles show a 50 % improvement in mobility on bare SiO_2 dielectric layer and an order of magnitude improvement in mobility on surface modified SiO_2 dielectric layer relative to those based on normal films.

5.3 Organic solar cells

It is well known that excitons formed in the active layer of organic solar cells usually have a migration distance less than 20 nm.[65] So in organic solar cells, the distance for excitons diffused to the interface of electron donors and acceptors should be smaller than 20 nm to ensure good light conversion efficiency. However, because the entropy of mixing is generally low for polymers, solid polymer blends tend to phase-separate at the macroscopic scale. Moreover, when a thin layer of immiscible polymers is deposited from solution, the resulting morphology strongly depends on various parameters, such as the individual solubility of the polymers in the solvent used, the interaction with the substrate surface, the layer thickness and the method of deposition, drying and annealing. Therefore, controlling of the lengths of phase separation in thin layers is important for organic solar cells to avoid large-scale phase separation. Kietzke et al. have reported that by using the blend of poly(9,9-dioctylfluorene-co-benzothiadiazole) (F8BT) and poly(9,9-dioctylfluorene-co-N,N-bis(4-butylphenyl)-N,Ndiphenyl-1,4-phenylenediamine) (PFB) nanoparticles as the active layer the phase separation in organic solar cells can be controlled on the nanoscale.[29,56]

5.4 Photonic crystals

If a photonic crystal is constructed from a material with sufficiently high refractive index, it can exhibit a photonic bandgap, a frequency range in which the mode density is zero and photons cannot propagate in any direction. Although this property is desirable since it would allow the inhibition of spontaneous emission and the ability to manipulate the flow of light, it is difficult to be achieved with organic semiconductors due to their relatively low refractive index. Instead, an organic photonic crystal would more likely have a partial bandgap, a frequency range where light can propagate in a limited number of directions. However, even in this case, the mode density for forbidden directions can be strongly modified by the photonic crystal. For frequencies just outside the partial bandgap, the mode density along these directions can be higher than in free space. This increase indicates that more optical modes are available to interact with electronic excitations. Thus, by combining organic semiconductors with photonic crystals, this enhanced interaction with light can be used to further improve optoelectronic properties.

One of the simplest ways of preparing photonic crystals is by the self assembly of monodisperse spheres, for instance, by the self-assembly of colloidal silica or polystyrene microspheres widely reported in the literatures. These spheres can then act as secondary templates for the fabrication of conjugated polymer inverse photonic crystals, where the interstitial voids of the sphere template have been filled with conjugated polymer. This approach has been successfully used in the preparation of poly(p-phenylenevinylene) (PPV) inverse photonic crystal films.[66]

6. Applications of OSN film in chemo-/biosensors

Although lots of applications of organic nanoparticles in chemo-/biosensors have been explored in recent year, most of them are carried out in solutions. Compared with solutions, solid-state samples can be more convenient for storage and transport which are highly desired for off-site laboratory analysis. Here we will introduce some applications of OSN film in chemical and biosensors.

6.1 Chemical sensors

Hydroxyl radical is one of the most important reactive oxygen species, which is recognized to play an important role in physiological and pathological processes of the organisms. In addition, hydroxyl radical is also involved in many chemical, environmental, and pharmaceutical processes such as semiconductor photocatalysis in aqueous solution, wastewater treatment, and tumor cell killing. By using a binary nanoparticle system combining PFO nanoparticles and MEH-PPV nanoparticles, a linear relationship between the concentration of hydroxyl radical and the intensity ratio (Band I to Band III) of PFO nanoparticles can be found in the deposited nanoparticle film.[67] The synergy between MEH-PPV NPs and PFO NPs are crucial to the response of free radicals in this kind of binary NP system. When exposed to free radicals, MEH-PPV NPs undergo molecular structure changes in the outer shell. As a result, a broad-sense polarity vector across the whole NP pointing from the weak-polarity core to the strong-polarity shell is established. Such a polarity vector will influence on the vibronic coupling among different electronic states of PFO molecules when the core-shell MEH-PPV NPs are adjacent to PFO NPs, which will change the relative PL emission intensity between bands I and III of PFO.

6.2 Biosensors

In general, conducting polymer nanoparticles are dispersed on the surface of the electrode to increase the area/volume ratio and to favor the adsorption of bio-molecules. By this means, uniform electrostatic adsorption of protein was enabled, thereby exhibiting higher signal-to-background ratios and shorter response times than electrochemically prepared films.[68] Taking advantage of conducting polymer nanoparticles, sufficient amounts of enzyme were firmly immobilized during the fabrication of a phosphate biosensor. The response time of the biosensors was about 6 s. A linear response was observed between 1.0 µM and 100 µM and the detection limit was determined to be about 0.3 µM.[69] Besides, an ascorbic acid sensor has been fabricated via the drop-casting of PANI nanoparticles onto a screen-printed carbon-paste electrode.[70] The PANI nanoparticles not only enhanced the catalytic reaction, but also allowed the detection of ascorbate at the reduced applied potential of 0 V and operation at neutral pH, avoiding the problem of sample interference.

7. Conclusions and prospects

Most of the organic semiconductors including both small molecules and conjugated polymers can be utilized for preparing OSNs, thus ensuring a very wide material selectivity for their applications. Various approaches have been reported for synthesizing OSNs. Besides reprecipitation method, additives such as surfactants are usually employed in other methods, which may be disadvantageous to the optical and electronic properties of OSNs. When depositing the OSN solution onto the substrate to form nanoparticle films, auxiliary underlayer and binders such as polymer matrix or surfactants are often used to improve the film quality. An exception is the vapor-driven self-assembly and solvent-evaporation induced self-assembly methods, which can prepare clean nanoparticle films and is highly desired for the optical and electronic applications.

Although OSNs have been proved to be effective building blocks in both optoelectronic devices and chemo/biosensors, a number of challenges and avenues of exploration remain. The interface between nanoparticles and the surroundings is crucial to its optical, electrical, and catalytic properties. So surface modification of OSNs can not only improve their contact properties but also endow them with a new function. However, the surface modification of OSNs is seldom reported yet.

The field of photonic crystals has recently provided a number of novel insights into the manipulation of light. These photonic properties have yet to be fully combined with the optoelectronic properties of OSNs, and the development of this area remains a very active area of research. Additionally, the ability to precisely control the morphology and alignment of OSNs is of importance to all fields of organic electronics.

In the field of electronic devices, OSN based OLEDs have been widely explored. However, OSN based OFETs still need to be paid more attention. By appropriate design, OFETs using OSNs as an active layer may be served as multifunctional optoelectronic devices.

With the great advantages of OSNs, they are believed to play an important role in more and more application fields and will provide new scientific insights in the coming years.

8. References

[1] J. Janata, M. Josowicz, *Nat. Mater.* 2003, 2, 19–24.
[2] J. E. Anthony, A. Facchetti, M. Heeney, S. R. Marder, X. Zhan, *Adv. Mater.* 2010, 22, 3876–3892.
[3] T. Sagawa, S. Yoshikawa, H. Imahori, *J. Phys. Chem. Lett.* 2010, 1, 1020–1025.
[4] L. Xia, Z. Wei, M. Wan, *J. Colloid Interface Sci.* 2010, 341, 1–11.
[5] X. Feng, F. Lv, L. Liu, H. Tang, C. Xing, Q. Yang, S. Wang, *ACS Appl. Mater. Interfaces* 2010, 2, 2429–2435. .
[6] B. -K. An, S.-K. Kwon, S.-D. Jung, S. Y. Park, *J. Am. Chem. Soc.* 2002, 124, 14410–14415.
[7] J. Chen, C. C. W. Law, J. W. Y. Lam, Y. Dong, S. M. F. Lo, I. D. Williams, D. Zhu, B. Tang, *Chem. Mater.* 2003, 15, 1535–1546.
[8] J. N. Clifford, T. D. M. Bell, P. Tinnefeld, M. Heilemann, S. M. Melnikov, J. Hotta, M. Sliwa, P. Dedecker, M. Sauer, J. Hofkens, E. K. L. Yeow, *J. Phys. Chem. B* 2007, 111, 6987–6991.

[9] Y. Zhou, G. Bian, L. Wang, L. Dong, L. Wang, J. Kan, *Spectrochim. Acta, Part A* 2005, *61*, 1841–1845.

[10] X. Xu, S. Chen, L. Li, G. Yu, C. Di, Y. Liu, *J. Mater. Chem.* 2008, *18*, 2555–2561.

[11] L. Wang, L. Dong, G. Bian, L. Wang, T. Xia, H. Chen, *Anal. Bioanal. Chem.* 2005, *382*, 1300–1303.

[12] H. Y. Kim, T. G. Bjorklund, S.-H. Lim, C. J. Bardeen, *Langmuir* 2003, *19*, 3941–3946.

[13] J. Jang, J. Ha, J. Cho, *Adv. Mater.* 2007, *19*, 1772–1775.

[14] T. Matsuya, K. Otake, S. Tashiro, N. Hoshino, M. Katada, T. Okuyama, *Anal. Bioanal. Chem.* 2006, *385*, 797–806.

[15] S.-J. Lim, B.-K. An, S. D. Jung, M.-A. Chung, S. Y. Park, *Angew. Chem. Int. Ed.* 2004, *43*, 6346–6350.

[16] F. Wang, M.-Y. Han, K. Y. Mya, Y. Wang, Y.-H. Lai, *J. Am. Chem. Soc.* 2005, *127*, 10350–10355.

[17] X. Xu, S. Chen, L. Li, G. Yu, C. Di, Y. Liu, *J. Mater. Chem.* 2008, *18*, 2555–2561.

[18] Y. Liu, Y. Tang, N. N. Barashkov, I. S. Irgibaeva, J. W. Y. Lam, R. Hu, D. Birimzhanova, Y. Yu, B. Z. Tang, *J. Am. Chem. Soc.* 2010, *132*, 13951–13953.

[19] S. Köstler, A. Rudorfer, A. Haase, V. Satzinger, G. Jakopic, V. Ribitsch, *Adv. Mater.* 2009, *21*, 2505–2510.

[20] E. H. Cho, M. S. Kim, D. H. Park, H. Jung, J. Bang, J. Kim, J. Joo, *Adv. Funct. Mater.* 2011, *21*, 3056–3063.

[21] X. Gong, T. Milic, C. Xu, J. D. Batteas, C. M. Drain, *J. Am. Chem. Soc.* 2002, *124*, 14290–14291.

[22] C. Huang, Y. Li, Y. Song, Y. Li, H. Liu, D. Zhu, *Adv. Mater.* 2010, *22*, 3532–3536.

[23] X. Zhang, Y. Wang, Y. Ma, Y. Ye, Y. Wang, K. Wu, *Langmuir* 2006, *22*, 344–348.

[24] K. L. Chen, M. Elimelech, *Langmuir* 2006, *22*, 10994–11001.

[25] A. M. Collins, S. N. Olof, J. M. Mitchels, S. Mann, *J. Mater. Chem.* 2009, *19*, 3950–3954.

[26] Y. S. Zhao, H. Fu, A. Peng, Y. Ma, Q. Liao, J. Yao, *Acc. Chem. Res.* 2010, *43*, 409–418.

[27] C. Szymanski, C. Wu, J. Hooper, M. A. Salazar, A. Perdomo, A. Dukes, J. McNeill, *J. Phys. Chem. B* 2005, *109*, 8543–8546.

[28] C. Wu, C. Szymanski, Z. Cain and J. McNeill, *J. Am. Chem. Soc.* 2007, *129*, 12904–12905.

[29] T. Kietzke, D. Neher, K. Landfester, R. Montenegro, R. Güntner, U. Scherf, *Nat. Mater.* 2003, *2*, 408–412.

[30] Y.-L. Chang, R. E. Palacios, F.-R. F. Fan, A. J. Bard, P. F. Barbara, *J. Am. Chem. Soc.* 2008, *130*, 8906–8907.

[31] J. E. Millstone, D. F. J. Kavulak, C. H. Woo, T. W. Holcombe, E. J. Westling, A. L. Briseno, M. F. Toney, J. M. J. Fréchet, *Langmuir* 2010, *26*, 13056–13061.

[32] T. Piok, S. Gamerith, C. Gadermaier, H. Plank, F. P. Wenzl, S. Patil, R. Montenegro, T. Kietzke, D. Neher, U. Scherf, et al. *Adv. Mater.* 2003, *15*, 800–804.

[33] N. A. A. Rahim, W. McDaniel, K. Bardon, S. Srinivasan, V. Vickerman, P. T. C. So, J. H. Moon, *Adv. Mater.* 2009, *21*, 3492–3496.

[34] J. Yang, J. Choi, D. Bang, E. Kim, E.-K. Lim, H. Park, J.-S. Suh, K. Lee, K.-H. Yoo, E.-K. Kim, Y.-M. Huh, S. Haam, *Angew. Chem. Int. Ed.* 2011, *50*, 441–444.

[35] J. H. Moon, R. Deans, E. Krueger and L. F. Hancock, *Chem. Commun.* 2003, 104–105.

[36] Z. Tian, A. D. Shallerz, A. D. Q. Li, *Chem. Commun.* 2009, 180–182.

[37] H. Kasai, H. S. Nalwa, H. Oikawa, S. Okada, H. Matsuda, N. Minami, A. Kakuta, K. Ono, A. Mukoh, H. Nakanishi, *Jpn. J. Appl. Phys.* 1992, *31*, L1132–L1134.

[38] K. Landfester, R. Montenegro, U. Scherf, R. Güntner, U. Asawapirom, S. Patil, D. Neher, T. Kietzke, *Adv. Mater.* 2002, *14*, 651–655.

[39] T. Asahi, T. Sugiyama, H. Masuhara, *Acc. Chem. Res.* 2008, *41*, 1790–1798.

[40] R. Yasukuni, T. Asahi, T. Sugiyama, H. Masuhara, M. Sliwa, J. Hofkens, F.C. De Schryver, M. Van der Auweraer, A. Herrmann, K. Müllen, *Appl. Phys. A* 2008, *93*, 5–9.

[41] M. C. Baier, J. Huber, S. Mecking, *J. Am. Chem. Soc.* 2009, *131*, 14267–14273.

[42] E. Hittinger, A. Kokil, C. Weder, *Angew. Chem. Int. Ed.* 2004, *43*, 1808–1811.

[43] J. W. Choi, M. G. Han, S. Y. Kim, S. G. Oh, S. S. Im, *Synth. Met.* 2004, *141*, 293–299.

[44] A. M. J. Henderson, J. M. Saunders, J. Mrkic, P. Kent, J. Gore, B. R. Saunders, *J. Mater. Chem.* 2001, *11*, 3037–3042.

[45] T. L. Kelly, M. O. Wolf, *Chem. Soc. Rev.* 2010, *39*, 1526–1535.

[46] J. H. Wosnick, J. H. Liao, T. M. Swager, *Macromolecules* 2005, *38*, 9287–9290.

[47] P. F. Grosso, R. E. Rutherford, D. E. Sargent, Jr. *J. Electrochem. Soc.* 1970, *117*, 1456–1459.

[48] C. T. Chu, B. Dunn, *Appl. Phys. Lett.* 1989, *55*, 492–494.

[49] B. Gao, G. Z. Yue, Q. Qiu, Y. Cheng, H. Shimoda, L. Fleming, O. Zhou, *Adv. Mater.* 2001, *13*, 1770–1773.

[50] K. Tada, M. Onoda, *Adv. Funct. Mater.* 2002, *12*, 420–424.

[51] R. Jagannathan, G. Irvin, T. Blanton, S. Jagannathan, *Adv. Funct. Mater.* 2006, *16*, 747–753.

[52] R. D. Deegan, O. Bakajin, T. F. Dupont, G. Huber, S. R. Nagel, T. A. Witten, *Nature* 1997, *389*, 827–829.

[53] C. Zheng, X. Xu, F. He, L. Li, B. Wu, G. Yu, Y. Liu, *Langmuir* 2010, *26*, 16730–16736.

[54] B.-K. An, S.-K. Kwon, S. Y. Park, *Angew. Chem. Int. Ed.* 2007, *46*, 1978 –1982.

[55] E. Fisslthaler, A. Blümel, K. Landfester, U. Scherfd, E. J. W. List, *Soft Matter* 2008, *4*, 2448–2453.

[56] T. Kietzke, D. Neher, M. Kumke, R. Montenegro, K. Landfester, U. Scherf, *Macromolecules* 2004, *37*, 4882–4890.

[57] J.-S. Heo, N.-H. Park, J.-H. Ryu, K.-D. Suh, *Adv. Mater.* 2005, *17*, 822–826.

[58] C. F. Huebner, J. B. Carroll, D. D. Evanoff, Jr., Y. Ying, B. J. Stevenson,J. R. Lawrence, J. M. Houchins, A. L. Foguth, J. Sperry, S. H. Foulger, *J. Mater. Chem.* 2008, *18*, 4942–4948.

[59] C. F. Huebner, S. H. Foulger, *Langmuir* 2010, *26*, 2945–2950.

[60] C. F. Huebner, R. D. Roeder, S. H. Foulger, *Adv. Funct. Mater.* 2009, *19*, 3604–3609.

[61] D. Tuncel, H. V. Demirb, *Nanoscale*, 2010, *2*, 484–494.

[62] K. Tada, M. Onoda, *Thin Solid Films* 2003, *438*, 365–368.

[63] S. Moynihan, D. Iacopino, D. O'Carroll, P. Lovera, G. Redmond, *Chem. Mater.* 2008, *20*, 996–1003.

[64] B. S. Ong, Y. Wu, P. Liu, S. Gardner, *Adv. Mater.* 2005, *17*, 1141–1144.

[65] H. Hoppe, N. S. Sariciftci, *Adv. Polym. Sci.* 2008, *214*, 1–86.

[66] M. Deutsch, Yu. A. Vlasov, D. J. Norris, *Adv. Mater.* 2000, *12*, 1176–1180.

[67] J. Wang, X. Xu, Y. Zhao, C. Zheng, L. Li, *J. Mater. Chem.* 2011, *21*, 18696–18703

[68] A. Morrin, O. Ngamna, A. J. Killard, S. E. Moulton, M. R. Smyth, G. G. Wallace, *Electroanalysis* 2005, *17*, 423–430.

[69] M. A. Rahman, D. S. Park, S. C. Chang, C. J. McNeil, Y. B. Shim, *Biosens. Bioelectron.* 2006, *21*, 1116–1124.

[70] A. Ambrosi, A. Morrin, M.R. Smyth, A.J. Killard, *Anal. Chim. Acta* 2008, *609*, 37–43.

10

Magnetic Properties and Size Effects of Spin-1/2 and Spin-1 Models of Core-Surface Nanoparticles in Different Type Lattices

Orhan Yalçın[1], Rıza Erdem[2] and Zafer Demir[3]
[1]*Department of Physics, Niğde University, Niğde*
[2]*Department of Physics, Akdeniz University, Antalya*
[3]*Institute of Graduate School of Natural and Applied Sciences, Niğde University, Niğde*
Turkey

1. Introduction

Dimension in the range of 1 to 100 nm, is called the nano regime. In recent years, nanoparticles/quantum dots are in a class of magnetic nanostructures (Aktaş et al., 2003, 2006; Kartopu & Yalçın, 2010). Nanoparticles (NPs) have been steadily interesting in Physics, Chemistry, Biology, Biomedicine, Spintronics, etc. As the dimensions of magnetic NPs decrease down to the nanometer scale, these core-surface NPs start to exhibit new and interesting physical properties mainly due to quantum size effects. Even the intrinsic physical characteristics of NPs are observed to change drastically compared to their macroscopic counterparts. The potential applications of NPs are very attractive for magneto-sensor, bio-sensor, magneto-electronics, data storage media, computer hard disks, microwave electronic devices, nano-transistors, etc. Especially, the studies of core-surface NPs are extremely important for technology because of transmission of data at high density to optical computer, nanorobot to assemble, compose rigid disk. The nanoparticles have relevance to thin film devices in the new breed of magnetoelectronics, spin-valve, spin-transistors, spin-dependence tunneling devices and etc. (Babin, et al., 2003). The hysteresis in fine magnetic particles applied to new technologies such as Magnetic Random Access Memory (MRAM).

In generally, a nanoparticle is divided into the inner, outer and intermediate regions. These zones are called core (C), surface (S) and core-surface (CS), respectively. The size effects of core-surface NPs are very important for technological and biomedicine applications (Fraerman et al., 2001; Pankhurst et al., 2003). Especially, superparamagnetic (single-domain) NPs are important for non surgecial interfere of human body. The ferromagnetic (FM) orders in magnetic systems were dominated as mono-domain (or single-domain) nanoparticles consisting of FM surface and antiferromagnetic (AFM) core regions which couple with each other (Rego & Figueiredo, 2001; Leite & Figueiredo, 2004). At the lower temperatures, the FM surface and AFM core are only ordered in the noninteracting (monodomain) NPs. Stoner-Wohlfarth (Stoner & Wohlfarth, 1948) and Heisenberg model (Heisenberg, 1928) to describe the fine structure were fistly used in detail. Magnetic evolutions with temperatures (Babin, et al., 2003; Szlaferek, 2004; Usov & Gudoshnikov, 2005), thermodynamic properties (Vargas et al., 2002) and experimental techniques (Wernsdorfer et al., 1995; Wernsdorfer et al., 2000) were performed by different type works for the core-surface NPs. A simple (Bakuzis & Morais, 2004) and the first atomic-scale models of the ferrimagnetic and heterogeneous systems in which the exchange energy plays a central role in determining the magnetization of the NPs, were studied (Kodama et al., 1996, 1999; Kodama & Berkowitz, 1999).

Ising models and real magnets have provided a rich and productive field for the interaction between theory and experiment over the past 86 years (Ising, 1925; Peierls, 1936). Ising models (Erdem, 1995; Keskin, & Erdem 1997; Erdem & Keskin, 2001; Erdem, 2009; Erdem, 2008; Chen & Levy, 1973) and thier variants such as Blume-Capel (Blume, 1966; Capel, 1966; Bakchich, et al., 1994), Blume-Emery-Griffiths (Blume, et al., 1971; Achiam, 1985; Hoston, & Berker, 1991; Bakkali, 1996; Goveas & Mukhopadhyay, 1997; Keskin, et al., 1999; Temizer, 2008) and mixed spin (Benayad & Dakhama 1997; Kaneyoshi, 1998; Albayrak, & Yigit, 2005; Albayrak, & Yigit, 2006; Albayrak, 2007; Albayrak, 2007; Deviren, et al., 2009) models were regarded as theoretical simplifications, designed to model the essential aspects of cooperative system (Kikuchi,1951) without detailed correspondence to specific materials.

In the scope of this chapter, we give a detailed analysis for both spin $S = 1/2$ and $S = 1$ Ising models of homegeneous and core-surface composite NPs to describe the magnetic properties of these particles. These models are based on the pair approximation in the Kikuchi version (Kikuchi, 1974; Keskin, 1986; Erdinç & Keskin, 2002; Yalçın, et al. 2008, Özüm, 2010; Çiftçi, 2011). Incorporating the pair correlations between the spins inside the NPs, we calculated the free energy and minimized with respect to pair variables to obtain the field-cooled magnetization. The field cooling magnetization (M) curves of homogeneous and composite NPs are given as a function of the reduced temperature with different radius and different type lattices. Hysteresis loops and coercive fields with their linear fit to the data were plotted as a function of radius and temperature of different NPs. We compared our result with other works (Kaneyoshi, 2005; Kodama, 1999; Usov & Gudoshnikov, 2005).

2. Theoretical model

2.1 Ising model

Ising model, which was introduced in the field of magnetism, is one of the most studied models in modern statistical physics. Although its greatest success during the last century has been in the theory of phase transitions, the model today is viewed as a mathematical structure which can represent a variaty of different physical phenomena. In this section, we give a brief summary for the basics of the model before its application to the nanoparticle (NP) magnetism.

Ising model is considered on a regular lattice where each interior site has the same number of nearest-neighbour sites. This is called the coordination number of the lattice and will be denoted by γ. The system under consideration is composed of the magnetic atoms (also called the spins) located at the lattice sites. It is assumed that, in the thermodynamic limit, boundary sites can be disregarded and that, with N sites, the number of nearest-neighbour site pairs is $N\gamma/2$. The standard Hamiltonian for the the simplest Ising model is given by

$$H\{S_i\} = -J\sum_{\langle ij\rangle} S_i S_j - h\sum_{\langle ij\rangle}(S_i + S_j), \text{ with } S_i = \pm 1,$$ (1)

where h is the external magnetic field at the site i and the summation is performed for nearest-neighbour sites. J is the exchange interaction between neighbouring sites $\langle ij\rangle$. Two distinctive cases corresponding to different signs of intersite interaction is considered, i.e., $J < 0$ (ferromagnetic (FM) coupling) and $J > 0$ (antiferromagnetic (AFM) coupling). The fractions of $S_i = \pm 1$ spins given by X_i are called the point (or state) variables. The X_i are normalized by $\sum_{i=1}^{2} X_i = 1$. The long-range order parameter in the model is called the magnetization (M) and it is defined by $M = X_1 - X_2$. From this definition and the normalization condition the point variables can be written as

$$X_1 = \frac{1}{2}(1 + M), \quad X_2 = \frac{1}{2}(1 - M).$$ (2)

On the other hand, Eq. (1) may be extended by allowing values $s = 0, \pm 1, \pm 2, ..., \pm S$ for the variables. It is then possible to consider higher order interactions such as $K\sum_{\langle ij\rangle} S_i^2 S_j^2$ or a chemical potential such as $\Delta\sum_i S_i^2$. These generalizations are regarded as extensions of the Blume-Emery-Griffiths (BEG) model (Blume et al., 1971). Recently, there have been many theoretical studies of mixed spin Ising systems. These are of interest because they have less translational symmetry than their single-spin counterparts since they consist of two interpenetrating inequivalent sublattices. The latter property is very important to study a certain type of ferrimagnetism, namely molecular-based magnetic materials which are of current interest (Kaneyoshi et al., 1998).

2.2 Pair approximation

In the pair approximation, we consider the pair correlations between the spins. Besides the point variables (X_i), we introduce new variables (Y_{ij}), indicating the average number of the states in which the first member of the nearest-neighbour pair is in state i and the second member in state j. These will be called the pair or bond variables. The bond variables are normalized by $\sum_{i,j=1}^{n} Y_{ij} = 1$ and related to the state varibales by the relations $X_i = \sum_{j=1}^{n} Y_{ij}$. Here n is the number of spin states in the given spin S model. The interaction energy E and entropy S_E can be written in terms of Y_{ij} as

$$\beta E = N\frac{\gamma}{2}\sum_{i,j}^{n}\eta_{ij}Y_{ij},$$ (3)

$$S_E = Nk\left((\gamma - 1)\sum_{i,j=1}^{n} X_i \ln(X_i) - \frac{\gamma}{2}\sum_{i,j=1}^{n} Y_{ij}\ln(Y_{ji})\right),$$ (4)

where $\beta = 1/kT$ (k Boltzmann's constant and T temperature). In Eq. (3), the parameters η_{ij} are called the bond energies for the spin pairs (i, j) and determined from Eq. (1). The free energy per site Φ can be found from

$$\Phi = \frac{\beta F}{N} = \frac{\beta}{N}(E - TS_E) \cdot \tag{5}$$

For the system at equilibrium, the minimization of Eq. (5) with respect to Y_{ij} ($\partial\Phi / \partial Y_{ij} = 0$) leads to the following set of self-consistent equations:

$$Y_{ij} = \frac{1}{Z}(X_i X_j)^{\bar{\gamma}} e^{-\beta\eta_{ij}} \equiv \frac{e_{ij}}{Z}, \tag{6}$$

where $\bar{\gamma} = (\gamma - 1)/\gamma$ and Z is the partition function:

$$Z = \exp(2\beta\lambda / \gamma) = \sum_{i,j=1}^{n} e_{ij} \cdot \tag{7}$$

In Eq. (7), λ is introduced to maintain the normalization condition. Applications of the above formulation to $S = 1/2$ and $S = 1$ Ising systems can be found in many works in the literature (Meijer et al., 1986; Keskin & Meijer, 1986; Keskin & Erdinç, 1995; Erdinç & Keskin, 2002). These applications are summerized for comparison in Table 1.

	$S = 1/2$	$S = 1$
Spin state variables (X_i)	X_1, X_2	X_1, X_2, X_3
Spin values (S_i)	$+1, -1$	$+1, 0, -1$
Bond variables $Y_{ij}(S_i, S_j)$	$Y_{11}(+1,+1), Y_{12}(+1,-1)$ $Y_{21}(-1,+1), Y_{22}(-1,-1)$	$Y_{11}(+1,+1), Y_{12}(+1,0), Y_{13}(+1,-1)$ $Y_{21}(0,+1), Y_{22}(0,0), Y_{23}(0,-1)$ $Y_{31}(-1,+1), Y_{32}(-1,0), Y_{33}(-1,-1)$
Normalization	$\sum_{i=1}^{2} X_i = 1, \ \sum_{i,j=1}^{2} Y_{ij} = 1$	$\sum_{i=1}^{3} X_i = 1, \ \sum_{i,j=1}^{3} Y_{ij} = 1$
Relations between point variables and pair variables	$X_i = \sum_{j=1}^{2} Y_{ij}$ $X_1 = Y_{11} + Y_{12}$ $X_2 = Y_{21} + Y_{22}$	$X_i = \sum_{j=1}^{3} Y_{ij}$ $X_1 = Y_{11} + Y_{12} + Y_{13}$ $X_2 = Y_{21} + Y_{22} + Y_{23}$ $X_3 = Y_{31} + Y_{32} + Y_{33}$
Avarage magnetization ($M \equiv \langle S_i \rangle$)	$M = X_1 - X_2$ $M = Y_{11} + Y_{12} - (Y_{21} + Y_{22})$	$M = X_1 - X_3$ $M = Y_{11} + Y_{12} + Y_{13}$ $-(Y_{31} + Y_{32} + Y_{33})$
Quadrupole moment $Q \equiv \langle Q \rangle \equiv \langle S_i^2 \rangle$	-----	$Q \equiv X_1 + X_3$ $Q = Y_{11} + Y_{12} + Y_{13} + Y_{31} + Y_{32} + Y_{33}$

Table 1. Comparison of the $S = 1/2$ and $S = 1$ Ising models under the pair approximation.

3. Magnetic properties of $S = 1/2$ and $S = 1$ Ising nanoparticles

The magnetic particles become single domain below a critical size in contrast with the usual multidomain structure of the bulk materials. Therefore, in the scope of this section, we study size effects and magnetic properties of monodomain NPs manifestations. We consider a noninteracting monodomain NP with Ising spins on both hexagonal and square lattices for any two-dimensional (2D) regular arrays which can also be extended to hexagonal closed packed (hcp) and simple cubic (sc) lattices for the three-dimensional (3D) case as in Fig.1. The shells and their numbers originate from the nearest-neighbor pair interactions for the hexagonal and square lattices in 2D. In this structure, number of shells for hexagonal and square lattices can be associated with radius (R) of the NPs. This behaviour can be seen explicitly in Fig. 2 for hexagonal lattice and in Fig. 3 for square lattice. The value of R includes number of shells and the size of a NP increases as the number of shells increses. Therefore, we have considered Ising spins in three parts that are core (C), core-surface (CS) and surface (S) within the NP. Each of these parts contain core spin number (N_C), core-surface spin number (N_{CS}) and surface spin number (N_S), respectively. The total number of spins (N) in a single NP involves core and surface spin numbers, i.e. $N = N_C + N_S$. The C and S spins interact ferromagnetically ($J < 0$) or antiferromagnetically ($J > 0$). The $S = 1/2$ and $S = 1$ Ising model Hamiltonians with dipol-dipol interaction (J) for a NP is given by

$$H = H_C + H_{CS} + H_S,$$ (8)

with

$$H_C = -J_C \sum_{\langle i,j \rangle} S_i S_j - h \sum_{\langle i,j \rangle} (S_i + S_j),$$

$$H_{CS} = -J_{CS} \sum_{\langle i,j \rangle} S_i \sigma_j,$$ (9)

$$H_S = -J_S \sum_{\langle i,j \rangle} \sigma_i \sigma_j - h \sum_{\langle i,j \rangle} (\sigma_i + \sigma_j),$$

where J_C, J_{CS} and J_S represent exchange interactions for C, CS and S atoms, respectively. If $J_C = J_{CS} = J_S$, the NP is known as a homegeneous NP. It is called a composite NP when $J_C \neq J_{CS} \neq J_S$, $J_C = J_{CS} \neq J_S$, $J_C \neq J_{CS} = J_S$ or $J_{CS} \neq J_C = J_S$. In Eqs. (9), S_i is called the core spin values and σ_i is the surface spin values. These variables take on the values ± 1 for $S = 1/2$ and $0, \pm 1$ for $S = 1$ Ising systems.

The interaction energies for $S = 1/2$ and $S = 1$ models of an Ising NP in 2D can be written shortly in term of Y_{ij} as

$$\beta E = \sum_{\langle i,j \rangle} (N_P^C \eta_{ij}^C + N_P^{CS} \eta_{ij}^{CS} + N_P^S \eta_{ij}^S) Y_{ij},$$ (10)

where the numbers of spin pairs for C, CS and S regions are defined by $N_P^C = (N_C \gamma_C / 2) - N_{CS}$, $N_P^{CS} = 2 N_{CS} \gamma_{CS} / 2$ and $N_P^S = N_S \gamma_S / 2$, respectively. Similarly γ_C, γ_{CS}, γ_S denote the coordination numbers for these regions. Since we consider the arrays of Ising spins for a structure made up of bigger particles in 2D, we choose $\gamma_C = 6$, $\gamma_{CS} = \gamma_S = 2$ for hexagonal lattice and $\gamma_C = 4$, $\gamma_S = 0$, $\gamma_{CS} = 2$ for square lattice, as depicted in Figs. 2 and 3, respectively. The values of these numbers for both suructures in 2D are given in Table 2. The expressions for the bond energies η_{ij}^C, η_{ij}^{CS} and η_{ij}^S of three regions are found using Eq. (9) for both models, as listed in Table 3.

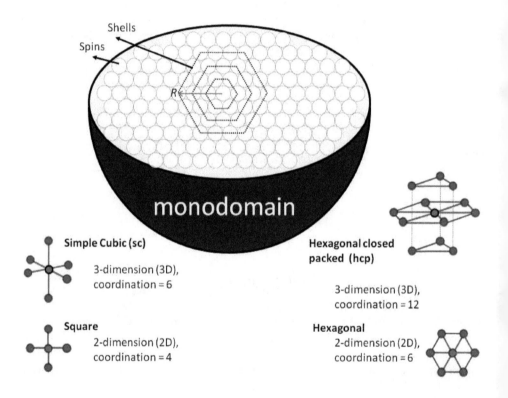

Fig. 1. A spherical monodomain magnetic NP spaced coherently in a form of 3D arrays. The shape of a single NP consists of the hexagonal lattice. The dashed lines displayed shells of spins in a 2D finite arrays. The radius of NP (R) includes shell numbers. The insets exhibit coordination numbers (γ) of hexagonal closed packed (hcp) and simple cubic (sc) lattices in 3D as well as hexagonal and square lattices in 2D structure.

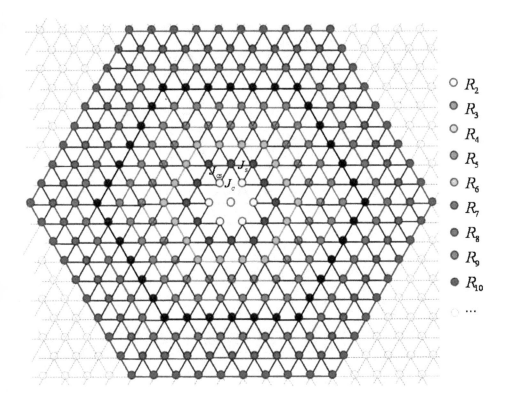

Fig. 2. Schematic representation of a NP on a hexagonal lattice in 2D exhibiting nine shells of
spins. Small full coloured circles correspond to ten radius of the NP. Solid grey lines are
number of the core-shell pairs. Solid coloured lines are number of shell pair (this line
corresponds to core and shell number for $R = 2$).

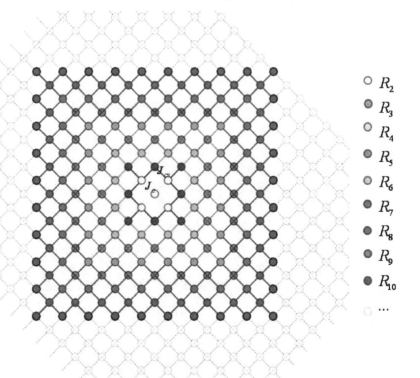

$\circ$ R_2

$\circ$ R_3

$\circ$ R_4

$\circ$ R_5

$\circ$ R_6

$\circ$ R_7

$\circ$ R_8

$\circ$ R_9

$\bullet$ R_{10}

$\circ$ $\cdots$

Fig. 3. Same as Fig. 2 but for the NP on square lattice in 2D.

Lattice Type	R	2	3	4	5	6	7	8	9	10
Hexagonal Lattice in 2D	N_C	7	19	37	61	91	127	169	217	271
	N_S	12	18	24	30	36	42	48	54	60
	N_{CS}	9	15	21	27	33	39	45	51	57
	N_P^C	12	42	90	156	240	342	462	600	756
	N_P^S	12	18	24	30	36	42	48	54	60
	N_P^{CS}	18	30	42	54	66	78	90	102	114
Square Lattice in 2D	N_C	5	13	25	41	61	85	113	145	181
	N_S	8	12	16	20	24	28	32	36	40
	N_{CS}	6	10	14	18	22	26	30	34	38
	N_P^C	4	16	36	64	100	144	196	256	324
	N_P^{CS}	12	20	28	36	44	52	60	68	76

Table 2. Numbers of the spins and spin pairs within the C, CS and S regions (Yalçın, et al., 2008).

Magnetic Properties and Size Effects of Spin-1/2 and Spin-1 Models of Core-Surface Nanoparticles in Different Type Lattices

211

Spin Model	Pair	Bond energy for Core (η_{ij}^{C})	Bond energy for Core-Surface (η_{ij}^{CS})	Bond energy for Surface (η_{ij}^{S})
$S = 1/2$ ($n = 2$)	η_{11}	$-J_C - 2h$	$-J_{CS}$	$-J_S - 2h$
	η_{12}	$+J_C$	$+J_{CS}$	$+J_S$
	η_{21}	$+J_C$	$+J_{CS}$	$+J_S$
	η_{22}	$-J_C + 2h$	$-J_{CS}$	$-J_S + 2h$
$S = 1$ ($n = 3$)	η_{11}	$-J_C - 2h$	$-J_{CS}$	$-J_S - 2h$
	η_{12}	$-h$	0	$-h$
	η_{13}	$+J_C$	$+J_{CS}$	$+J_S$
	η_{21}	$-h$	0	$-h$
	η_{22}	0	0	0
	η_{23}	$+h$	0	$+h$
	η_{31}	$+J_C$	$+J_{CS}$	$+J_S$
	η_{32}	$+h$	0	$+h$
	η_{33}	$-J_C + 2h$	$-J_{CS}$	$-J_S + 2h$

Table 3. Bond energies for $S = 1/2$ and $S = 1$ models.

Using Eq. (6) we obtain four self-consistent equations of Y_{ij} for $S = 1/2$ model of core-surface NPs:

$$\begin{aligned}
Y_{11} &= \frac{1}{Z}(X_1 X_1)^{\bar{\gamma}} \exp\left[-\beta\left(N_P^C \eta_{11}^C + N_P^{CS} \eta_{11}^{CS} + N_P^S \eta_{11}^S\right)\right] \equiv \frac{e_{11}}{Z}, \\
Y_{12} &= \frac{1}{Z}(X_1 X_2)^{\bar{\gamma}} \exp\left[-\beta\left(N_P^C \eta_{12}^C + N_P^{CS} \eta_{12}^{CS} + N_P^S \eta_{12}^S\right)\right] \equiv \frac{e_{12}}{Z}, \\
Y_{21} &= \frac{1}{Z}(X_2 X_1)^{\bar{\gamma}} \exp\left[-\beta\left(N_P^C \eta_{21}^C + N_P^{CS} \eta_{21}^{CS} + N_P^S \eta_{21}^S\right)\right] \equiv \frac{e_{21}}{Z}, \\
Y_{22} &= \frac{1}{Z}(X_2 X_2)^{\bar{\gamma}} \exp\left[-\beta\left(N_P^C \eta_{22}^C + N_P^{CS} \eta_{22}^{CS} + N_P^S \eta_{22}^S\right)\right] \equiv \frac{e_{22}}{Z}.
\end{aligned} \tag{11}$$

Similarly, nine self-consistent equations of Y_{ij} for $S = 1$ model of these particles are

$$Y_{11} = \frac{1}{Z}(X_1 X_1)^{\bar{\gamma}} \exp\left[-\beta\left(N_P^C \eta_{11}^C + N_P^{CS} \eta_{11}^{CS} + N_P^S \eta_{11}^S\right)\right] \equiv \frac{e_{11}}{Z},$$

$$Y_{12} = \frac{1}{Z}(X_1 X_2)^{\bar{\gamma}} \exp\left[-\beta\left(N_P^C \eta_{12}^C + N_P^{CS} \eta_{12}^{CS} + N_P^S \eta_{12}^S\right)\right] \equiv \frac{e_{12}}{Z},$$

$$Y_{13} = \frac{1}{Z}(X_1 X_3)^{\bar{\gamma}} \exp\left[-\beta\left(N_P^C \eta_{13}^C + N_P^{CS} \eta_{13}^{CS} + N_P^S \eta_{13}^S\right)\right] \equiv \frac{e_{13}}{Z},$$

$$Y_{21} = \frac{1}{Z}(X_2 X_1)^{\bar{\gamma}} \exp\left[-\beta\left(N_P^C \eta_{21}^C + N_P^{CS} \eta_{21}^{CS} + N_P^S \eta_{21}^S\right)\right] \equiv \frac{e_{21}}{Z},$$

$$Y_{22} = \frac{1}{Z}(X_2 X_2)^{\bar{\gamma}} \exp\left[-\beta\left(N_P^C \eta_{22}^C + N_P^{CS} \eta_{22}^{CS} + N_P^S \eta_{22}^S\right)\right] \equiv \frac{e_{22}}{Z}, \qquad (12)$$

$$Y_{23} = \frac{1}{Z}(X_2 X_3)^{\bar{\gamma}} \exp\left[-\beta\left(N_P^C \eta_{23}^C + N_P^{CS} \eta_{23}^{CS} + N_P^S \eta_{23}^S\right)\right] \equiv \frac{e_{23}}{Z},$$

$$Y_{31} = \frac{1}{Z}(X_3 X_1)^{\bar{\gamma}} \exp\left[-\beta\left(N_P^C \eta_{31}^C + N_P^{CS} \eta_{31}^{CS} + N_P^S \eta_{31}^S\right)\right] \equiv \frac{e_{31}}{Z},$$

$$Y_{32} = \frac{1}{Z}(X_3 X_2)^{\bar{\gamma}} \exp\left[-\beta\left(N_P^C \eta_{32}^C + N_P^{CS} \eta_{32}^{CS} + N_P^S \eta_{32}^S\right)\right] \equiv \frac{e_{32}}{Z},$$

$$Y_{33} = \frac{1}{Z}(X_3 X_3.)^{\bar{\gamma}} \exp\left[-\beta\left(N_P^C \eta_{33}^C + N_P^{CS} \eta_{33}^{CS} + N_P^S \eta_{33}^S\right)\right] \equiv \frac{e_{33}}{Z}.$$

Eqs. (11) and (12) are solved numerically using Newton-Raphson method and normalized magnetization (M) is easily calculated for both $S = 1/2$ and $S = 1$ models of homegeneous and core-surface composite NPs. Results are shown as the magnetization curves and hysteresis loops in Figs. 4–9.

4. Result and discussions

4.1 Magnetization

The evolution of normalized magnetization (M) as a function of the reduced temperature ($k_B T / J_0$) and particle size dependence of the transition temperature T_C from FM to paramagnetic (PM) phases for homogeneous and composite Ising NPs are shown in Figs. 4 and 5, respectively. The magnetization curves in Fig. 4 are plotted for $S = 1/2$ and $S = 1$ models of homogeneous NPs using the FM core ($J_0 = 1, J_C = 1$), FM surface ($J_S = J_0$) and FM core-surface ($J_{CS} = J_0$) interactions and the curves in Fig. 5 are obtained for both models of the composite NPs based on FM core ($J_C = J_0$), FM surface ($J_S = J_0$) and AFM core-surface ($J_{CS} = -J_0$) interactions. In the plots, different values for the applied magnetic field are considered ($h = 0.0$-0.1). The solid curves in the figures correspond to hexagonal lattice while dotted ones denote the square lattice. As seen from the figures, the changes in the magnetization with the reduced temperature point out an interesting aspect for NPs on the hexagonal and square lattices in 2D. The magnetization curves are decreasing from one (1) to zero (0) value while the reduced temperature is increasing (Figs. 4(a), 4(b), 5(a), 5(b)). These decreases terminate at the phase transition temperature (or Curie temperature, T_C) from FM phase to PM phase for $h = 0.0$, seen in Figs. 4(c) and 5(c). To show the size dependence of the critical temperature we plot $\sqrt{T_C}$ vs R in Figs. 4(d) and 5(d). All critical temperature values follow a linear increase with the particle radius. With increase in the particle radius it approaches to the Crue temperatures of the bulk materials. This is consistent with the mean-field approximation for the magnetic structure of Heisenberg NP (Usov & Gudoshnikov, 2005). On the other hand, it is interesting that composite $S = 1/2$ and $S = 1$ Ising NPs show smaller transition temperatures than their corresponding homogeneous NPs. This can easily be seen by comparing the same coloured fits in Figs. 4(d) and 5(d).

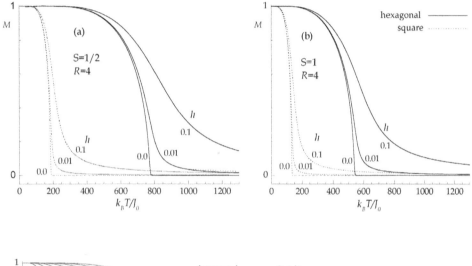

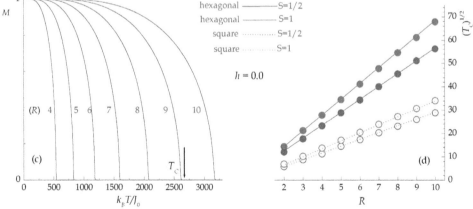

Fig. 4. Normalized magnetization (M) vs. reduced temperature ($k_B T / J_0$) and particle size dependence of the transition temperature T_C from FM to PM phases for homogeneous $S = 1/2$ and $S = 1$ Ising NPs on the hexagonal and square lattices. $J_0 = J_C = J_{CS} = J_S = 1$ and $h = 0.0$-0.1.

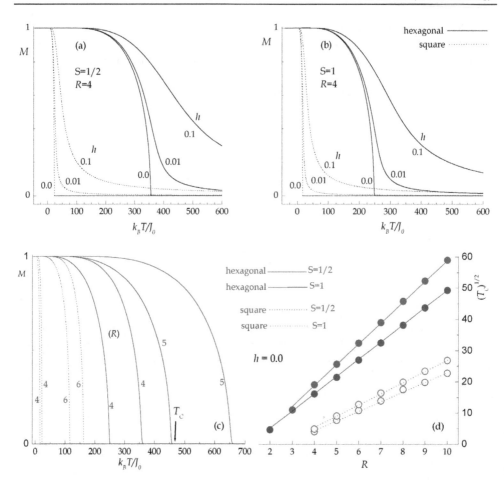

Fig. 5. Same as Fig. 4 but for the core-surface composite NPs with $J_0 = J_C = J_S = 1$, $J_{CS} = -J_0 = -1$.

4.2 Hysteresis loops

The magnetic field evolution of normalized magnetization (or hysteresis loops) for the homegeneous $S = 1/2$ and $S = 1$ Ising NPs which has different particle sizes and their corresponding coercive field vs. R^{-2} variation are given Figs. 6 and 7, respectively. We consider a FM coupling in core $(J_C = J_0)$, surface $(J_S = J_0)$ and core-surface $(J_{CS} = J_0)$ regions with $J_0 = 1$ on the hexagonal and square lattice structures. The hysteresis curves of small diameters, namely with radius $R = 2, 4, 5$ in Figs. 6(a)-6(d), are approximately the same. These behaviours are called superparamegnetic (SP) regime. However, the loops strongly depend on the size of NP. The hysteresis curves of high diamater values change sharply, as also shown in Figs. 6(a)-6(d). Moreover, the hysteresis curves for this type of NPs are broadening while the diamater of NPs is increasing so that it approaches to bulk materials. The size dependence of the coercive fields h_C is determined from the hysteresis loops in Fig. 7. In Fig. 7, the full red and blue circles correspond to the curves obtained for $R = 2-9$ in the Figs. 6(a) and 6(c), respectively. Similarly, the open red and blue circles correspond to the curves obtained for $R = 3-11$ and $R = 4-10$ in Figs. 6(b) and 6(d), respectively. The straight solid and dotted lines are the results from a linear fit to the calculated data. From this fit, it is obvious that the coercive field (h_C) depends linearly on $1/R^2$.

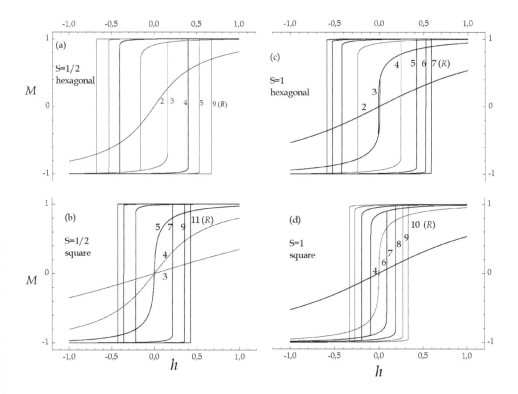

Fig. 6. (a) Hysteresis loops of a homegeneous $S = 1/2$ Ising NP on the hexagonal lattice for various sizes. (b) Same as Fig. 6(a) but for NP on the square lattice. (c) Hysteresis loops of a homegeneous $S = 1$ Ising NP on the hexagonal lattice for different sizes. (d) Same as Fig. 6(c) but for NP on the square lattice. $J_0 = J_C = J_{CS} = J_S = 1$ and $T = 300 J_0/k_B$.

Magnetic hysteresis loops of composite $S = 1/2$ and $S = 1$ Ising NPs on the hexagonal and square lattice (in 2D) structures for various values of particle sizes are shown in Fig. 8. The exchange interactions in the C and S regions are FM, i.e. $J_0 = J_C = J_S$, while the coupling between C and S is an AFM exchange constant $J_{CS} = -J_0$ for each type of NP. From the figure, it is clear that the hysteresis loops strongly depend on the particle size. The loops for the $S = 1/2$ and $S = 1$ Ising NPs on the hexagonal lattice change suddenly in low radius values while those for the $S = 1/2$ and $S = 1$ Ising NPs on the square lattice in high radius values.

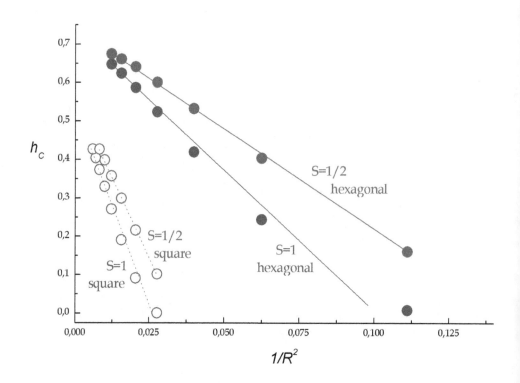

Fig. 7. The coercive field (h_C) plotted as a function of R^{-2} for the hysteresis loops of the homegeneous NP in Fig. 6.

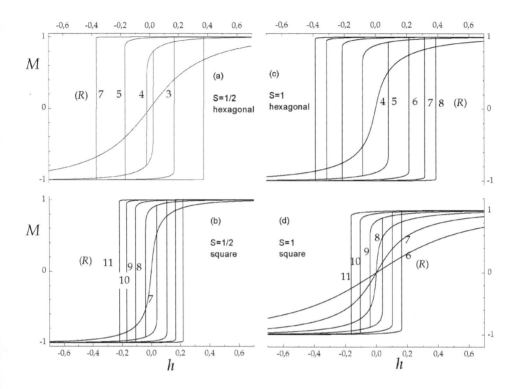

Fig. 8. Same as Fig. 6 but for the composite NP. $J_0 = J_C = J_S = 1$, $J_{CS} = -J_0$.

Finally, the evolutions of hysteresis loops and their coercive field according to the
temperature of composite Ising NPs are seen to change monotically as the temperature
increases, illustrated in Fig. 9(a) and 9(b), respectively. Since the loops for both models of
NPs on the hexagonal and square lattices display the same behaviour we have drawn only
the loops of $S = 1/2$ Ising NP on the hexagonal lattice. In this case, hysteresis for the NP is
in superparamagnetic (SP) regime at $700 J_0 / k_B$. But, the loops for the temperature regime
between $150 J_0 / k_B$ - $600 J_0 / k_B$ belong to the FM phase (Fig. 9(a)). The tempereture
dependence of the coercivity (h_C) are determined from the hysteresis loops of Fig. 9(a), as
given in Fig. 9(b).

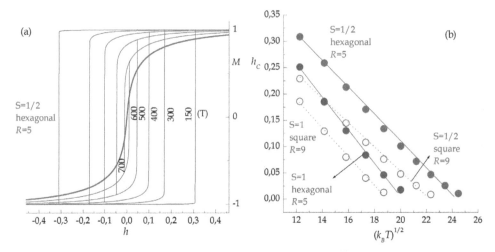

Fig. 9. (a) Temperature dependence of the hysteresis loops for the $S = 1/2$ Ising NP on the hexagonal lattice exhibiting five shells of spins ($R = 5$). (b) The coercive field (h_C) plotted as function of $(k_B T)^{1/2}$ for two models of NP on both structures studied above. $J_0 = J_C = J_S = 1$, $J_{CS} = -J_0$.

5. Conclusion

In the scope of this chapter, we have focused on the magnetic properties with size effects for homogeneous and core-surface composite NPs which have Ising spins (1, 1/2) on 2D lattice structures (hexagonal, square). The transition for all NPs corresponds to a second-order phase transition in the absence of magnetic field ($h \approx 0$). The spin disorder can be caused by lower coordination of the surface atoms in core-surface NPs broken exchange interactions that produce spin-glass (SG) like state of spatially disordered spin in the surface captions with inhomogeneous surface effects (Kodama, 1999; Kaneyoshi, 2005). Our theoretical observations are scrutinized below briefly.

i.   All critical temperature ($\sqrt{T_C}$) values of both types of Ising NPs on 2D lattice structures follow a linear increase with the particle size. With increase in the NP size it approaches to the Crue temperature of the bulk materials. These results agree with the mean-field magnetic structure of Heisenberg NPs (Usov & Gudoshnikov, 2005).

ii.   From the hysteresis loops for the homogeneous $S = 1/2$ and $S = 1$ Ising NPs which have different sizes and corresponding coercive field (h_C) vs. R^{-2} variations, it is clearly seen that the coercivity strongly depends on the particle size. Due to the superparamegnetic regime the hysteresis curves of small diameters are almost independent of each other while the curves of big diameters sharply change. This shows that the NP approaches to bulk materials.

iii. The hysteresis loops at different temperatures show a monotonic change in the coercive field of composite Ising NPs on 2D lattice structures. This property probably is an important aspect in the future high-density magnetic data storage.

6. Acknowledgements

One of us (Orhan Yalçın) would like to express his gratitude to "The Scientific and
Technological Research Council of Turkey" (TÜBİTAK) for financial support (Grant No.
107T635) during the this work.

7. References

Achiam, Y. (1985). Critical Relaxation of the One-Dimensional Blume-Emery-Griffiths
    Model. *Physical Review B*, Vol.31, pp.260-265, ISSN:1095-3795.
Aktaş, B.; Tagirov, L. & Mikailov, F. (October, 2006). *Magnetic Nanostructures,* Springer Series
    in materials science, Vol. 94, ISBN 978-3-540-49334-1.
Aktaş, B.; Tagirov, L. & Mikailov, F. (2004). *Nanostructures Magnetic Materials and Their
    Applications.*, Kluwer Academic Publisher. Nato Science Series. Mathematics,
    Physics and Chemistry. Vol. 143. ISBN 1-4020-2004-X.
Albayrak, E. (2007). Mixed Spin-2 and Spin-5/2 Blume–Emery–Griffiths Model. *Physica A,*
    Vol. 375, pp. 174-184, ISSN:0378-4371.
Albayrak, E. (2007). The Critical and Compensation Temperatures for the Mixed Spin-3/2
    and Spin-2 Ising Model. *Physica B,* Vol. 391, pp. 47-53, ISSN: 0921-4526.
Albayrak, E. & Yigit, A. (2005). The Critical Behavior of the Mixed Spin-1 and Spin-2 Ising
    Ferromagnetic System on the Bethe lattice. *Physica A,* Vol. 349, pp. 471-486,
    ISSN:0378-4371.
Albayrak, E. & Yigit, A. (2006). Mixed Spin-3/2 and Spin-5/2 Ising System on the Bethe
    Lattice. *Physics Letters A,* Vol. 353, pp. 121–129, ISSN: 0375-9601.
Bakkali, A.; Kerouad, M. & Saber, M. (1996). The Spin-3/2 Blume-Emery-Griffiths Model.
    *Physica A,*Vol. 229, No.3-4, pp.563-573, ISSN:0378-4371.
Babin, V.; Garstecki, P. & Holyst, R. (2003). Multiple Photonic Band Gaps in the Structures
    Composed of Core-Shell Particles. *Journal of Applied Physics,* Vol. 94, pp. 4244, ISSN:
    1089-7550.
Benayad, N. & Dakhama, A. (1997). Magnetic Properties of the Mixed-Spin Ising
    Ferromagnet with a Ferrimagnet Surface. *Physical Review B,* Vol. 55, No.18, pp.
    12276-12289, ISSN:1095-3795.
Bakchich, A.; Bekhechi, S. & Benyoussef, A. (1994). Multicritical Behavior of the
    Antiferromagnetic Spin-3/2 Blume-Capel Model. *Physica A,* Vol. 210, pp. 415-423,
    ISSN:0378-4371.
Bakuzis, A.F. & Morais, P.C. (2004). Magnetic nanoparticle systems: an Ising model
    approximation *Journal of Magnetism and Magnetic Materials,* Vol.272-276, pp. e1161-
    e1163 ISSN: 0304-8853.
Blume, M. (January 1966). Theory of the First-Order Magnetic Phase Change in UO_2, *Physical
    Review,* Vol. 141, No.2, pp. 517-524, ISSN 1094-1622.
Blume, M.; Emery, V. J. & Griffiths, R. B. (September 1971). Ising Model for the λ Transition
    and Phase Separation in He^3-He^4 Mixtures, *Physical Review A,* Vol. 4, No.3, pp.
    1071-1077, ISSN 1094-1622.
Capel, H. W., (1966). On the Possibility of First-Order Phase transitions in Ising Systems of
    Triplet Ions with Zero-Field Splitting. Physica Vol. 32, pp. 966-988.
Chen, H.H. & Levy, P.M. (1973). High Temperature Series Expansions for a Spin-1 Model of
    Ferromagnetism, *Physical Review B,* Vol. 7, pp. 4284–4289, ISSN: 1538-4446.

Çiftçi, N. (July, 2011). *Magnetic properties of a monodomain nanoparticle with dipole-quadropole interaction*, Master of Science Thesis. Thesis Supervisor, R. Erdem. Gaziosmanpaşa University, Turkey.

Deviren, B.; Keskin, M.& Canko, O. (2009). Kinetics of a Mixed Spin-1/2 and Spin-3/2 Ising Ferrimagnetic Model. *Journal of Magnetism and Magnetic Materials*, Vol. 321, pp. 458–466,. ISSN: 0304-8853.

Deviren, B.; Keskin, M. & Canko, O. (2009). Magnetic Properties of an Anti-ferromagnetic and ferrimagnetic Mixed Spin-1/2 and Spin-5/2 Ising Model in the Longitudinal Magnetic Field within the Effective-Field Approximation. *Physica A*, Vol. 388, pp. 1835-1848, ISSN:0378-4371.

Erdem, R. (2009). Frequency Dependence of the Complex Susceptibility for a Spin-1 Ising Model. *Journal of Magnetism and Magnetic Materials*, Vol. 321, pp. 2592–25595,. ISSN: 0304-8853.

Erdem, R. & Keskin, M. (2001). Dynamics of a Spin-1 Ising System in the neighborhood of Equilibrium States. *Physical Review E*, Vol. 64, pp. 0261102-1-9, ISSN 1550-2376.

Erdem, R. (2008). Magnetic Relaxation in a Spin-1 Ising Model near the second-order Phase Transition Point. *Journal of Magnetism and Magnetic Materials*, Vol. 320, pp. 2273–2278,. ISSN: 0304-8853.

Erdem, R. (September, 1995) *A study of the dynamics of a spin-1 Ising model with bilinear and biquadratic interactions*. Science Thesis. Thesis Supervisor, M. Keskin. Gaziosmanpaşa University, Turkey.

Erdinç, A. & Keskin, M. (May 2002). Equilibrium and Nonequilibrium Behavior of the Spin-1 Ising Model in the Quadrupole Phase. *Physica A*, Vol. 307, No. 3-4, pp. 453-468, ISSN:0378-4371.

Fraerman, A. A.; Gusev, S.A.; Nefedov, I.M.; Nozdrin, Y.N.; Karetnikova, I.R.; Mazo L.A.; Sapozhnikov, M.V.; Shereshevsky, I.A. & Suhodoev, L.V. (2001). *Journal of Physics Condensed Matter*, Vol. 13, pp. 683-689. ISSN 1361-648X.

Goveas, N. & Mukhopadhyay, G. (1997). Study of Blume-Emery-Griffiths Model by a Modified Bethe-Peierls Method. *Physica Scripta, Vol.* 56, pp.661-666, ISSN: 0031-8949.

Heisenberg, W. (1928). Theory of ferromagnetism *Zeitschrift für Physik*. Vol. 49, pp. 619-636.

Hoston, W. & Berker, A. N. (1991). Multicritical Phase Diagrams of the Blume-Emery-Griffiths Model with Republisive Biquadratic Coupling. *Physical Review Letters*, Vol.67, pp.1027-1030. ISSN:0031-9007.

Ising, E. (1925). Beitrag zur Theorie des Ferromagnetismus. *Zeitschrift für Physik*, Vol. 31, pp. 253-258.

Kaneyoshi, T.; Nakamura, Y. & Shin, S. (1998). A Diluted Mixed Spin-2 and Spin-5/2 Ferrimagnetic Ising System; A Study of a Molecular-Based Magnet. *Journal of Physics: Condensed Matter*, Vol.10, pp. 7025-7035. ISSN: 1361-648X.

Kartopu G. & Yalçın O., (February, 2010). Electrodeposited Nanowires and their Applications, In: *Electrodeposited Nanowires and Their Applications*, Nicolate Lupu, pp. 113-140. ISBN 978-953-7619-88-6.

Keskin, M.; Ekiz, C. &Yalcin, O. (1999). Stable, Metastable and Unstable Solutions of the Blume-Emery-Griffiths Model. *Physica A*, Vol. 267, pp. 392-405, ISSN:0378-4371.

Keskin, M. (1986). A Model for Quenching Via Hidden Variables Non-Equilibrium
Behaviour of a system with Two Range Order Parameters II. Influence of a
Magnetic Field. *Physica A*, Vol. 135, pp. 226-236, ISSN:0378-4371.

Keskin, M. & Erdem, R. (1997). Dynamic Behavior of a Spin-1 Ising Model. I. Relaxation of
Order Parameters and the 'Flatness' Property of Metastable States. *Journal of
Statistical Physics*, Vol. 89, pp. 1035-1046, ISSN: 1572-9613.

Keskin, M. & Erdinç, A. (1995). The Spin-1 Ising Model on the Body-Centered Cubic Lattice
Using the Pair Approximation. *Turkish Journal of Physics*, Vol.19, pp. 88-100, ISSN
1010-7630.

Keskin, M. & Meijer, P. H. E. (December 1986). Dynamics of a Spin-1 Model with the Pair
Correlation. *Journal of Chemical Physics*, Vol. 85, pp.7324-7333. ISSN: 1520-5207.

Kikuchi, R. (1951). A Theory of Cooperative Phenomena. *Physical Review*, Vol. 81, pp. 988-
1002, ISSN: 1536-6065.

Kaneyoshi, T. (2005). Phase Diagrams of a Nanoparticle Described by the Transverse Ising
Model. *Physica Status Solidi (b)* Vol.242, pp. 2938-2948, ISSN: 1521-3951.

Kikuchi, R. (1974). Superposition Approximation and Natural İteration Calculation in
Cluster-Variation Method. *Journal of Chemical Physics*, Vol. 60, pp. 1071, ISSN:1089-
7690.

Kodama, R.H. (1999)., Magnetic Nanoparticles. *Journal of Magnetism and Magnetic Materials*,
Vol.200, pp. 359–372, ISSN:0304-8853.

Kodama, R.H. & Berkowitz, A.E. (March 1999). Atomic-Scale Magnetic Modeling of Oxide
Nanoparticles. *Physical Review B*, *Vol.59*, No. 9, pp. 6321-6336, ISSN:0163-1829.

Kodama, R.H.; Berkowitz, A.E.; McNiff Jr. E.J. & Foner S. (July 1996). Surface Spin Disorder
in $NiFe_2O_4$ Nanoparticles. *Physical Review Letters*, Vol.77, No.2, pp.394-397,
ISSN:1079-7114.

Leite, V. S. & Figueiredo, W. (2004). Monte Carlo Simulations of Antiferromagnetic Small
Particles. *BrazilianJournal of Physics*, Vol. 34, No. 2a, pp.452-454, ISSN 0103-9733.

Meijer, P. H. E.; Keskin, M. & Bodegom, E. (October 1986). A Simple Model for the
Dynamics Towards Metastable States. *Journal of Statistical Physics*, Vol. 45, No1-2.
pp. 215-232. ISSN: 1572-9613.

Özüm, S. (July, 2010). *A study with pair approximation of spin- Ising model of noninteracting
nanoparticles with quadratic and crystal field interactions*, Master of Science Thesis.
Thesis Supervisor, Yalçın, O. Bozok University, Turkey.

Peierls, R. (1936). On Isings Model of Ferromagnetism. *Proceedings of the Cambridge
Philosophical Society,.*Vol. 32, pp. 477-481, ISSN: 1469-8064.

Pankhurst, Q. A.; Connolly, J.; Jones, S.K. & Dobson, J. (2003). Applications of magnetic
nanoparticles in biomedicine. *Journal of Physics D Applied Physics*, Vol. 36,pp. R167,
ISSN:1361-6463.

Rego, L. G. C. & Figueiredo, W. (September 2001). Magnetic Properties of Nanoparticles in
the Bethe-Peierls Aproximation. *Physical Review B*, Vol. 64, pp. 144424-1-7, ISSN:
1538-4446.

Stoner, E. C. & Wohlfarth, E. P. (1948). A Mechanism of Magnetic Hysteresis in
Heterogeneous Alloys. *Philosophical Transactions of the Royal Society of London Series
A*, Vol. 240,pp. 599-642, ISSN:0261-0523.

Szlaferek, A. (May 2004). Model Exchange-Spring Nanocomposite Magnetic Grains. *Physica
Status Solidi B*, Vol. 241, pp. 1312-1315, ISSN: 1521-3951.

Temizer, U.; Kantar, E.; Keskin, M. & Canko, O. (2008). Multicritical Dynamical Phase Diagrams of the Kinetic Blume–Emery–Griffiths Model with Repulsive Biquadratic Coupling in an Oscillating Field. *Journal of Magnetism and Magnetic Materials*, Vol. 320, pp. 1787–1801, ISSN: 0304-8853.

Usov, N. A. & Gudoshnikov, S. A. (2005). Magnetic Structure of a Nanoparticle in Mean-Field Approximation. *Journal of Magnetism and Magnetic Materials*, Vol.290, pp. 727-730, ISSN: 0304-8853.

Vargas, P.; Altbir, D.; Knobel, M. & Laroze, D. (May 2002). Thermodynamics of Two-Dimensional Magnetic Nanoparticles. *Europhysics Letters*, Vol.58, No. 4, pp. 603-609, ISSN:1286-4854.

Wernsdorfer, W.; Hasselbach, K.; Mailly, D.; Barbara, B.; Benoit, A.; Thomas, L. & Suran, G. (1995). DC-SQUID Magnetization Measurements of Single Magnetic Particles. *Journal of Magnetism and Magnetic Materials*, Vol.145, pp. 33, ISSN:0304-8853

Wernsdorfer, W.; Mailly, D. & Benoit, A. (May 2000). Single Nanoparticle Measurement Techniques. *Journal of Applied Physics*, Vol. 87, No. 9, pp. 5094-5096, ISSN: 1089-7550.

Yalçın, O.; Erdem, R. & Övünç, S. (2008). Spin-1 Model of Noninteracting Nanoparticles. *Acta Physica Polonica A*, Vol. 114, No.4, pp. 835-844, ISSN: 0587-4246.

Thermal Conductivity
of Nanoparticles Filled Polymers

Hassan Ebadi-Dehaghani and Monireh Nazempour
Shahreza Branch, Islamic Azad University
Iran

1. Introduction

Thermal conductivity of polymers is an important thermal property for both polymer applications and processing. Polymers typically have intrinsic thermal conductivity much lower than those for metals or ceramic materials, and therefore are good thermal insulators. Further enhancement of this thermal insulating quality can be achieved by foaming polymers. In other applications which require higher thermal conductivity, such as in electronic packaging and encapsulations, satellite devices, and in areas where good heat dissipation, low thermal expansion and light weight are needed, polymers reinforced with fillers, organic or inorganic, are becoming more and more common in producing advanced polymer composites for these applications (Hodgin & Estes, 1999; Tavman, 2004; Lee & Eun, 2004; Liu & Mather, 2004; Ishida & Heights, 1999; Frank & Phillip, 2002; Hermansen, 2001; Ishida, 2000). Most polymeric materials are processed and fabricated at elevated temperatures, often above their melting temperatures. This process may be long and expensive because of the low thermal conductivity of polymers. Subsequently, the cooling process or annealing may also be controlled by heat transport properties of polymers, which eventually affect the physical properties of the materials. One example is crystalline polymers, for which the structural and morphological features may be significantly changed with the speed of cooling. Careful consideration in designing polymer processing is vital to achieve desired properties.

For one-dimensional and rectilinear heat flow, the steady-state heat transfer in polymeric materials can be described by the Fourier's law of heat conduction:

$$q = -k\frac{dt}{dx} \tag{1}$$

where q is the heat flux (i.e., the heat transfer rate per unit area normal to the direction of flow), x is the thickness of the material, dT/dx is the temperature gradient per unit length, and the proportionality constant k is known as the thermal conductivity. The units for thermal conductivity k are expressed as $W/(m\ K)$ in SI units, $Btu\ in./(ft^2\ h\ °F)$ in English units, and $cal/(cm\ s\ °C)$ in cgs units. The corresponding units for heat flux are expressed as $W/(m^2)$, $Btu/(ft^2\ h)$, and $cal/(cm^2\ s)$, respectively.

Heat transfer involves the transport of energy from one place to another by energy carriers. In a gas phase, gas molecules carry energy either by random molecular motion (diffusion) or by an overall drift of the molecules in a certain direction (advection). In liquids, energy can be transported by diffusion and advection of molecules. In solids, phonons, electrons, or photons transport energy. Phonons, quantized modes of vibration occurring in a rigid crystal lattice, are the primary mechanism of heat conduction in most polymers since free movement of electrons is not possible (Majumdar, 1998). In view of theoretical prediction, the Debye equation is usually used to calculate the thermal conductivity of polymers (Han & Fina, 2010).

$$\lambda = \frac{C_p v l}{3} \qquad (2)$$

where C_p is the specific heat capacity per unit volume; v is the average phonon velocity; and l is the phonon mean free path.

For amorphous polymers, l is an extremely small constant (i.e. a few angstroms) due to phonon scattering from numerous defects, leading to a very low thermal conductivity of polymers (Agari et al., 1997). Table 1 displays the thermal conductivities of some polymers (T'Joen et al., 2009), (Hu et al., 2007) and (Speight, 2005).

Material	Thermal Conductivity at 25°C (W/m K)
Low density polyethylene (LDPE)	0.30
High density polyethylene (HDPE)	0.44
Polypropylene (PP)	0.11
Polystyrene (PS)	0.14
Polymethylmethacrylate (PMMA)	0.21
Nylon-6 (PA6)	0.25
Nylon-6.6 (PA66)	0.26
Poly(ethylene terephthalate) (PET)	0.15
Poly(butylene terephthalate) (PBT)	0.29
Polycarbonate (PC)	0.20
Poly(acrylonitrile-butadiene-styrene) copolymer (ABS)	0.33
Polyetheretherketone (PEEK)	0.25
Polyphenylene sulfide (PPS)	0.30
Polysulfone (PSU)	0.22
Polyphenylsulfone (PPSU)	0.35
Polyvinyl chloride (PVC)	0.19
Polyvinylidene difluoride (PVDF)	0.19
Polytetrafluoroethylene (PTFE)	0.27
Poly(ethylene vinyl acetate) (EVA)	0.34
Polyimide, Thermoplastic (PI)	0.11
Poly(dimethylsiloxane) (PDMS)	0.25
Epoxy resin	0.19

Table 1. Thermal conductivities of some polymers (T'Joen et al., 2009), (Hu et al., 2007) and (Speight, 2005).

3. Thermal conductivity – measurement and modeling

3.1 Methods for thermal conductivity measurements

Several methods, as reviewed elsewhere (Tritt & Weston, 2004) and (Rides et al., 2009), have been proposed and used for measurement of the thermal conductivity of polymers and composites. Classical steady-state methods measure the temperature difference across the specimens in response to an applied heating power, either as an absolute value or by comparison with a reference material put in series or in parallel to the sample to be measured. However, these methods are often time consuming and require relatively bulky specimens.

Several non steady-state methods have also been developed, including hot wire and hot plate methods, temperature wave method and laser flash techniques (Nunes dos Santos, 2007). Among these, laser-flash thermal diffusivity measurement is widely used, being a relatively fast method, using small specimens (Nunes dos Santos, 2007), (Nunes dos Santos, 2005) and (Gaal et al., 2004). In this method, the sample surface is irradiated with a very short laser pulse and the temperature rise is measured on the opposite side of the specimen, permitting calculation of the thermal diffusivity of the material, after proper mathematical elaboration. The thermal conductivity k is then calculated according to Eq. (3):

$$k = a C_p \rho \tag{3}$$

where a, C_p and ρ are the thermal diffusivity, heat capacity and density, respectively.

Differential scanning calorimetry (DSC) methods may also be used, applying an oscillary (Marcus & Blaine, 1994) or step temperature profile (Merzlyakov & Schick, 2001) and analyzing the dynamic response.

Significant experimental error may be involved in thermal conductivity measurements, due to difficulties in controlling the test conditions, such as the thermal contact resistance with the sample, leading to accuracy of thermal conductivity measurements typically in the range of 5–10%. In indirect methods, such as those calculating the thermal conductivity from the thermal diffusivity, experimental errors on density and heat capacity values will also contribute to the experimental error in the thermal conductivity.

3.2 Modeling of thermal conductivity in composites

Several different models developed to predict the thermal conductivity of traditional polymer composites are reviewed elsewhere (Bigg, 1995), (Zhou et al., 2007), (Zeng et al., 2009) and (Wang et al., 2008). The fundamentals are recalled in this section.

The two basic models representing the upper bound and the lower bound for thermal conductivity of composites are the rule of mixture and the so-called series model, respectively. In the rule of mixture model, also referred to as the parallel model, each phase is assumed to contribute independently to the overall conductivity, proportionally to its volume fraction (Eq. (4)):

$$k_c = k_p \Phi_p + k_m \Phi_m \tag{4}$$

where k_c, k_p, k_m are the thermal conductivity of the composite, particle, matrix, respectively, and Φ_p, Φ_m volume fractions of particles and matrix, respectively. The parallel model maximizes the contribution of the conductive phase and implicitly assumes perfect contact between particles in a fully percolating network. This model has some relevance to the case of continuous fiber composites in the direction parallel to fibers, but generally results in very large overestimation for other types of composites.

On the other hand, the basic series model assumes no contact between particles and thus the contribution of particles is confined to the region of matrix embedding the particle. The conductivity of composites accordingly with the series model is predicted by Eq. (5):

$$k_c = \frac{1}{(\phi_m+k_m)+(\phi_p+k_p)} \tag{5}$$

Most of the experimental results were found to fall in between the two models. However, the lower bound model is usually closer to the experimental data compared to the rule of mixture (Ebadi-Dehaghani et al., 2011; Bigg, 1995), which brought to a number of different models derived from the basic series model, generally introducing some more complex weighted averages on thermal conductivities and volume fractions of particles and matrix. These so-called second-order models including equations by Hashin and Shtrikman, Hamilton and Crosser, Hatta and Taya, Agari, Cheng and Vachon as well as by Nielsen (Bigg, 1995), (Zhou et al., 2007) and (Okamoto & Ishida 1999), appear to reasonably fit most of the experimental data for composites based on isotropic particles as well as short fibers and flakes with limited aspect ratio, up to loadings of about 30% in volume.

In the case of the geometric mean model, the effective thermal conductivity of the composite is given by:

$$K_c = K_m^{\phi_m} + K_f^{\phi_f} \tag{6}$$

Lewis and Nielsen modified the Halpin-Tsai equation (Nielsen et al., 1994) to include the effect of the shape of the particles and the orientation or type of packing for a two-phase system.

$$K_c = K_m \left[\frac{1+AB\phi_f}{1-B\psi\phi_f}\right] \tag{7}$$

Where

$$B = \frac{\frac{K_f}{K_m}-1}{\frac{K_f}{K_m}-A} \quad \psi = 1 + \left(\frac{1-\phi_{max}}{\phi_{max}^2}\right)\phi_f \tag{8}$$

The values of A and Φ_{max} were given for many geometric shapes and orientations (Weidenfeller et al., 2004).

This model appears to reasonably fit most of the experimental data for composites based on isotropic particles as well as short fibers and flakes with limited aspect ratio, up to loading of about 30% in volume. For higher loadings, the Nielsen's model appear to best fit the rapid increase of thermal conductivity above 30 vol.%, thanks for the introduction of the maximum packing factor into the fitting equation, despite the evaluation of maximum packing factor in real composites may present difficulties due to particle size distribution and particle dispersion in the matrix. However, the basic assumption of separated particles in the effective medium approach is not valid in principle for highly filled composites, where contacts are likely to occur, possibly leading to thermally conductive paths (Tavman, 1996).

Maxwell, using potential theory, obtained an exact solution for the conductivity of randomly distributed and non-interacting homogeneous spheres in a homogeneous medium:

$$K_c = K_m \left[\frac{K_f + 2K_m + 2\phi_f(K_f - K_m)}{K_f + 2K_m - \phi_f(K_f - K_m)} \right] \tag{9}$$

Other theoretical models have attempted to explain the thermal conductivity of two-phased composites. Some of these models, such as those by Bruggeman, Botcher, De Loor, and Ce Wen Nan et al., equations 6 to 9 respectively, have been used for prediction of thermal conductivity of carbon nanotube composites (Bruggeman, 1935; Böttcher, 1952; deLoor, 1956; Nan et al., 2004).

$$K_c = \frac{K_m}{(1 - \phi_f)^3} \tag{10}$$

$$K_c = \frac{K_m}{(1 - \phi_f)} \tag{11}$$

$$K_c = \frac{K_m(1 + \phi_f)}{1 - 2\phi_f} \tag{12}$$

$$K_c = \frac{K_m \left[3 + \phi_f \left(\frac{K_f}{K_m} \right) \right]}{3 - 2\phi_f} \tag{13}$$

In order to take into account fluctuations in thermal conductivity in the composites, Zhi et al. (Zhi et al., 2009) proposed the concept of heat-transfer passages, to model the conduction in regions where interparticle distance is low, applying the series model to "packed-belt" of conductive particles.

Even though these macroscopic approaches may be of interest from the engineering point of view, they deliver little or no information about the physical background of the observed behavior. As an example, very limited interpretation is given to the rapidly increasing conductivity with filler content above a certain filler loading (typically above 30 vol.%), or why the experimental results are so far away from the upper bound conductivity, even for highly percolated systems.

Attempts to model thermal conductivity taking into account the interfacial thermal resistance between conductive particles and matrix have been reported by several research groups (Nan et al., 1997), (Every et al., 1992), (Dunn & Taya, 1993), (Lipton & Vernescu, 1996) and (Torquato & Rintoul, 1995) and applied particles with different geometries and topologies, including aligned continuous fibers, laminated flat plates, spheres, as well as disoriented ellipsoidal particles. In general, these models provided an improved fit with experimental data for ceramic based composites than models not accounting for interface thermal resistance. These approaches generally assume conductive particles to be isolated in the matrix and take into account the thermal resistance in heat transfer between conductive particle and matrix, also known as Kapitza resistance, from the name of the discoverer of the temperature discontinuity at the metal–liquid interface. A very simple proof of thermal interfacial resistance is the fact that a thermal conductivity lower than the reference matrix was experimentally found with some composites containing particles with thermal conductivity higher than the matrix (Nan et al., 1997) and (Every et al., 1992). This phenomenon is explained by the very low efficiency of heat transfer between particles and matrix, so that the higher thermal conductivity of the filler cannot be taken into advantage and the composite behaves like a hollow material, thus reducing its conductivity compared to the dense reference matrix. Evaluation of the effective thermal conductivity of composite polymers by considering the filler size distribution law was investigated by Holotescu et al (Holotescu et al., 2009).

They presented an empirical model for the effective thermal conductivity (ETC) of a polymer composite that includes dependency on the filler size distribution – chosen as the Rosin-Rammler distribution. The ETC is determined based on certain hypotheses that connect the behavior of a real composite material A, to that of a model composite material B, filled with mono-dimensional filler. The application of these hypotheses to the Maxwell model for ETC is presented. The validation of the new model and its characteristic equation was carried out using experimental data from the reference. The comparison showed that by using the size distribution law a very good fit between the equation of the new model (the size distribution model for the ETC) and the reference experimental results is obtained, even for high volume fractions, up to about 50%.

4. Crystallinity and temperature dependence

Polymer crystallinity strongly affects their thermal conductivity, which roughly varies from 0.2 W/m K for amorphous polymers such as polymethylmethacrylate (PMMA) or polystyrene (PS), to 0.5 W/m K for highly crystalline polymers as high-density polyethylene (HDPE) (Hu et al., 2007). The thermal conductivity of semi-crystalline polymers is reported to increase with crystallinity. As an example, the thermal conductivity of polytetrafluoroethylene (PTFE) was found to increase linearly with crystallinity at 232 °C (Price & Jarratt, 2002).

However, there is a large scatter in the reported experimental data of thermal conductivity of crystalline polymers, even including some contradictory results. It should be noticed that the thermal conductivities of polymers depend on many factors, such as chemical constituents, bond strength, structure type, side group molecular weight, molecular density distribution, type and strength of defects or structural faults, size of intermediate range order, processing conditions and temperature. Furthermore, due to the phonon scattering at the interface between the amorphous and crystalline phase and complex factors on crystallinity of polymer, the prediction of the thermal conductivity vs. crystallinity presents a significant degree of complexity (Han & Fina, 2010).

Semicrystalline and amorphous polymers also vary considerably in the temperature dependence of the thermal conductivity. At low temperature, semicrystalline polymers display a temperature dependence similar to that obtained from highly imperfect crystals, having a maximum in the temperature range near 100 K, shifting to lower temperatures and higher thermal conductivities as the crystallinity increases (Greig & Hardy, 1981) and (Yano

& Yamaoka, 1995), while amorphous polymers display temperature dependence similar to that obtained for inorganic glasses with no maximum, but a significant plateau region at low temperature range (Reese, 1969). The thermal conductivity of an amorphous polymer increases with increasing temperature to the glass transition temperature (T_g), while it decreases above T_g (Zhong et al., 2001) and (Dashora & Gupta, 1996). The study of the thermal conductivity of some amorphous and partially crystalline polymers (PE, PS, PTFE and epoxy resin) as a function of temperature in a common-use range (273–373 K) indicates that the conductivity of amorphous polymers increases with temperature and that the conductivity is significantly higher in crystalline than amorphous regions (Kline, 1961).

From the general overview given in the preceding, it appears that very limited thermal conductivity is usually characteristic of polymers. On the other hand, there are many reasons to increase thermal conductivity of polymer-based materials in various industrial applications including circuit boards in power electronics, heat exchangers, electronics appliances and machinery. This justifies the recent significant research efforts on thermally conductive composite materials to overcome the limitations of traditional polymers.

5. Fillers for thermally conductive composites

Many applications would benefit from the use of polymers with enhanced thermal conductivity. For example, when used as heat sinks in electric or electronic systems, composites with a thermal conductivity approximately from 1 to 30 W/m K are required (King et al., 1999). The thermal conductivity of polymers has been traditionally enhanced by the addition of thermally conductive fillers, including graphite, carbon black, carbon fibers, ceramic or metal particles (see Table 2) (Pierson, 1993), (Wypych, 2000), (Fischer, 2006),

Material	Thermal Conductivity at 25 °C (W/m K)
Graphite	100∼400 (on plane)
Carbon black	6∼174
Carbon Nanotubes	2000∼6000
Diamond	2000
PAN-based Carbon Fibre	8∼70 (along the axis)
Pitch-based Carbon Fibre	530∼1100 (along the axis)
Copper	483
Silver	450
Gold	345
Aluminum	204
Nickel	158
Boron Nitride	250∼300
Aluminum nitride	200
Beryllium oxide	260
Aluminum oxide	20∼29

Table 2. Thermal conductivities of some thermally conductive fillers (Pierson, 1993), (Wypych, 2000), (Fischer, 2006), (Wolff & Wang, 1993) and (Kelly, 1981).

(Wolff & Wang, 1993) and (Kelly, 1981). It is worth noticing that significant scatter of data are typically reported for thermal conductivity of fillers. This is caused by several factors, including filler purity, crystallinity, particle size and measurement method. It is also important to point out that some materials, typically fibers and layers, are highly anisotropic and can show much higher conductivity along a main axis or on a plane, compared to perpendicular direction.

High filler loadings (>30 vol.%) are typically necessary to achieve the appropriate level of thermal conductivity in thermally conductive polymer composites, which represents a significant processing challenge. Indeed, the processing requirements, such as possibility to be extruded and injection molded, often limit the amount of fillers in the formulation and, consequently, the thermal conductivity performance (King et al., 2008). Moreover, high inorganic filler loading dramatically alters the polymer mechanical behavior and density. For these reasons, obtaining composites having thermal conductivities higher than 4 W/m K and usual polymer processability is very challenging at present (Han & Fina, 2010).

5.1 Carbon-based fillers

Carbon-based fillers appear to be the best promising fillers, coupling high thermal conductivity and lightweight. Graphite, carbon fiber and carbon black are well-known traditional carbon-based fillers. Graphite is usually recognized as the best conductive filler because of its good thermal conductivity, low cost and fair dispersability in polymer matrix (Causin et al., 2006) and (Tu & Ye, 2009). Single graphene sheets constituting graphite show intrinsically high thermal conductivity of about 800 W/m K (Liu et al., 2008) or higher (theoretically estimated to be as high as 5300 W/m K (Veca et al., 2009) and (Stankovich et al., 2006)), this determining the high thermal conductivity of graphite, usually reported in the range from 100 to 400 W/m K. Expanded graphite (EG), an exfoliated form of graphite with layers of 20–100 nm thickness, has also been used in polymer composites (Ganguli et al., 2008), for which the thermal conductivity depends on the exfoliation degree (Park et al., 2008), its dispersion in matrix (Mu & Feng, 2007) and the aspect ratio of the EG (Kalaitzidou et al., 2007).

Thermal conductivity of exfoliated graphite nanocomposites was investigated by Fukushima et al. (Fukushima et al., 2006). Since the late 1990's, research has been reported where intercalated, expanded, and/or exfoliated graphite nanoflakes could also be used as reinforcements in polymer systems. The key point to utilizing graphite as a platelet nanoreinforcement is in the ability to exfoliate graphite using Graphite Intercalated Compounds (GICs). Natural graphite is still abundant and its cost is quite low compared to the other nano–size carbon materials, the cost of producing graphite nanoplatelets is expected to be ~$5/lb. This is significantly less expensive than single wall nanotubes (SWNT) (>$45000/lb) or vapor grown carbon fiber (VGCF) ($40–50/lb), yet the mechanical, electrical, and thermal properties of crystalline graphite flakes are comparable to those of SWNT and VGCF. The use of exfoliated graphite flakes (xGnP) opens up many new applications where electromagnetic shielding, high thermal conductivity, gas barrier resistance or low flammability are required. A special thermal treatment was developed to exfoliate graphite flakes for the production of nylon and high density polypropylene nanocomposites. X-ray diffraction (XRD), scanning electron microscopy (SEM) and transmission electron microscopy (TEM) were used to assess the degree of exfoliation of the graphite platelets and the morphology of the nanocomposites. The thermal conductivity of these composites was investigated by three different methods, namely, by DSC, modified hot wire, and halogen flash lamp methods. The addition of small amounts of exfoliated graphite flakes showed a marked improvement in thermal and electrical conductivity of the composites.

Carbon fiber, typically vapor grown carbon fiber (VGCF), is another important carbon-based filler. Polymer/VGCF composites have been reviewed by Tibbetts et al. (Tibbetts et al., 2007). Since VGCF is composed of an annular geometry parallel to the fiber axis, thermal conductive properties along the fiber axis are very different from the transverse direction (estimated up to 2000 W/m K in the axial direction vs. 10–110 W/m K in the transverse direction (Chen & Ting, 2002) and (Zhang et al, 2000)), directly affecting the thermal conductivity of aligned composites (Mohammed & Uttandaraman, 2009) and (Kuriger et al., 2002).

Carbon black particles are aggregates of graphite microcrystals and characteristic of their particle size (10–500 nm) and surface area (25–150 m^2/g) (Pierson, 1993). Carbon black is reported to contribute to electrical conductivity rather than thermal conductivity (Wong et al., 2001), (Abdel-Aal et al., 2008) and (King et al., 2006).

5.2 Metallic fillers

The filling of a polymer with metallic particles may result in both increase of thermal conductivity and electrical conductivity in the composites. However, a density increase is also obtained when adding significant metal loadings to the polymer matrix, thus limiting applications when lightweight is required. Metallic particles used for thermal conductivity improvement include powders of aluminum, silver, copper and nickel. Boudenne et al studied the thermal behavior of polypropylene filled with copper particles (Kumlutaş et al., 2003). In this work thermal conductivity, diffusivity, effusivity and specific heat of polypropylene matrix filled with copper particles of two different sizes were investigated. A parallel study of the evolution of the electrical conductivity was also carried out. The highest heat transport ability was observed for the composites filled with the smaller particles. The Agari's model provides a good estimation of the thermal conductivity of composites for all filler concentrations. Polymers modified with the inclusion of metallic particles include polyethylene (Kumlutaş et al., 2003), polypropylene (Boudenne et al., 2005), polyamide (Tekce et al., 2007), polyvinylchloride and epoxy resins (Mamunya et al., 2002), showing thermal conductivity performance depending on the thermal conductivity of the metallic fillers, the particle shape and size, the volume fraction and spatial arrangement in the polymer matrix. Thermal conductivity of metal powder-polymer feedstock for powder injection moulding was studied by Kowalski et al. (Kowalski et al., 1999).

Thermal conductivity of a powder injection molding feedstock (mixture of metal powders and polymers) in solid and molten states has been measured by using the laser flash method. The filler material was 316L stainless steel powder and its content in the mixture amounted 60% by volume. An attempt has been made to employ two most promising existing mathematical models (theoretical Maxwell- and semi-theoretical Lewis & Nielsen model) to calculate the thermal conductivity of the mixture (see section 1.3.2). Comparison of the experimental and calculated results has revealed that the Lewis & Nielsen model predicts better than Maxwell model the thermal conductivity of the feedstock. As the difference between the calculated (Maxwell model) and the measured results amounts to 15–85%, it is suggested that it can only be used for preliminary assessment of the thermal conductivity of so highly filled composite material. If accurate thermal conductivity data are required (as in case of numerical simulation of the powder injection moulding process), measurement of this property has to be performed if meaningful simulation results are to be expected.

5.3 Ceramic fillers

Ceramic powder reinforced polymer materials have been used extensively as electronic materials. Being aware of the high electrical conductivity of metallic particles, several ceramic materials such as aluminum nitride (AlN), boron nitride (BN), silicon carbide (SiC) and beryllium oxide (BeO) gained more attention as thermally conductive fillers due to their high thermal conductivity and electrical resistivity (Nu et al., 2008) and (Ishida & Rimdusit, 1998). Thermal conductivities of composites with ceramic filler are influenced by filler packing density(Ohashi et al., 2005), particle size and size distribution (Yu et al., 2002) and (Mu et al., 2007), surface treatment (Gu et al., 2009) and mixing methods (Zhou et al., 2007). Models and theories for predicting the thermal conductivity of polymer composites were discussed. Effective Medium Theory (EMT), Agari model and Nielsen model respectively are introduced and are applied as predictions for the thermal conductivity of ceramic particle filled polymer composites. Thermal conductivity of experimentally prepared Si_3N_4/epoxy composite and some data cited from the literature are discussed using the above theories. Feasibility of the three methods as a prediction in the whole volume fraction region of the filler from 0 to 1 was evaluated for a comparison. As a conclusion: both EMT and Nielsen model can give a well prediction for the thermal conductivity at a low volume fraction of the filler; Agari model give a better prediction in the whole range, but with larger error percentage (He et al., 2007).

6. Nanocomposites for thermal conductivity

Polymer nanocomposites are commonly defined as the combination of a polymer matrix and additives that have at least one dimension in the nanometer range. The additives can be one-dimensional (examples include nanotubes and fibres), two-dimensional (which include layered minerals like clay), or three-dimensional (including spherical particles). Over the past decade, polymer nanocomposites have attracted considerable interests in both academia and industry, owing to their outstanding mechanical properties like elastic stiffness and strength with only a small amount of the nanoadditives. This is caused by the large surface area to volume ratio of nanoadditives when compared to the micro- and macro-additives. Other superior properties of polymer nano-composites include barrier resistance, flame retardancy, scratch/wear resistance, as well as optical, magnetic, thermal conductivity and electrical properties. Polymer based nanocomposites can be obtained by the addition of nanoscale particles which are classified into three categories depending on their dimensions: nanoparticles, nanotubes and nanolayers. The interest in using nanoscaled fillers in polymer matrices is the potential for unique properties deriving from the nanoscopic dimensions and inherent extreme aspect ratios of the nanofillers (Mai et al., 2006). Kumar et al. (Kumar et al., 2009) summarized six interrelated characteristics of nanocomposites over conventional micro-composites: (1) low-percolation threshold (about 0.1–2 vol.%), (2) particle–particle correlation (orientation and position) arising at low-volume fractions (less than 0.001), (3) large number density of particles per particle volume (10^6 to 10^8 particles/μm^3), (4) extensive interfacial area per volume of particles (10^3 to 10^4 m^2/ml), (5) short distances between particles (10–50 nm at 1–8 vol.%) and (6) comparable size scales among the rigid nanoparticles inclusion, distance between particles, and the relaxation volume of polymer chains (Han & Fina, 2010).

Different nanoparticles have been used to improve thermal conductivity of polymers. As a few examples, HDPE filled with 7 vol.% nanometer size expanded graphite has a thermal conductivity of 1.59 W/m K, twice that of microcomposites (0.78 W/m K) at the same volume content (Ye et al., 2006). Poly(vinyl butyral) (PVB), PS, PMMA and poly(ethylene vinyl alcohol) (PEVA) based nanocomposites with 24 wt.% boron nitride nanotubes (BNNT) have thermal conductivities of 1.80, 3.61, 3.16 and 2.50 W/m K, respectively (Zhi et al., 2009). Carbon nanofiber was also reported to improve the thermal conductivity of polymer composites (Sui et al., 2008) and (Elgafy & Lafdi, 2005). However, the most widely used and studied nanoparticles for thermal conductivity are certainly carbon nanotubes (either single wall-SWCNT or multiwall-MWCNT), which have attracted growing research interest. Indeed, CNT couples very high thermal conductivity with outstanding aspect ratio, thus forming percolating network at very low loadings.

Droval and co-workers (Droval et al., 2006) investigated the effect of boron nitride (BN), talc ($Mg_3Si_4O_{10}$ $(OH)_2$), aluminum nitride (AlN) and aluminum oxide (Al_2O_3) particles, and their impact on thermal properties. Lewis and Nielson, Cheng and Vachon, Agari and Uno models were used to predict the evolution of thermal conductivity with filler content and were found to describe correctly thermal conductivity. Only BN shows a real exponential increase of conductivity over 20% v/v filler. Consequently, in best conditions introducing 30% v/v of BN allows the thermal conductivity to be multiplied by six.

A technology has been developed for making carbon-ceramic composite refractories by combining carbon fibers as reinforcing component with a mixture matrix, which allows one to make refractory components of various sizes and geometry, including thin-walled large constructions (Chernenko et al., 2009). The heat resistance of these composite refractories increases with the bulk silicization during ceramic production on a carbon-carbon substrate. The degree of silicization is determined by the volume of the open microporosity of transport type, which is formed by pyrolysis of a polymer coke-forming matrix in the initial carbon plastic. The transport micropores are produced by a modification of the phenol-formaldehyde resin additive treatment, which does not give rise to coke on pyrolysis. As a result, the content of open pores in the carbon framework attains 55%, which enables one to make a silicized composite refractory of density up to 2.7 g/cm^3 with a compressive strength of 250 – 300 MPa, bending strength 120–140, and tensile strength 60–80 MPa, elastic modulus 120–140 GPa, linear expansion coefficient 3.5×10^{-6} – 4.5×10^{-6} K^{-1}, and thermal conductivity 6 – 8 W/(m K). These refractories are widely used in various branches of industry. Thermal conductivity of particle filled polyethylene composite materials was investigated by Kumlutas et al. (Kumlutas et al., 2003). In this study, the effective thermal conductivity of aluminum filled high-density polyethylene composites is investigated numerically as a function of filler concentration. The obtained values are compared with experimental results and the existing theoretical and empirical models. The thermal conductivity is measured by a modified hot-wire technique. For numerical study, the effective thermal conductivity of particle-filled composite was calculated numerically using the micro structural images of them. By identifying each pixel with a finite difference equation and accompanying appropriate image processing, the effective thermal conductivity of composite material is determined numerically. As a result of this study, numerical results, experimental values and all the models are close to each other at low particle content. For particle content greater than 10%, the effective thermal conductivity is exponentially formed. All the models fail to predict thermal conductivity in this region. But, numerical results give satisfactory values in the whole range of aluminum particle content.

6.1 Nanocomposites using inorganic fillers

Thermally conductive polymer nanocomposites based on polypropylene has been studied (Vakili et al., 2011; Ebadi-Dehaghani et al., 2011). In this study three nanocomposite containing 5 to 15 wt% of ZnO and $CaCO_3$ nanoparticles prepared by extrusion were used. The thermal conductivity (TC) of compression moulded polypropylene (PP) and PP filled nanoparticles was studied using thermal conductivity analyser (TCA). The effect of nanoparticle content and crystallinity on thermal conductivity was investigated using conventional methods like SEM, XRD and DSC. The incorporation of nanoparticles improved crystallinity and thermal conductivity simultaneously. The experimental TC values of PP nanocomposites with different level of nanoparticles concentration showed a linear increase with an increase in crystallinity.

6.1.1 Differential Scanning Calorimetry (DSC)

DSC measurements were investigated by conventional differential scanning calorimeter Labsys TG (Setaram Instumentation, Caluire, France). A pellet of extruded sample, with a weight of 8-10 mg, was placed into an alumina pan in the presence of air as the furnace atmosphere. Measurements were performed from ambient temperature up to 200°C with heating rate of 10°C/min. The DSC results for pure PP and nanocomposites, The Tm (peak temperature of melting) and ΔHm (enthalpy of melting), are listed in Table 3.

The degree of crystallinity of a specimen can be calculated from the melting heat of crystallization according to the following equation:

$$X_c = \frac{\Delta H_m}{\Delta H_0 (1 - W_f)} \times 100$$

(14)

Where w_f is the weight fraction of nanofiller and $\Delta H_0 = 207.1$ Jg^{-1} is the melting heat of 100% crystalline PP (Bai et al., 1999).

The DSC results indicated that the addition of both nanoparticles to the PP caused only a marginal effect on melting temperature (T_m) and no correlation of the results with the filler concentration could be established. The calculated degree of crystallinity of the PP phase increased with increasing content of both nanoparticles, indicating that the nanofillers nucleated the crystallization process. (Frormann et al., 2008) This implies that the existence of nanoparticles facilitates the crystallization of PP and this effect becomes more evident with higher nanoparticle content (Zhao & Li, 2006). Similar results for PP/CaCO3 nanocomposites, PP/carbon nanotube composites and PP/nanoclay composites (Han & Fina , 2011; Frormann et al., 2008; Vakili et al., 2011) have been reported. However there are some contradicting results in the literature (Zhao & Li, 2006).

	Neat PP	PP+5wt% nanofiller	PP+10wt% nanofiller	PP+15wt% nanofiller
T_m (°C)	167.8	168.5	168.8	167.0
ΔH_m (Jg^{-1})	78.0	91.2	108.9	107.7
X_c (%)	37.7	44.0	52.6	52.0

a)

	Neat PP	PP+5wt% nanofiller	PP+10wt% nanofiller	PP+15wt% nanofiller
T_m (°C)	167.8	168.6	168.1	168.4
ΔH_m (Jg^{-1})	78.0	81.4	85.8	104.6
X_c (%)	37.7	39.3	41.4	50.5

b)

Table 3. Crystallization parameters of neat PP and nanocomposites. a) PP/ZnO nanocomposite. b) PP/CaCO$_3$ nanocomposite.

6.1.2 Thermal conductivity measurement

Thermal conductivity was measured using a TCA Thermal Conductivity Analyser (TCA-200LT-A, Netzsch, Selb, Germany) with the guarded heat flow meter method. Each compression molded sample (30cm×30cm sheets with 10 mm thickness) was placed between two heated surface controlled at different temperatures with a flow of heat from the hot to the cold surface. When thermal equilibrium was attained thermal conductivity data were taken within an accuracy of 3%. Fig. 1 compares the effect of the nanoparticles' content on the thermal conductivity of the nanocomposites. As seen, the value of thermal conductivity increased with an increase in the nanoparticle concentration up to 64% and 82% for CaCO$_3$ and ZnO respectively. The increase in TC for ZnO nanoparticles is more than CaCO$_3$ due to the nature of nanofiller and also crystallinity degree regarding to DSC results (Table 3). These increases in TC for both nanoparticles are higher than that of reported values for CNF in a PP matrix (Frormann et al., 2008).

The values obtained from the experimental study of PP nanocomposites were compared with several TC models (Figs 2.a and 2.b). As seen the experimental results were found to fall in between the Series and Parallel models. However, the lower bound model (series) is usually closer to the experimental data.

Maxwell, Lewis & Nielson, Bruggeman, Bottcher and De Loor models predict fairly well thermal conductivity values up to 10 wt% for PP/ZnO nanocomposites (Fig. 2.a). In the concentration of 15 wt% no model could predict well the TC values and all of the mentioned models underestimated the TC values of nanocomposite whereas in the case 5 wt% all models overestimated the TC values.

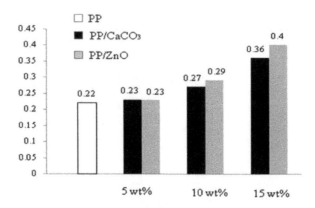

Fig. 1. Comparing the effect of nanoparticles on the TC of PP.

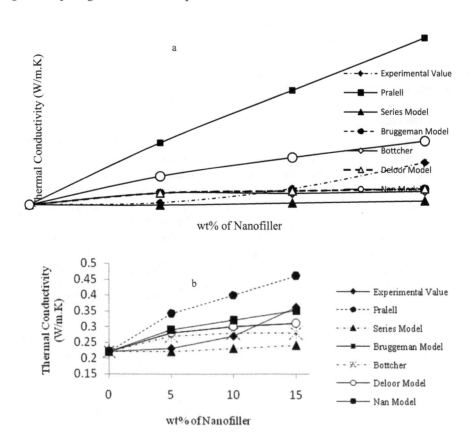

Fig. 2. Comparing the experimental TC values vs nanoparticle content with theoretical models. a) PP/ZnO nanocomposite b) PP/CaCO$_3$ nanocomposite.

This fact can be attributed to the intrinsic thermal conductivity of both nanoparticles and their large surface area which even at lower loadings of nanofillers they are still effective to transfer heat through the samples (Frormann et al., 2008). At a higher volume fraction, this effect becomes stronger. Fig. 2.b the values obtained from the experimental study for PP/CaCO$_3$ nanocomposites are compared with a number of TC models. As seen the Ce Wen Nan model predicts fairly well the thermal conductivity values up to 15 wt%. For the concentration of 10 wt% all the models predict the TC values well. In the case of 15 wt% other models underestimated the TC values of nanocomposites except for the Ce Wen Nan model, whereas for 5 wt% all models overestimated the TC value. The predicted TC values by the models depend on the nature of nanofiller and their relative concentrations (Weidenfeller et al., 2004; Frormann et al., 2008).

The TC improvement in PP/ZnO nanocomposite is greater than that of PP/calcium carbonate nanocomposites. This fact can be attributed to intrinsic thermal conductivity of the ZnO nanoparticles. Several models have been used for prediction of TC in the nanocomposites (see section 3.2). In the PP/ZnO nanocomposites TC values correlated well with the values predicted by Series, Maxwell, Lewis & Nielson, Bruggeman and De Loor models up to 10 wt%.

7. Conclusions

As electronic devices tend to become slimmer and more integrated, heat management become a central task for device design and application. Similar issues are faced in several other applications, including electric motors and generators, heat exchangers in power generation, automotive, etc. Metallic materials are widely used as heat dissipation materials, but there have been many attempts to replace the metallic materials with highly thermally conductive polymer based composites due to their lightweight, corrosion resistance, easy processing and lower manufacturing cost.

Thermally conductive polymer based composites are tentatively prepared by the incorporation of thermally conductive fillers. The outstanding thermal conductivity of mentioned fillers makes them a promising candidates to obtain highly thermally conductive polymer based composites.

PP nanocomposites were prepared by melt extrusion in a twin screw extruder. The introduction of nanoparticles resulted in an increase in crystallinity. Scanning electron microscopy (SEM) indicated a good dispersion of the nanofillers within the PP matrix that might enhance the thermal conductivity of the nanocomposites even at lower nanofiller loadings owing to enhanced filler-matrix interaction. The thermal conductivity of PP/ZnO nanocomposites had an increase of 82% at 15 wt% concentration comparing to that of pure PP, while for PP/CaCO$_3$ nanocomposite with same level of nanoparticle content it was 64%, so it is concluded that ZnO nanoparticles had more intrinsic potential to improve thermal conductivity of PP comparing to CaCO$_3$ nanoparticles regarding to its nature and crystallinity.

The thermal conductivity was increased from K=0.22 W/mK for pure PP by 64% for the sample with 15 wt% of CaCO$_3$ nanoparticles. These results for both nanocomposites (PP/ZnO and PP/CaCO$_3$) are higher than the values which reported for CNF in a PP matrix

Frormann L, Iqbal A, Abdullah S.A. 2008. The measured values were also compared with various models in the investigated range of nanofiller concentration. The Series, Maxwell, Lewis & Nielson, Bruggeman, Bottcher, De Loor and Ce Wen Nan models predicted fairly well the thermal conductivity values for the samples containing more than 5 wt% of nanoparticles. The experimental TC values of PP nanocomposites showed a linear increase with an increase in concentration and crystallinity.

8. Abbreviations

ABS   poly(acrylonitrile-butadiene-styrene) copolymer
AlN   aluminum nitride
BeO   beryllium oxide
BN   boron nitride
BNNT   boron nitride nanotubes
CNT   carbon nanotube
C_p heat capacity
DSC   differential scanning calorimetry
DWCNT double-walled carbon nanotube
EG   expanded graphite
EPDM   ethylene propylene diene rubber
EVA   poly(ethylene vinyl acetate)
GNP   graphite nanoplatelet
HDPE   high density polyethylene
K thermal conductivity (in some figures taken from literature referred as K_e = effective thermal conductivity)
k_c thermal conductivity of composite
k_m thermal conductivity of matrix
k_p thermal conductivity of particle
l phonon mean free path
L length parameter
LDPE   low density polyethylene
MWCNT   multi-walled carbon nanotube
PA6   polyamide 6
PA66   polyamide 6-6
PBT   poly(butylene terephthalate)
PC   polycarbonate
PDMS   poly(dimethylsiloxane)
PE   polyethylene
PEEK   polyetheretherketone
PET   poly(ethylene terephthalate)
PEVA   poly(ethylene vinyl alcohol)
PI   polyimide
PMDA   pyromellitic dianhydride
PMMA   polymethylmethacrylate
PP   polypropylene
PPS   polyphenylene sulfide
PPSU   polyphenylsulfone

PS   polystyrene
PSU   polysulfone
PTFE   polytetrafluoroethylene
PU   polyurethane
PVB   poly(vinyl butyral)
PVC   polyvinyl chloride
PVDF   polyvinylidene difluoride
R_k interfacial resistance
SiC   silicon carbide
SWCNT single-walled carbon nanotube
T_g glass transition temperature
v average phonon velocity
VGCF   vapor grown carbon fiber
A thermal diffusivity
ρ density of the material
Φ_m volume fractions of matrix
Φ_p volume fractions of particles
w_f weight fraction of particles

9. References

Abdel-Aal N.; El-Tantawy F.; Al-Hajry A. & Bououdina M. (2008), Epoxy resin/plasticized carbon black composites. Part I. Electrical and thermal properties and their applications, *Polymer Composites,* vol.29 pp. 511–517.

Agari Y., Ueda A., Omura Y. & Nagai S. (1997), Thermal diffusivity and conductivity of PMMA–PC blends, *Polymer,* vol.38, pp. 801–807.

Bai F.; Li F.; Calhoun B.; Quirk R.P. & Cheng S.Z.D. (1999), Polymer Handbook, 4th edition. In: *John Wiley &Sons, Inc.,* New York.

Bigg D.M. (1995), Thermal conductivity of heterophase polymer compositions, *Advanced Polymer Science,* vol.119, pp. 1–30.

Böttcher C. (1952), Theory of Electric Polarization, In: *Elsevier.* Amsterdam, Netherlands.

Boudenne A.; Ibos L.; Fois M.; Majeste J.C. & Géhin E. (2005), Electrical and thermal behavior of polypropylene filled with copper particles, *Composites: part A,* vol.36, pp. 1545–1554.

Bruggeman D. (1935), Dielectric constant and conductivity of mixtures of isotropic materials. *Annual Physics,* vol.24, pp. 636–645.

Causin V.; C. Marega;A. Marigo; G. Ferrara & A. Ferraro (2006), Morphological and structural characterization of polypropylene/conductive graphite nanocomposites, *European Polymer Journal,* vol.42, pp. 3153–3161.

Chen Y. & Ting J. (2002), Ultra high thermal conductivity polymer composites, *Carbon,* vol.40, pp. 359–362.

Chernenko N. M.; Yu. N.; Beilina A. & Sokolov I. (2009), TECHNOLOGICAL ASPECTS OF HEAT RESISTANCE IN CARBON-CERAMIC COMPOSITE REFRACTORIES. *Refractories and Industrial Ceramics,* Vol. 50, No. 2.

Dashora P. & Gupta G. (1996), On the temperature dependence of the thermal conductivity of linear amorphous polymers, *Polymer,* vol.37, pp. 231–234.

DeLoor G. (1956), Dielectric Proberties of heterogeneous mixtures. PhD Thesis, University of Leiden, NL. Leiden.

Droval G.; Feller J.F., Salagnac P. & Glouannec P. (2006), Thermal conductivity enhancement of electrically insulating syndiotactic poly(styrene) matrix for diphasic conductive polymer composites. *Polymers for Advanced Technology*, vol.17, pp.732–745.

Dunn M.L. & Taya M. (1993), The effective thermal conductivity of composites with coated reinforcement and the application to imperfect interfaces, *Journal of Applied Physics.* Vol.73, pp. 1711–1722.

Elgafy A. & Lafdi K. (2005), Effect of carbon nanofiber additives on thermal behavior of phase change materials, *Carbon,* vol. 43, pp. 3067–3074.

Every A.G.; Tzou Y.; Hasselman D.P.H. & Ray R. (1992), The effect of particle size on the thermal conductivity of ZnS/diamond composites, *Acta Metallurgica et Materialia*, vol.40, pp. 123–129.

Ebadi-Dehaghani H.; Reiszadeh M.; Chavoshi A. & Nazempour M. (2011), A study on the effect of zinc oxide (ZnO) and calcium carbonate ($CaCO_3$) nanoparticles on the thermal conductivity (TC) of polypropylene. *Composites: part B* (revised)

Fischer J.E. (2006), Carbon nanotubes: structure and properties. Gogotsi Y., *Carbon nanomaterials*, Taylor and Francis Group, New York, pp. 51–58.

Frank H. R. & Phillip D. S. (2002), Enhanced Boron Nitride Composition and Polymer Based High Thermal Conductivity Molding Compound, EP 0 794 227 B1.

Frormann L.; Iqbal A. & Abdullah S.A. (2008), Thermo-viscoelastic behavior of PCNF-filled polypropylene nanocomposites. *Journal of Applied Polymer Science*; 107: 2695-2703.

Fukushima H.; Drzal L. T.; Rook B. P. & Rich M. J. (2006), Thermal conductivity of exfoliated Graphite nanocomposites. *Journal of Thermal Analysis and Calorimetry*, Vol. 85, pp.235–238.

Gaal P.S.; Thermitus M.A. & Stroe D.E. (2004), Thermal conductivity measurements using the flash method, *Journal of Thermal Analysis and Calorimetry*, vol.78, pp. 185–189.

Ganguli S.; Roy A.K. & Anderson D.P. (2008), Improved thermal conductivity for chemically functionalized exfoliated graphite/epoxy composites, *Carbon*, vol.46, pp. 806–817.

Greig D. & Hardy N.D. (1981), The thermal conductivity of semicrystalline polymers at very low temperatures, *Journal of Physics Colloq*, vol.42, pp. 69–71.

Gu J.; Zhang Q.; Dang J.; Zhang J. & Yang Z. (2009), Thermal conductivity and mechanical properties of aluminum nitride filled linear low-density polyethylene composites, *Polymer Engineering Science,* vol.49, pp. 1030–1034.

Han Z. & Fina A. (2011). Thermal conductivity of carbon nanotubes and their polymer nanocomposites: A review, *Progress in Polymer Science*, vol.36, Issue 7, pp.914-944.

He H.; Fu R.; Han Y.; Shen Y. & Song X (2007),Thermal conductivity of ceramic particle filled polymer composites and theoretical predictions. Journal of Material Science, vol.42, pp.6749–6754. DOI 10.1007/s10853-006-1480-y.

Hermansen R. D. (2001), Room-Temperature Stable, One-Component, Thermally-Conductive, Flexible Epoxy Adhesives, EP 0 754 741 B1.

Hodgin M. J. & Estes R. H. (1999). Proceedings of the Technical Programs, *NEPCO WEST Conference*, pp. 359–366.Anaheim, CA.

Holotescu S. & Stoian F. D. (2009), Evaluation of the effective thermal conductivity of composite polymers by considering the filler size distribution law. *Journal of Zhejiang University SCIENCE A*. vol. 10(5) pp.704-709.

Hu M.; Yu D. & Wei J. (2007), Thermal conductivity determination of small polymer samples by differential scanning calorimetry. *Polymer Testing*, vol.26, pp.333-337.

Ishida H. (2000), Surface Treated Boron Nitride for Forming A Low Viscosity High Thermal Conductivity Polymer Based on Boron Nitride Composition and Method, US Patent, 6,160,042.

Ishida H. & Heights S. (1999), Composition for Forming High Thermal Conductivity Polybenzoxazine-Based Material and Method, US Patent 5,900,447.

Ishida H. & Rimdusit S. (1998), Very high thermal conductivity obtained by boron nitride-filled polybenzoxazine, *Thermochimica Acta*, vol.320, pp. 177–186.

Kalaitzidou K.; Fukushima H. & Drzal L.T. (2007), Multifunctional polypropylene composites produced by incorporation of exfoliated graphite nanoplatelets, *Carbon*, vol.45, pp. 1446–1452.

Kelly B.T. (1981), Physics of graphite, Applied Science Publishers, Barking (UK).

King J.A.; Barton R.L., Hauser R.A. & Keith J.M. (2008), Synergistic effects of carbon fillers in electrically and thermally conductive liquid crystal polymer based resins, *Polymer Composites*, vol.29, pp. 421–428.

King J.A.; Morrison F.A.; Keith J.M.; Miller M.G.; Smith R.C.; Cruz M.; Neuhalfen A.M. & Barton R.L. (2006), Electrical conductivity and rheology of carbon-filled liquid crystal polymer composites, *Journal of Applied Polymer Science*, vol.101, pp. 2680–2688.

King J.A.; Tucker K.W.; Vogt B.D.; Weber E.H. & Quan C. (1999), Electrically and thermally conductive nylon 6,6, *Polymer Composites*, vol.20, pp. 643–654.

Kline D.E. (1961), Thermal conductivity studies of polymers, *Journal of Polymer Science*, vol.50, pp. 441–450.

Kowalski L.; Duszczyk J. & Katgerman L(1999), Thermal conductivity of metal powder-polymer feedstock for powder injection moulding. *Journal of Material Science*, vol.34, pp.1– 5.

Kumar A.P.; Depan D.; Tomer N.S. & Singh R.P. (2009), Nanoscale particles for polymer degradation and stabilization—trends and future perspectives, *Progress Polymer Science*, vol.34, pp. 479–515.

Kumlutas D.; Tavman I. H. & Coban M.T. (2003). Thermal conductivity of particle filled polyethylene composite materials. *Composites Science and Technology*, vol.63, pp.113–117.

Kuriger R.J.; Alam M.K.; Anderson D.P. & Jacobsen R.L. (2002), Processing and characterization of aligned vapor grown carbon fiber reinforced polypropylene, *Composites: PartA*, vol.33, pp. 53–62.

Lee H. & Eun S. (2004). In Composites 2004 Convention and Trade Show, *American Composites Manufacturers Association*, Tampa, Florida.

Lipton R. & Vernescu B. (1996), Composites with imperfect interface, *Proc R Soc Lond A*, vol.452, pp. 329–358.

Liu C. & Mather T. (2004), ANTEC 2004, *Society of Plastic Engineers*, pp. 3080–3084.

Liu Z., Guo Q., Shi J., Zhai G. & Liu L. (2008), Graphite blocks with high thermal conductivity derived from natural graphite flake, *Carbon*, vol.46, pp. 414–421.

Mai Y. & Yu Z. (2006), Polymer nanocomposites. Woodhead publishing limited, Cambridge England.

Majumdar A. (1998), Microscale transport phenomena. Rohsenow W.M.; Hartnett J.R. & Cho Y.I., *Handbook of heat transfer,* (3rd ed.), McGraw-Hill, New York, pp. 8.1–8.8.

Mamunya Y.P.; Davydenko V.V.; Pissis P. & Lebedev E.V. (2002), Electrical and thermal conductivity of polymers filled with metal powders, *European Polymer Journal,* vol.38, pp. 1887–1897.

Marcus S.M. & Blaine R.L. (1994), Thermal conductivity of polymers, glasses and ceramics by modulated DSC, *Thermochimica Acta,* vol.243, pp. 231–239.

Merzlyakov M. & Schick C. (2001), Thermal conductivity from dynamical response of DSC, *Thermochimica Acta,* vol.377, pp. 183–191.

Mohammed H.A. & Uttandaraman S. (2009), A review of vapor grown carbon nanofiber/polymer conductive composites, *Carbon,* vol.47, pp. 2–22.

Momentive Performance Materials, Boron nitride finds new applications in thermoplastic compounds. (2008), *Plastics, Additives and Compounding,* (May/June), pp. 26–31.

Mu Q.; Feng S. & Diao G. (2007), Thermal conductivity of silicone rubber filled with ZnO, *Polymer Composites,* vol.28, pp. 125–130.

Mu Q. and Feng S. (2007), Thermal conductivity of graphite/silicone rubber prepared by solution intercalation, *Thermochimica Acta,* vol.462, pp. 70–75.

Nan C.W.; Birringer R.; Clarke D.R. & Gleiter H. (1997), Effective thermal conductivity of particulate composites with interfacial thermal resistance, *Journal of Applied Physics,* vol. 10, pp. 6692–6699.

Nan C.W.; Liu G.; Lin Y. & Li M. (2004), Interface effect on thermal conductivity of carbon nanotube composites. *Applied Physics Letter,* vol. 85, pp.3549-3555.

Nielsen E.; Landel R.F. (1994), Mechanical Properties of Polymers and Composites. 2nd edition, Marcel Dekker Inc., New York.

Nunes dos Santos W., Mummery P. & Wallwork A. (2005), Thermal diffusivity of polymers by the laser flash technique, *Polymer Testing,* vol.24, pp. 628–634.

Nunes dos Santos W. (2007), Thermal properties of polymers by non-steady-state techniques, *Polymer Testing,* vol.26, pp. 556–566.

Ohashi M.; Kawakami S. & Yokogawa Y. (2005), Spherical aluminum nitride fillers for heat-conducting plastic packages, *Journal of American Ceramic Society,* vol.88, pp. 2615–2618.

Okamoto S. & Ishida H. (1999), A new theoretical equation for thermal conductivity of two-phase systems, *Journal of Applied Polymer Science,* vol.72, pp. 1689–1697.

Park S.H.; Hong C.M.; Kim S. & Lee Y.J. (2008), Effect of fillers shape factor on the performance of thermally conductive polymer composites, *ANTEC Plastics - Annual Technical Conference Proceedings 2008,* pp. 39–43.

Price D.M. & Jarratt M. (2002), Thermal conductivity of PTFE and PTFE composites, *Thermochimiuca Acta,* vol.392-393, pp. 231–236.

Pierson H.O. (1993), Handbook of carbon, graphite, diamond and fullerenes: properties. Processing and applications, Noyes Publications, New Jersey.

Reese W. (1969), Thermal properties of polymers at low temperatures, *Journal of Macromolecular Science A,* vol.3, pp. 1257–1295.

Rides M.; Morikawa J.; Halldahl L.; Hay B.; Lobo H.; Dawson A. & Allen C. (2009), Intercomparison of thermal conductivity and thermal diffusivity methods for plastics, *Polymer Testing,* vol.28, pp. 480–489.

Speight J.G. (2005), Lange's handbook of chemistry (16th ed.), McGraw-Hill, New York pp. 2794–2797.

Stankovich S.; Dikin D.A.; Dommett G.H.B.; Kohlhaas K.M.; Zimney E.J.; Stach E.A.; Piner R.D.; Nguyen S.T. & Ruoff R.S., Graphene-based composite materials, *Nature,* vol.442 (2006), pp. 282–286.

Sui G.; Jana S.; Zhong W.H.; Fuqua M.A. & Ulven C.A. (2008), Dielectric properties and conductivity of carbon nanofiber/semi-crystalline polymer composites, *Acta Material,* vol.56, pp. 2381–2388.

Tavman I.H. (1996), Thermal and mechanical properties of aluminum powder-filled high-density polyethylene composites, *Journal of Applied Polymer Science,* vol. 62, pp. 2161–2167.

Tavman I. H. (2004). in Nanoengineered Nanofibrous Materials, *NATO Science Series II, Mathematics, Physics and Chemistry,* Guceri S. I.; Gogotsi Y. & Kuzentsov V., Kluwer Academic Book Publication, Dordrecht, Netherlands, Vol. 169, p. 449.

Tekce H.S.; Kumlutas D. & Tavman I.H. (2007), Effect of particle shape on thermal conductivity of copper reinforced polymer composites, *Journal of Reinforced Plastics and Composites,* vol.26. pp. 113–121.

Thostenson E.K.; Li C. & Chou T.W. (2005), Nanocomposites in context, *Composite Science Technology,* vol.65, pp. 491–516.

Tibbetts G.G.; Lake M.L.; Strong K.L. & Rice B.P. (2007), A review of the fabrication and properties of vapor-grown carbon nanofiber/polymer composites, *Composite Science Technology,* vol.67, pp. 1709–1718.

T'Joen C.; Park Y.; Wang Q.; Sommers A.; Han X. & Jacobi A. (2009), A review on polymer heat exchangers for HVAC&R applications, *International Journal of Refrigeration,* vol.32, pp. 763–779.

Torquato S. & Rintoul M.D. (1995), Effect of interface on the properties of composite media, *Physical Review Letters,* vol.75, pp. 4067–4070.

Tritt T.M. & Weston D. (2004), Measurement techniques and considerations for determining thermal conductivity of bulk materials. Tritt T.M., *Thermal conductivity theory, properties, and applications,* Kluwer Academic/Plenum Publishers, New York, pp. 187–203.

Tu H. & Ye L. (2009), Thermal conductive PS/graphite composites, *Polymer Advanced Technology,* vol.20, pp. 21–27.

Vakili M.H.; Ebadi-Dehaghani H.; Haghshenas-Fard M. (2011), Crystallization and Thermal Conductivity of $CaCO_3$ Nanoparticle Filled Polypropylene. *Journal of Macromolocular Science: Part B.* Physics, vol.50, pp.1637–1645.

Veca M.L.; Meziani M.J.; Wang W.; Wang X.; Lu F.; Zhang P.; Lin Y., Fee R.; Connell J.W. & Sun Y. (2009), Carbon nanosheets for polymeric nanocomposites with high thermal conductivity, *Advanced Material,* vol. 21, pp. 2088–2092.

Wang J.; Carson J.K.; North M.F. & Cleland D.J. (2008), A new structural model of effective thermal conductivity for heterogeneous materials with cocontinuous phases, *International Journal of Heat and Mass Transfer,* vol.51, pp. 2389–2397.

Weidenfeller B.; Hofer M & Schilling F. R. (2004), Thermal conductivity, thermal diffusivity, and specific heat capacity of particle filled polypropylene. *Composites: Part A,* vol.35, pp. 423–429.

Wolff S. & Wang M.J. (1993), Carbon black science & technology (2nd ed.), Marcel Dekker, New York.

Wong Y.W.; Lo K.L. & Shin F.G. (2001), Electrical and thermal properties of composite of liquid crystalline polymer filled with carbon black, *Journal of Applied Polymer Science*, vol.82, pp. 1549-1555.

Wypych G. (2000), Handbook of fillers: physical properties of fillers and filled materials, ChemTec Publishing, Toronto.

Yano O. & Yamaoka H. (1995), Cryogenic properties of polymers, *Progress Polymer Science*, vol.20, pp. 585-613.

Ye C.M.; Shentu B.Q. & Weng Z.X. (2006), Thermal conductivity of high density polyethylene filled with graphite, *Journal of Applied Polymer Science*, vol.101, pp. 3806-3810.

Yu S.; Hing P. & Hu X. (2002), Thermal conductivity of polystyrene–aluminum nitride composite, *Composites: Part A*, vol.33, pp. 289-292.

Zhang X.; Fujiwara S. & Fujii M. (2000), Measurements of thermal conductivity and electrical conductivity of a single carbon fiber, *International Journal of Thermophysics*, vol.21, pp. 965-980.

Zhao H. & Li R.K.Y. (2006), A study on the photo-degradation of zinc oxide (ZnO) filled polypropylene nanocomposites. *Polymer;* vol.47, pp.3207-3217.

Zeng J.; Fu R.; Agathopoulos S.; Zhang S.; Song X. & He H. (2009), Numerical simulation of thermal conductivity of particle filled epoxy composites, *Journal* of *Electronic Packaging*, vol.131, pp.1-7.

Zhi C.; Bando Y.; Terao T.; Tang C.; Kuwahara H. & Golberg D. (2009), Towards thermoconductive, electrically insulating polymeric composites with boron nitride nanotubes as fillers, *Advanced Functional Materials*, vol.19, pp. 1857-1862.

Zhong C.; Yang Q. & Wang W. (2001), Correlation and prediction of the thermal conductivity of amorphous polymers, *Fluid Phase Equilibria*, vol.181, pp. 195-202.

Zhou H.; Zhang S. & Yang M. (2007), The effect of heat-transfer passages on the effective thermal conductivity of high filler loading composite materials, *Composites Science and Technology*, vol.67, pp. 1035-1040.

Zhou W.; Qi S.; An Q.; Zhao H. & Liu N. (2007), Thermal conductivity of boron nitride reinforced polyethylene composites, *Materials Research Bulletin*, vol.42, pp. 1863-1873.

Permissions

The contributors of this book come from diverse backgrounds, making this book a truly international effort. This book will bring forth new frontiers with its revolutionizing research information and detailed analysis of the nascent developments around the world.

We would like to thank Dr. Abbass A. Hashim, for lending his expertise to make the book truly unique. He has played a crucial role in the development of this book. Without his invaluable contribution this book wouldn't have been possible. He has made vital efforts to compile up to date information on the varied aspects of this subject to make this book a valuable addition to the collection of many professionals and students.

This book was conceptualized with the vision of imparting up-to-date information and advanced data in this field. To ensure the same, a matchless editorial board was set up. Every individual on the board went through rigorous rounds of assessment to prove their worth. After which they invested a large part of their time researching and compiling the most relevant data for our readers. Conferences and sessions were held from time to time between the editorial board and the contributing authors to present the data in the most comprehensible form. The editorial team has worked tirelessly to provide valuable and valid information to help people across the globe.

Every chapter published in this book has been scrutinized by our experts. Their significance has been extensively debated. The topics covered herein carry significant findings which will fuel the growth of the discipline. They may even be implemented as practical applications or may be referred to as a beginning point for another development. Chapters in this book were first published by InTech; hereby published with permission under the Creative Commons Attribution License or equivalent.

The editorial board has been involved in producing this book since its inception. They have spent rigorous hours researching and exploring the diverse topics which have resulted in the successful publishing of this book. They have passed on their knowledge of decades through this book. To expedite this challenging task, the publisher supported the team at every step. A small team of assistant editors was also appointed to further simplify the editing procedure and attain best results for the readers.

Our editorial team has been hand-picked from every corner of the world. Their multi-ethnicity adds dynamic inputs to the discussions which result in innovative outcomes. These outcomes are then further discussed with the researchers and contributors who give their valuable feedback and opinion regarding the same. The feedback is then collaborated with the researches and they are edited in a comprehensive manner to aid the understanding of the subject.

Apart from the editorial board, the designing team has also invested a significant amount of their time in understanding the subject and creating the most relevant covers. They scrutinized every image to scout for the most suitable representation of the subject and create an appropriate cover for the book.

The publishing team has been involved in this book since its early stages. They were actively engaged in every process, be it collecting the data, connecting with the contributors or procuring relevant information. The team has been an ardent support to the editorial, designing and production team. Their endless efforts to recruit the best for this project, has resulted in the accomplishment of this book. They are a veteran in the field of academics and their pool of knowledge is as vast as their experience in printing. Their expertise and guidance has proved useful at every step. Their uncompromising quality standards have made this book an exceptional effort. Their encouragement from time to time has been an inspiration for everyone.

The publisher and the editorial board hope that this book will prove to be a valuable piece of knowledge for researchers, students, practitioners and scholars across the globe.

List of Contributors

Fei Duan
School of Mechanical and Aerospace Engineering, Nanyang Technological University, Singapore

Mirosław R. Dudek and Marcin Kośmider
Institute of Physics, University of Zielona Góra, Zielona Góra, Poland

Nikos Guskos
Solid State Physics Section, Department of Physics, University of Athens, Panepistimiopolis, Greece
Institute of Physics, West Pomeranian University of Technology, Szczecin, Poland

Hon-Man Liu
National Taiwan University, Department of Radiology, Taiwan

Jong-Kai Hsiao
Tzu-Chi University, Department of Radiology, Taiwan

Peter Siffalovic, Eva Majkova, Matej Jergel, Karol Vegso, Martin Weis and Stefan Luby
Institute of Physics, Slovak Academy of Sciences, Slovakia

Wei Yu, Huaqing Xie and Lifei Chen
Shanghai Second Polytechnic University, P. R. China

Giovanni Filippone and Domenico Acierno
Dept. of Materials and Production Engineering, University of Naples Federico II, Italy

Nickolay N. Kononov
Prokhorov General Physics Institute, Russian Academy of Sciences, Moscow, Russia

Sergey G. Dorofeev
Faculty of Chemistry, Moscow State University, Moscow, Russia

Jelena Tamulienė
Vilnius University, Institute of Theoretical Physics and Astronomy, Vilnius, Lithuania

Rimas Vaišnoras
Vilnius Pedagogical University, Vilnius, Lithuania

Goncal Badenes
Institut de Ciences Fotoniques ICFO, Barcelona, Spain

Mindaugas L. Balevičius
Vilnius University, Vilnius, Lithuania

Xinjun Xu and Lidong Li
School of Materials Science and Engineering, University of Science and Technology Beijing, Beijing, P. R. China

Orhan Yalçın
Department of Physics, Niğde University, Niğde, Turkey

Rıza Erdem
Department of Physics, Akdeniz University, Antalya, Turkey

Zafer Demir
Institute of Graduate School of Natural and Applied Sciences, Niğde University, Niğde, Turkey

Hassan Ebadi-Dehaghani and Monireh Nazempour
Shahreza Branch, Islamic Azad University, Iran